DC Power Supplies
Power Management
and Surge Protection
for Power
Electronic Systems

D0075131

DC Power Supplies
Power Management
and Surge Protection
for Power
Electronic Systems

Nihal Kularatna

CRC Press
Taylor & Francis Group
Boca Raton   London   New York

CRC Press is an imprint of the
Taylor & Francis Group, an **informa** business

CRC Press
Taylor & Francis Group
6000 Broken Sound Parkway NW, Suite 300
Boca Raton, FL 33487-2742

Printed in the United States of America on acid-free paper
Version Date: 2011916

International Standard Book Number: 978-0-415-80247-5 (Hardback)

Library of Congress Cataloging-in-Publication Data

DC power supplies : power management and surge protection for power electronic
 systems / editor, Nihal Kularatna.
       p. cm.
   "A CRC title."
   Includes bibliographical references and index.
   ISBN 978-0-415-80247-5 (hardcover : alk. paper)
   1. Electric circuits--Direct current. 2. Power electronics. I. Kularatna, Nihal. II. Title:
Direct current power supplies.

  TK454.15.D57D37 2012
  621.381'044--dc23   2011032873

**Visit the Taylor & Francis Web site at
http://www.taylorandfrancis.com**

**and the CRC Press Web site at
http://www.crcpress.com**

This book is dedicated to Sir Arthur C Clarke (the scientist who predicted satellite communication in the year 1945) and Prof John Robinson Pierce (who named the transistor) … they inspired me …

With loving thanks to my wife Priyani and daughters Dulsha and Malsha and their families, who tolerate my addiction to tech writing and electronics.

Contents

5   Control Loop Design of DC-to-DC Converters

6   Power Management

7   Off-the-Line Switching Power Supplies

8   Rechargeable Batteries and Their Management

Preface

In mid-1976, I was a young EE graduate starting my first industry job at the area control center of the Department of Civil Aviation in Sri Lanka. When I met my senior engineer, B. L. Ramanayake, on my first day of work, he took me to the Alcatel KLB-5 Message Switching System, a monster TTL IC-based real-time computer system (no microprocessor-based systems were there at that time), and showed me a massive 30-volume documentation set. He said when the system fails, you have read these documents and repair the fault. The system was a large set of equipment cabinets full of basic TTL families and core memories, with just double-sided PCBs, linear power supplies, and wire-wrapped back planes supported by a 20 kVA UPS system with a massive battery bank. As a fresh 22-year-old graduate, I was stunned, since I have not taken a course on computer science, and thought of the amount of data to be referred to in a repair attempt.

Within a year, after five months of training in a subsidiary company of Alcatel in Paris, when I started attempting fault diagnosis on this monster system I realized that what Ramanayake said on my first day of work was absolutely true. For a period of more than five years, I detailed logic circuits, analog power supplies, UPS systems, and applied the hard-earned practical know-how in my repair attempts. The instrumentation available was only a 300 MHz Tektronix scope, multimeter, and a PCB tester. At every fault appearing in my system, I felt that I was not a true engineer until I was able to fix it.

I was also lucky to work as a commissioning engineer installing solid-state VHF omniranges (VORs) and distance-measuring equipment (DME) for air navigation during the same period to learn a lot more analog- and mixed-signal design approaches in real-world navaids. From 1982 to 1985, I accepted a job in Saudi Arabia to work as an electronics engineer doing maintenance on Ericsson-AXE 10 digital exchanges with eight-bit microprocessors in fault-tolerant time-sharing designs dealing with telephone switching. As an engineer who had never worked in the telecom area, I was to learn the telecom jargon soon … but the applicable fundamentals I learned in my time in aviation helped me fast-track into the application of real-time processor systems in digital telephony.

In summary, during my first 10 years it was a massive exposure and a challenge to learn the fundamentals and apply them in scientific fault diagnosis. Later, this exposure was extremely helpful in my 16-year-long research/research-management career, at the

Arthur C. Clarke Institute for Modern Technologies (ACCIMT) back in Sri Lanka, to creatively apply fundamentals in power electronic circuit designs.

To summarize my 25 years of early career, I simply learned that practically applying Ohm's and Kirchoff's laws, equivalent circuits, device characteristics, and a reasonable amount of simple math is all an electronic circuit designer will have to use creatively. Once you start enjoying this world of circuit design, even during your vacation time you can think of electronic circuits, with no stress in your brain!

With the inspiration given to me by Sir Arthur C. Clarke (who predicted satellite communication in 1945, writing a paper to *Wireless World* magazine) and Prof. John Robinson Pierce (a former executive director of Bell Labs, and the engineer who named the transistor), I was able to document my hard-earned knowledge in six published books, with this my seventh attempt, while completing a 10-year spell as a university academic.

In this work I have attempted to summarize the information in several hundreds of published papers (authored by subject experts) to guide the people who intend to learn power electronics (PE) and the design of PE circuits as applicable to modern-day electronic systems. I am very thankful to all my readers who have encouraged me to keep my book writing in a continuum. If you point out any errors here, I will be very thankful to you.

Enjoy power electronics, a serious enabling technology in the world of greener electronic systems.

Nihal Kularatna
School of Engineering
The University of Waikato

Acknowledgments

I am really grateful to Prof. John Robinson Pierce, the engineer who named the transistor, for his great advice during a dinner at a U.S. West Coast restaurant in the early 1990s, on how to write a technical book. One of his two simple guidelines was to summarize the contents of the book on a back of a business card before the project begins!

The subject of this book can be summarized as "essentials of the power electronics circuits applicable to low-power systems, including modern portable devices." During my 35-year career as an electronics engineer, I was exposed to many great applications developed by experts in their fields using many analog- and mixed-signal circuits and the power conversion techniques. Many new ICs, passive devices, and software enter the marketplace almost every week, and the *knowledge half-life* of electronics and affiliated fields keeps dropping toward less than three to four years.

I thought of developing this book based on my 25 years of industry and research experiences, mixed with the lessons I learned from teaching full-time in two universities in New Zealand for 10 years. Undergraduate students, with little or no real-life design experience, tend to learn only from the guidance we give them as academics. If we direct them to "theory-only" texts and limited laboratory experiences, we limit their opportunities to develop adequate hands-on abilities. If we can give them more projects, with *burning-and-learning* types of power electronics laboratory and project experiences, as well as reading new-material collections that can *bridge the gap between theory and practice*, they tend to learn faster with confidence to face real engineering.

For practicing engineers, time limitations to complete projects do not let them search new publications frequently to keep abreast of new developments and research directions. If a reference-type book can be there on the bookshelf, it can help one save time and do further referencing easily. However, such books also will get outdated quickly!

This work is aimed at fulfilling both tasks, and for my opportunities and exposure to real-world PE environments, I first acknowledge all my students and the junior members of staff (particularly at my Sri Lankan workplace from 1985 to 2002, Arthur C. Clarke Institute for Modern Technologies—ACCIMT) and thank them for those wonderful learning opportunities I received.

The attempt in this book is to provide a reasonable link between the theoretical knowledge domain and the valuable practical information domain from the perspective of technology developers. The broad approach in this work is to *understand complete power electronic solutions and appreciate their interfacing aspects*, embracing many mixed-signal circuitries in complete systems.

In this work, a large amount of published material from industry and academia has been used, and the work and organizations that deserve strong acknowledgments are:

Many published textbooks for the material in Chapters 1 and 2, which provide a review of power supply fundamentals

Many published articles in *Power Electronics Technology* (formerly *PCIM*) magazine

Industry magazines such as *EDN, Electronic Design*, and *Test & Measurement World*, and many IEEE/IEE research publications

In this exercise I am very thankful for the tireless assistance given by the following to create the figures and help with word processing tasks and the like:

ACCIMT staff members from Sri Lanka, Chandrika Weerasekera and Jayathu Fernando, who spent a lot of their private time coordinating with me across different time zones

My students Lasantha Tilakaratne, Kosala Kankanamge, and Sisira James at the University of Waikato

My special gratitude is extended to Associate Prof. Patrick Hu (University of Auckland) for authoring Chapter 5, and to my PhD student Kosala Kankanamge for coauthoring Chapter 4.

For the copyright permission for certain contents in the book, I am very thankful to Sam Davis, chief editor of the *Power Electronics Technology Magazine*, Penton Publications, USA, and of IEEE, USA. And thank you too to all parties who helped me with copyright permissions for the contents of Chapter 3 in my previous CRC book published in 2008.

It was a pleasant experience to work with the staff of the CRC Press for this book project, and I am particularly thankful to Nora Konopka, the publisher, for her understanding and the support to get the project moving smoothly. Vakili Jessica, Brittany Gilbert, and the other members of the editorial/sales staff are gratefully acknowledged for their support in collecting my manuscript in several stages, allowing me to balance my time between work, family, and book writing. I am very grateful to Michael Davidson and the production staff for their assistance in solving problems in the proof of the work.

Also, I am very thankfully acknowledging the encouragement given by the readers of my previous six books, and for some of those works continuing into reprints and e-books even after 10 years from the original publication time.

Last but not least, my special thanks go to my loving wife, Priyani, who made this work possible by taking over my family commitments—looking after the family needs and taking the full responsibility of planning the wedding of my second daughter, Malsha, to Kasun in Sri Lanka, while I was spending time on my laptop finalizing the book's manuscript in 2010, during which year we became grandparents, too.

Thank you all once again,
Nihal Kularatna

About the Author

Former CEO of the Arthur C. Clarke Institute for Modern Technologies (ACCIMT) in Sri Lanka, **Nihal Kularatna** is an electronics engineer with more than 35 years of experience in professional and research environments. He is the author of two Electrical Measurement Series books for the IEE (London), titled *Modern Electronic Test and Measuring Instruments* (1996) and *Digital and Analogue Instrumentation: Testing and Measurement* (2003/2008), and two Butterworth (USA) titles, *Power Electronics Design Handbook: Low Power Components and Applications* (1998) and *Modern Component Families and Circuit Block Design* (2000). He coauthored *Essentials of Modern Telecommunications Systems* for Artech House Publishers (2004).

Electronic Circuit Design: From Concept to Implementation was his first CRC Press publication in 2008.

From 1976 to 1985 he worked as an electronics engineer responsible for navigational aids and communications projects in civil aviation and digital telephone exchange systems. In 1985 he joined the ACCIMT as a research and development engineer and earned a principal research engineer status in 1990; he was appointed CEO/Director of the ACCIMT in 2000. From 2002 to 2005 he was a senior lecturer at the Department of Electrical and Electronic Engineering, University of Auckland, New Zealand. He has participated in many specialized training programs with equipment manufacturers, universities, and other organizations in the United States, United Kingdom, France, and Italy.

A Fellow of the IEE (London), a Senior Member of IEEE (USA), and an honors graduate from the University of Peradeniya, Sri Lanka, during his research career in Sri Lanka, he was the winner of a Presidential Award for Inventions (1995), Most Outstanding Citizens Award (1999, Lions Club), and a TOYP Award for academic accomplishment (Jaycees) in 1993. He has contributed more than 70 papers to academic and industry journals and international conference proceedings. He is currently employed as a senior lecturer in the School of Engineering, University of Waikato, New Zealand.

Nihal is currently active in research in supercapacitor applications, transient propagation, power conditioning and smart sensor systems. Recently a US patent was granted for his low frequency supercapacitor circulation technique for significantly improving the end to end efficiency of linear regulators. His hobby is gardening cacti and succulents.

Contributors

Derek Bowers is a Fellow of Analog Devices Inc., USA and has developed numerous analog ICs including the XFET®. He was responsible for developing many op-amps, voltage references, data converters, surround sound chips and has contributed to numerous patents. He is a Fellow of IET and a Senior Member of IEEE.

Patrick Hu is with the Department of Electrical and Computer Engineering, University of Auckland, NZ. With over 26 years of a career in China and New Zealand, he has published over 80 papers and holds 6 patents in the area of inductive power transfer. He is the author of *Wireless/Contactless Power Supply—Inductively Coupled Resonant Converter Solutions* [2009] and a Senior Member of IEEE.

Kosala Kankanamge-Gunawardane graduated in 2005 with a BSc (Honours) degree in Electronics and Telecommunication Engineering from University of Moratuwa, Sri Lanka. Presently she is pursuing a PhD at the University of Waikato, New Zealand, developing supercapacitor-based high efficiency linear regulators.

1

Review of Fundamentals Related to DC Power Supply Design and Linear Regulators

1.1 Introduction

All electronic circuits require a clean and constant voltage DC power supply. However, the energy source available for the system may be a commercial AC supply, a battery pack, or a combination of the two. In some special cases, this energy source may be another DC bus within the system or the universal serial bus (USB) port of a laptop. In a successful total system design exercise, the power supply should not be considered as an afterthought or the final stage of the design process, because it is the most vital part of a system for reliable performance under worst-case circumstances. Another serious consideration in system design is the total weight and the volume, and this can be very much dependent on the power supply and the power management system. Also, it is important for design engineers to keep in mind that the power supply design may entail many analog design concepts.

Most power supply design issues are due to resource and component limitations within the power supply and the power management system. Nonideal components—particularly, passives, commercial limitations to allocate sufficient backup energy storage within the battery pack, unexpected surges and transients from the commercial AC supply, and the fast load current transients—can create extreme and unexpected conditions within the system unless the power management system adequately addresses all the possible worst cases at an early design stage. Many product design experts choose to have the power supply and the power management system designed at an early stage with estimated parameters, with the actual system blocks powered from the system power supply. This approach may help minimize late-stage disasters in a large design project.

In the 1960s and early 1970s, power supplies were linear designs with efficiencies in the range of 30%–50%. With the introduction of switching techniques in the 1980s, this rose to 60%–80%. In the mid-1980s, power densities were about 50 W/in^3. With the introduction of resonant converter techniques in the 1990s, this was increased to

100 W/in³ [1, 2]. When high-speed and power-hungry processors were introduced during the mid-1990s, much attention was focused on transient response, and industry trends were to mix linear and switching systems to obtain the best of both worlds. Low-dropout (LDO) regulators were introduced to power noise-sensitive and fast transient loads in many portable products. In the late 1990s, power management and digital control concepts and many advanced approaches were introduced into the power supply and overall power management [3].

In this chapter we consider simple fundamentals related to an unregulated DC power supply and simple calculations to select the important components, with simple linear regulator concepts gradually extending to a discussion on LDOs. Because of space constraints, for detailed theoretical aspects and deeper design considerations, the reader is referred to the many useful references cited herein.

1.2 Simple Unregulated DC Power Supply and Estimating the Essential Component Values

In a DC power supply derived from the commercial AC source we can have two fundamental approaches: (a) transformer isolated and (b) nontransformer isolated. Transformer-isolated power supplies are safer, but bulky due their its line frequency transformer. This was the case for older electronic equipment where size was not a major concern, and still this approach is used in safety-critical applications or where common mode noise (discussed in Chapter 8) is a serious concern. Classic example in the modern scenario is the high-fidelity music systems. When the regulatory requirements for electrical isolation are covered by the DC-DC converters followed by the unregulated DC power supply, direct rectification and smoothing are used. One common example is the desktop computer power supplies (known as the "silver box"). In these incoming lines, voltage is directly rectified and filtered by a smoothing capacitor rated above the value of peak value of the AC line voltage, which is either 165 VDC (for 120 V, 60 Hz systems) or 325 VDC (for 230 V, 50 Hz). Figure 1.1 indicates the concept.

Given the case in Figure 1.1(c), if the peak voltage of the waveform of line frequency, f, appearing at the input of the bridge rectifier is V_{peak}, the peak-to-peak ripple voltage at the output could be approximated by,

$$\Delta V_{p-p} \approx \frac{I_L}{2fC} \tag{1.1}$$

Considering the capacitor is ideal, and the forward voltage drop for each of diodes is V_D, the approximate output DC voltage, V_{DC}, will be,

$$V_{DC} \approx V_{peak} - 2V_D - \frac{I_L}{4fC} \tag{1.2}$$

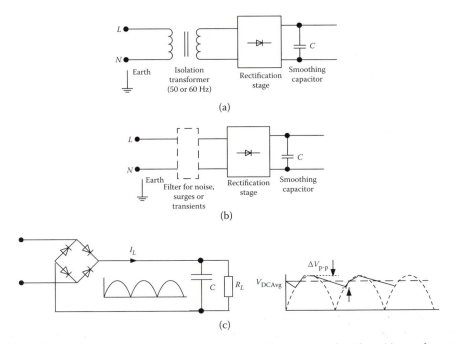

FIGURE 1.1 Unregulated power supply derived from the commercial AC line: (a) transformer isolated; (b) nontransformer isolated direct rectification; (c) estimating the size of the smoothing capacitor.

These approximate relationships allow us to estimate the approximate design parameters for an unregulated DC power supply, with or without a transformer. In the case of an ideal transformer with a turns ratio of n, and the input AC line RMS voltage is V_{rms},

$$V_{DC} \approx n\sqrt{2}V_{rms} - 2V_D - \frac{I_L}{4fC} \tag{1.3}$$

Given the practical situations of path resistances, diode dynamic resistances, the equivalent series resistance (ESR) of the smoothing capacitor and the like, and any losses in the transformer, an unregulated DC power supply will have a load regulation curve as per Figure 1.2.

1.3 Linear Regulators

Given the case of Figure 1.2, an unregulated DC power supply needs further improvements to make the output DC voltage constant at different load current. Historically, with the availability of semiconductor components such as diodes and transistors, linear regulator techniques were developed to solve this load regulation issue. In the following sections, we discuss the linear regulator techniques in a brief manner to highlight the important considerations in linear regulator designs.

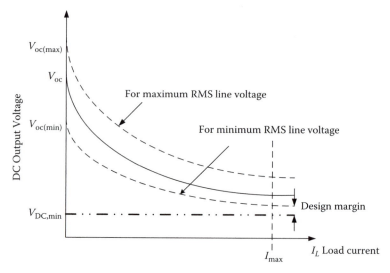

FIGURE 1.2   Load regulation curve of a typical unregulated DC power supply.

In designing a regulated DC power supply, the designer should consider the output voltage changes due to three important situations: (1) output voltage variations due to load current changes (usually depicted in a load regulation curve), (2) output voltage variations due to input source voltage fluctuations (line regulation curve), and (3) output voltage variations due to temperature variations. To illustrate this let us take a very simple case of a shunt regulator based on a simple zener/avalanche-type diode. In designing a simple power converter of this kind, let us start from some specifications as below:

    Unregulated input voltage range: 7 to 9 V DC
    Regulated output voltage: 5 V
    Maximum output current: 10 mA
    Output resistance of the unregulated input source: 10 Ω

The most simple solution could be the circuit in Figure 1.3 with a single resistor and a zener diode, where the regulated output is available at the terminals of the zener diode. Given such a simple specification, if we are to develop this circuit from simple and basic calculations based on available commercial components, we can develop this circuit using a zener diode such as BZX84C5V1from ON Semiconductor. If we consider the overall circuit in Figure 1.3(a), and the equivalent circuit for the diode in Figure 1.3(b), we can design this circuit to achieve the approximate specifications given above. As the maximum load current expected is 10 mA, we can allow the diode to carry about 2 mA under full load situation. The diode we have chosen above has a nominal zener voltage of 5.1 at 5 mA, and the data sheet indicates a resistance of 60 Ω. With reference to Figure 1.3(a), we can write the following relationships:

$$V_{out} = V_Z + I_Z r_Z \approx V_z \qquad (1.4)$$

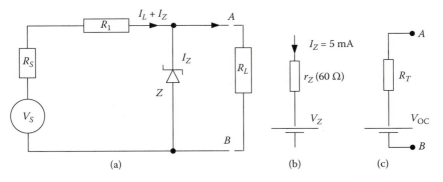

FIGURE 1.3 Simple shunt regulator: (a) basic circuit; (b) simplified equivalent circuit for a breakdown diode such as a zener or an avalanche diode; (c) simplified representation of the power supply using Thevenin's equivalent circuit.

and

$$V_s = V_Z + I_Z r_Z + (R_S + R_1)(I_L + I_Z) \tag{1.5}$$

Based on the design conditions selected in the above paragraph, R_1 can be estimated by applying the worst-case condition of lowest source voltage of 7 V and the case of zener diode taking up the whole current of 12 mA (when no load is connected),

$$R_1 = \frac{(7-5.1)}{12} - R_s - r_z \approx 88\,\Omega,$$

giving a close enough E12 range resistor of 82 Ω.

When the input is at the maximum possible value and the circuit is on no-load condition, worst-case zener current occurs. This worst-case zener current is

$$I_{Z,worst} = \frac{9-5.1}{10+60+82} \approx 26\,\text{mA}$$

This gives a worst-case zener dissipation of $(5.1+0.060*26)*26 \approx 173\text{mW}$, which is well within the data sheet limit. Under this condition the output voltage is approximately $5.1 + 0.060 * 26$, which is approximately 6.7 V, and under worst-input voltage and maximum load current, output voltage is approximately $5.1 + 0.060 * 2$), which is 5.22 V. This clearly shows that the output voltage can vary over a wide range, around an approximate value of nominal 5 V.

Given the variables in Figure 1.3(a), using basic circuit theory we can estimate the approximate Thevenin's equivalent circuit parameters of the simple shunt regulator, as shown in Figure 1.3(c). We can derive the Thevenin's equivalent circuit parameters as:

$$V_{OC} = V_Z \left[\frac{1}{1+\frac{r_z}{R_s+R_1}} \right] + V_s \left[\frac{1}{1+\frac{R_s+R_1}{r_z}} \right] \tag{1.5}$$

and

$$R_o = \frac{r_z}{1 + \frac{r_z}{R_s + R_l}}$$ (1.6)

From these relationships, we can clearly see that the impact of input source voltage variations can be minimized by keeping the value of zener impedance, r_z, much smaller than the value of $(R_s + R_1)$, and the same criteria applies to minimization of load regulation. However, practical limitations of available diodes make these circuits useful only in very low current circuits. Also one major disadvantage of this kind of a circuit is the very high no-load power dissipation.

Based on the same simple example of the shunt regulator, we can develop a relationship for output voltage fluctuations in the form of,

$$\Delta V_o = k_1 \Delta I_L + k_2 \Delta V_s + k_3 \Delta T$$ (1.7)

where the coefficient k_1 represents the Thevenin resistance, R_o, of the circuit, $k2$ represents the coefficient representing line regulation, and k_3 represents the temperature coefficient of the power supply. If you take the simple example in Figure 1.3(a) for a regulated power supply, when the load current is I_L, output voltage V_o can be written as

$$V_o = V_{OC} - R_o I_L$$ (1.8)

For the simplest case of Figure 1.3(a),

$$V_o = V_z \left[\frac{1}{1 + \frac{r_z}{R_s + R_l}} \right] + V_s \left[\frac{1}{1 + \frac{R_s + R_l}{r_z}} \right] - \frac{r_z}{1 + \frac{r_z}{R_s + R_l}} I_L$$ (1.9a)

This gives,

$$\Delta V_o = \left[\frac{1}{1 + \frac{r_z}{R_s + R_l}} \right] \Delta V_z + \left[\frac{1}{1 + \frac{R_s + R_l}{r_z}} \right] \Delta V_S - \frac{r_z}{1 + \frac{r_z}{R_s + R_l}} \Delta I_L$$ (1.9b)

assuming that r_z is a constant value.

For this case,

$$k_{1.} = -\frac{r_z}{1 + \frac{r_z}{R_s + R_l}}; \quad k_2 = \frac{1}{1 + \frac{R_s + R_l}{r_z}}.$$

If the temperature effects on the breakdown voltage can be simplified by $V_z(T) = V_{z,nom}(1+k_T\Delta T)$ where T is the absolute temperature, k_T is the temperature coefficient of the zener, related to the nominal zener voltage, $V_{z,nom}$, at a specified temperature.

$$k_3 = \frac{1}{1+\frac{r_z}{R_s+R_1}}k_T \qquad (1.10)$$

The above discussion leads us to consider selection of devices with appropriate data sheet parameters to get the best possible specifications. For example, if we can select a diode with low r_z compared to the total input path resistance of $(R_S + R_1)$, line regulation coefficient, k_2, can be minimized.

While the above discussion leads us an example to developing suitable input-output relationships for the regulated power supply, if one requires better output regulation with higher output current capability, more advanced circuit configurations are required. Figure 1.4 indicates two examples of more improved shunt regulator circuits. In Figure 1.4(a), R_1 and Z_1 act the same as in Figure 1.3, where R_2 and Z_2 act as a pre-regulator. In effect the preregulator provides a lower source resistance to the pair R_1 and Z_1. Relationships for this circuit can be developed using approximations applied to Equations (1.5) and (1.6). For example, if the preregulator can be developed with the useful relationships as applied to Figure 1.3(a), the second stage sees very approximate Thevenin equivalent circuit with

$$V_{OC,prereg} \approx V_{z2}; \quad \text{and} \quad R_{o,prereg} \approx r_{Z2} \qquad (1.11)$$

In effect this creates a condition where the regulator stage sees a near constant input and with a lower source resistance. If we can select the diode Z_1 suitably, we get a more precisely regulated output. More discussion on this kind of circuits is in [1].

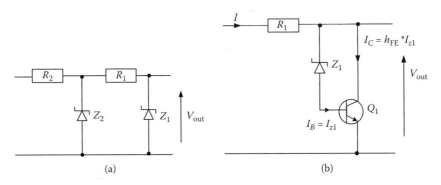

(a)   (b)

FIGURE 1.4 Improved shunt regulator circuits: (a) a circuit with a better load and line regulation; (b) a circuit with higher current capability.

When we need to achieve a higher output current capability, a transistor can be added to the basic circuit in Figure 1.3(a), resulting in the case of Figure 1.4(b). For this case, when the transistor is kept in the active mode,

$$V_{out} \approx V_Z + V_{BE} \tag{1.12}$$

As the load sees the collector and emitter of the transistor, maximum load current is given by

$$I_{L,max} \approx (h_{FE}I_{Z1}) \approx \left[\frac{V_s - V_Z - V_{BE}}{R_S + R_1} \right]\left[\frac{h_{FE}}{h_{FE} + 1} \right] \tag{1.13}$$

This clearly indicates that it improves the capability by a factor nearly equal to the gain of the transistor.

1.3.1 Series Regulators

In all the above shunt regulator circuits, when the output current is zero the transistor/ zener dissipates lot of heat, and these circuits are generally used for low current requirements. In general, series regulator concepts were more attractive to industrial applications and consumer electronics, except for their drawback of low efficiency. Figure 1.5(a) depicts a very simple open-loop-type linear regulator. In general, the output regulated voltage V_{Out} can be approximated by,

$$V_{Out} = V_Z - V_{BE} \tag{1.14}$$

Under maximum load condition, if we need to maintain the zener diode voltage at the breakdown value, with a minimum current of $I_{Z,min}$, current through resistor R_1 under maximum load current will be

$$I_{R1} \approx \frac{I_{L,max}}{h_{FE}} + I_{Z,min} \tag{1.15}$$

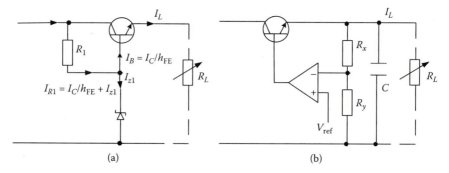

(a)   (b)

FIGURE 1.5 The basic linear regulator configurations: (a) an open-loop type; (b) a closed-loop type based on an op-amp in the control loop.

If the load is removed, the base current gets driven into the zener diode and under that situation, worst-case current through the resistor R_1 will be,

$$I_{R1,worstcase} \approx \frac{V_{in,max} - V_z}{R_1 + r_z} \approx I_{z,max} \qquad (1.16)$$

If you neglect the zener diode internal resistance,

$$I_{R1,worstcase} \approx I_{z,max} \approx \frac{I_{L,max}}{h_{FE}} + I_{Z,min} \qquad (1.17)$$

In this situation, approximate Thevenin's equivalent circuit parameters will be

$$V_{oc} \approx \frac{V_Z \cdot R_1 - V_{in} r_Z}{R_1 + r_Z} - V_{BE} \qquad (1.18)$$

and

$$R_o \approx (R_1 \,/\!/\, r_Z) + r_e \qquad (1.19)$$

where r_e is one of the transistor T model circuit parameters.

$$V_o = -\left(\frac{R_1 \cdot r_Z}{R_1 + r_Z} + r_e \right) I_L - \left(\frac{r_Z}{R_1 + r_Z} \right) V_{in} + \left(\frac{R_1}{R_1 + r_Z} V_z - V_{BE} \right) \qquad (1.20a)$$

and

$$I_E = I_L = \frac{I_S}{\alpha} e^{\frac{V_{BE}}{V_T}} \qquad (1.20b)$$

which gives

$$V_{BE} = V_T \ln\left(\alpha \cdot \frac{I_L}{I_S} \right) = \frac{kT}{q} \ln\left(\alpha \cdot \frac{I_L}{I_S} \right) \qquad (1.20c)$$

where I_S is the saturation current and V_T is the thermal voltage of the B-E junction. Considering the temperature coefficient of the zener, k_T,

$$V_o = -\left(\frac{R_1 \cdot r_Z}{R_1 + r_Z} + r_e \right) I_L - \left(\frac{r_Z}{R_1 + r_Z} \right) V_{in} + \left(\frac{R_1}{R_1 + r_Z} V_z (1 + k_T T) - \frac{kT}{q} \ln\left(\alpha \cdot \frac{I_L}{I_S} \right) \right) \qquad (1.21)$$

Given the above simplified analysis, we see that the regulated output has a severe dependency on the value of the V_{BE} and the impact of the base current variations under load conditions. This leads to a case of coefficients in Equation (1.7) given by,

$$k_1 \approx -\left(\frac{R_1 \cdot r_z}{R_1 + r_z} + r_e \right) \quad k_2 \approx -\left(\frac{r_z}{R_1 + r_z} \right) \quad k_3 \approx \left(\frac{k_T R_1}{R_1 + r_z} \right) V_z - \frac{k}{q} \ln \left(\alpha \cdot \frac{I_L}{I_S} \right) \quad (1.22)$$

where I_S is the saturation current and V_Z is the nominal voltage of the Zener diode.

Given the above simplified analysis, we see that the regulated output has a severe dependency on the value of the load current also, in addition to the zener diode's temp performance.

Having discussed the basic behavior of an open-loop series regulator, we can use Figure 1.5(b) to illustrate the basic elements of a closed-loop linear regulator, which can minimize some of these issues. Similar to the case in open-loop series regulator, the output is regulated by controlling the voltage drop across the series-pass element, a power transistor biased in the linear region. The control circuit compares the sample of the output voltage with a reference source, and changes the on-resistance of the series-pass power transistor.

If we consider the op-amp to be ideal, and the reference voltage to be constant at V_{ref}, as far as the op-amp could maintain its basic function, V_{out} can be given by

$$V_{out} = V_{ref} \left(1 + \frac{R_x}{R_y} \right) \quad (1.23)$$

Given this simple relationship, we see that the circuit behaves much better than the previous cases of linear regulators, as far as the op-amp, and the reference sources are considered ideal. If a nonideal op amp with an open-loop gain of A_{OL} and input resistance between the inverting and noninverting inputs is R_{in}, we can develop the following relationship for an output load current of I_L as

$$V_{Out} = \left(1 + \frac{R_Y}{R_X} \right) V_{ref} - V_{BE} \left(\frac{1}{A_{OL}} \left(1 + \frac{R_X}{R_Y} \right) \frac{(1 + R_X // R_Y)}{R_{in}} \right) \quad (1.24)$$

The power dissipation in the linear regulator is a function of the difference between the input and the output voltage, output load current, and power consumed by control circuits. The power dissipation in the series pass device contributes largely to lower the efficiency of linear regulators compared to switching regulators. Efficiency of a linear regulator can be approximated by

$$\eta \approx \frac{V_{out} I_L}{V_{in} I_{in}} \approx \frac{V_{out} I_L}{V_{in} (I_L + I_{Control})} \approx \frac{V_{Out}}{V_{in} (1 + \frac{I_{Control}}{I_L})} \quad (1.25)$$

$I_{Control}$ is the current drawn by the control circuits, referred to the input side. This current is sometimes called the ground pin current (particularly in cases such as the three terminal regulators). This indicates to us that the best possible theoretical efficiency in a linear regulator is given by

$$\eta_{max} \approx \frac{V_{out}}{V_{in}} \qquad (1.26)$$

The major advantages of linear regulators in comparison with switching regulators are their (a) low noise, (b) transient response to load current fluctuations (output current slew rate), (c) design simplicity, and (d) low cost. However, due to the low efficiency of these circuits they are not attractive to high power requirements with wide differential voltage between the input and output sides. The following sections discuss the specifics of the essential components of a series linear regulator circuit.

1.3.1.1 Series Pass Device

There are many different options for a series pass device of a linear regulator circuit, either in the form of a discrete design or in the monolithic IC form. Table 1.1 compares the characteristics of these options, as applicable to integrated circuits. In discrete form of circuits, the current capability could be much higher than the values in Table 1.1, but requires large heat sink to keep the series pass device within safe temperature limits. Two possible circuit configurations are given in Figure 1.6.

1.3.1.2 Control Circuits

The control circuit samples the output voltage through a resistive divider, and uses this feedback signal to control an error amplifier's output to control the resistance of the series pass device. Control circuit characteristics directly affect system bandwidth and the achievable DC regulation. The voltage reference is used for comparison of the output voltage in the control circuit, and primarily governs the steady-state accuracy of the device. The control circuit can be based on an op-amp or a circuit designed with discrete components. This directly governs the output transient response and the stability of the output. More on this will be discussed later. In general the designer should be conscious of the power consumption of the control circuits to get the best efficiency.

Any overcurrent or thermal protection needs to be incorporated into the control circuits, and Figure 1.7 indicates a general block diagram.

1.3.1.3 The Output Capacitor

The bulk capacitance at the output maintains the output during transients. The output capacitor is required in order for the design to meet the specified transient requirements. As with any control system, the voltage loop has a finite bandwidth and cannot instantaneously respond to a change in load conditions. The supply rail for many of today's microprocessors cannot vary more than ±100 mV while handling load transients of the order of 5 A with 20 ns rise and fall times. This translates to a current slew rate of 250 A/μs [2].

TABLE 1.1   Pass Transistor Configurations and Their Comparison in Linear Regulators

Configuration	Single NPN	Darlington NPN	Single PNP	PNP/NPN combination	p-MOSFET
Minimum dropout voltage	$\approx 1\,V$	$\approx 2\,V$	$\approx 0.1\,V$	$\approx 1.5\,V$	$\approx R_{DS}(on)I_L$
Load current capability	$<1\,A$	$>1\,A$	$<1\,A$	$>1\,A$	$>1\,A$
Output impedance	Low	Low	High	High	High
Bandwidth	Wide	Wide	Narrow	Narrow	Narrow
Effect of load capacitance	Immune	Immune	Sensitive	Sensitive	Sensitive

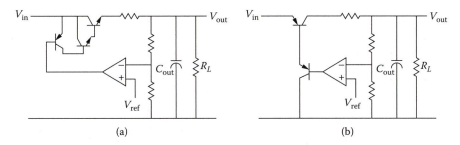

FIGURE 1.6 Two possible linear regulator configurations: (a) with an NPN Darlington pair for higher output current capability; (b) with a PNP series pass transistor.

In order to keep the output voltage within the specified tolerance, sufficient capacitance must be provided to source the increased load current throughout the initial portion of the transient period. During this time, charge is removed from the capacitor, and its voltage decreases until the control loop can catch the error and correct for the increased current demand. The amount of capacitance used must be sufficient to keep the voltage drop within specifications. Design considerations in the selection of the capacitor value are detailed in [3].

1.3.1.4 Voltage References

1.3.1.4.1 *Voltage Reference Fundamentals*

A wide variety of voltage references are available today. The most common ones are based on the action of either a zener diode or a bandgap cell with additional circuitry included to obtain good temperature stability. Although discrete zener diodes are available in voltage ratings as low as 1.8 V to as high as 200 V, with power-handling capabilities in excess of 100 W, their tolerance and temperature characteristics are unsuitable for many

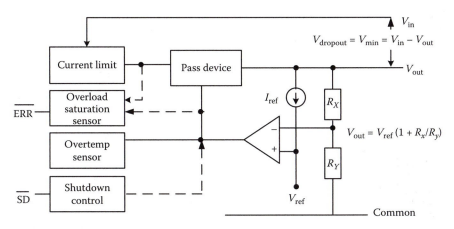

FIGURE 1.7 Generalized block diagram of a linear regulator with additional protection and external control inputs.

applications. Therefore discrete zener diode based references have additional circuitry to improve performance. One commonly used version is the temperature-compensated zener diode, particularly for voltages above 5 V.

The operation of a bandgap reference is based on specific characteristics of diodes operating at the same current but at different current densities. Bandgap references are available with output voltage ratings of about 1.2 V to 10 V. The principal advantage of these devices is their ability to provide stable low voltages such as 1.2, 2.5, or 5 V. However, bandgap references of 5 V and higher tend to have more noise than equivalent zener-based references. This is due to the fact that in bandgap references, higher voltages are obtained by amplification of the 1.2 V bandgap voltage by an internal amplifier. Their temperature stability is also below that of zener-based references.

In the commercial domain of semiconductors, there are several options today for voltage references such as (a) zener diodes, (b) buried zener diodes, (c) bandgap-based devices, and (d) XFET™ and FGA™. A comprehensive account of these technologies is available in [4].

1.3.1.4.2 Reverse-Biased Diode-Based Voltage References

The most common and simple way to achieve a reference source is to use a reverse-biased diode, or a zener diode as it is commonly called, where it enters into a voltage breakdown region. A zener diode has two distinctly different breakdown mechanisms: zener breakdown and avalanche breakdown. The zener breakdown voltage decreases as the temperature increases creating a negative temperature coefficient (TC). The avalanche breakdown voltage increases with temperature (positive TC). This is illustrated in Figure 1.8. The zener effect dominates usually below 5 V, and the avalanche effect dominates above 6 V. By the use of additional diode (in forward-biased mode) in series with an avalanche-type diode, it is possible to achieve a better temperature stability in a reference circuit [5].

1.3.1.4.3 Bandgap References

This is one of the popular solutions to achieve a very stable reference source in a regulator circuit. The concept behind this circuit is to have two base emitter junctions operating at

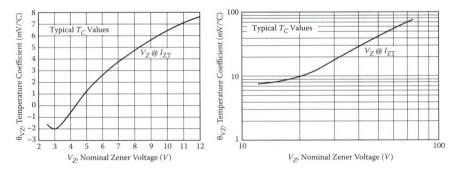

FIGURE 1.8 Typical example of temperature characteristics of reverse-biased silicon diodes: (a) Zener breakdown; (b) avalanche breakdown. (From Motorola-BZX series.)

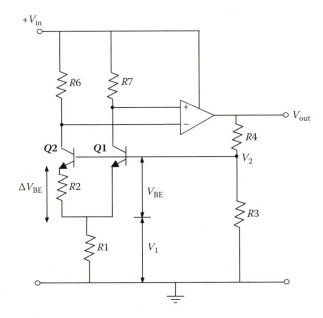

FIGURE 1.9 The circuit diagram of a bandgap reference.

different current densities [4] where temperature compensation can be easily maintained. A circuit diagram of a bandgap reference is shown in Figure 1.9. This circuit, developed by Paul Brokaw, is called Brokaw bandgap circuit. Transistors Q_1 and Q_2 are operating at the same current but at different current densities. This is achieved by fabricating Q2 with a larger emitter area than Q_1. Therefore the base-emitter voltages of the two transistors are different. This difference is dropped across R_2. Extrapolated to absolute zero, V_{BE} is equal to 1.205 V, the bandgap voltage of silicon, and has a predictable, negative temperature coefficient of –2 mV/°C. By adding a voltage to V_{BE} that has a positive temperature coefficient, a bandgap reference can, at least theoretically, generate a constant voltage at any temperature.

The base-emitter voltage difference is given by

$$\Delta V_{BE} = \frac{kT}{q} \ln\left(\frac{J_1}{J_2}\right) \tag{1.27}$$

where J_1 and J_2 are the current densities of transistors Q_1 and Q_2 respectively. Since the sum of the two transistor currents flows through R_1, the voltage across R_1 can be expressed as

$$V_1 = 2\left(\frac{R_1}{R_2}\right)\Delta V_{BE} \tag{1.28}$$

Also,

$$V_2 = V_{BE} + V_1 \tag{1.29}$$

Using Equations (1.28) and (1.29),

$$V_2 = V_{BE} + 2\left(\frac{R_1}{R_2}\right)\Delta V_{BE} \tag{1.30}$$

Therefore V_2 is the sum of V_{BE} and the scaled ΔV_{BE}. It is shown that if ratio of the emitter areas of the two transistors is eight, the temperature coefficients of V_{BE} and ΔV_{BE} cancel each other. The op-amp raises the bandgap voltage V_2 to a higher voltage at the output of the reference. There are many variations of this basic circuit in commercial bandgap references by Analog Devices Inc., USA, and readers can refer to their application notes for details.

Bandgap references typically provide voltages ranging from 1.2 V to 10 V. The advantage of bandgap references is their ability to provide voltages below 5 V. The greatest appeal of bandgap devices is the ability to function with operating currents from milliamps down to microamps. Commercial IC bandgap references have additional features such as multiple calibrated voltages. Because most bandgap references are constructed in monolithic form, they are relatively inexpensive. However, their temperature coefficient could be sometimes inferior to that of temperature-compensated zener-based references. This is due to second-order dependencies of ΔV_{BE} on temperature.

1.3.1.4.4 Buried-Zener References

The above two types of common reference sources have their own advantages and disadvantages. Another development to compete with disadvantages of these types was the buried-zener reference where some process improvements were used to get a lower noise and improved stability. Figures 1.10(a) and 1.10(b) depict some comparison of the device structure in relation to a regular zener diode. The device comes with a heating element to stabilize the temperature as shown in the commercial example of LM199 from National Semiconductor. Due to the difference in construction, it has achieved far superior performance, which can be summarized by,

- Very low initial error, between 0.01% and 0.05%
- Ultralow temperature coefficient, from 0.05 to 10 ppm/°C

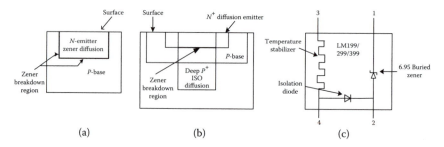

FIGURE 1.10 Buried zener diode: (a) ordinary zener diode; (b) buried version of the zener diode; (c) schematic showing the heating element for temperature stabilization in LM199/299/399.

- Ultralow noise level of less than 10 µV peak to peak, in the frequency band of 0.1 to 10 Hz
- Long-term stability of typically less than 25 ppm/1000 hours

More details can be found in [4], with historical developments occurring in Silicon Valley, USA. This will also provide more details on other state-of-the-art devices such as the XFET™ and Intersil/Xicor FGA™ types. Reference [5] provides some comparison of zener device families and bandgap families commonly available. Appendix A provides an overview of XFET™.

1.3.1.4.5 Quality Measures of Voltage References

An ideal voltage reference would have the exact specified voltage, and it would not vary with time, temperature, input voltage or load conditions. However, as it is not possible to fabricate such ideal references, manufacturers provide specifications informing the user of the device's important quality parameters.

1.3.1.4.5.1 Output Voltage Error   This is the initial untrimmed accuracy of the reference at 25°C at a specified input voltage. This is specified in millivolts or a percentage. Some references provide pin connections for trimming their initial accuracy with an external potentiometer.

1.3.1.4.5.2 Temperature Coefficient   The temperature coefficient of a reference is its average change in output voltage as a function of temperature compared with its value at 25°C. This is specified in ppm/°C or mV/°C.

1.3.1.4.5.3 Line Regulation   This is the change in output voltage for a specified change in input voltage. Usually specified in %/V or µV/V of input change, line regulation is a measure of the reference's ability to handle variations in supply voltage.

1.3.1.4.5.4 Load Regulation   This is the change in output voltage for a specified change in load current. Specified in µV/mA, %/mA, or ohms of DC output resistance, load regulation includes any self-heating effects due to changes in power dissipation with load current.

1.3.1.4.5.5 Long-Term Stability   This is the change in the output voltage of a reference as a function of time. Specified in ppm/1000 hrs at a specific temperature, long-term stability is difficult to quantify. As a result, manufacturers usually provide only typical specifications based on device data collected during the characterization process.

1.3.1.4.5.6 Noise   Although the above are the most important quality parameters of a voltage reference, noise is particularly of importance in certain applications such as A/D or D/A converters. In such applications, the noise from the reference should be less than 10% of the LSB value of the converter. Therefore the higher the resolution of the converter, the lower should be the noise generated from the reference. Noise depends on the operating current of the reference, and is generally specified over a particular bandwidth and for a particular current. The specified bandwidths are 0.1–10 Hz (low-frequency noise) and 10 Hz–10 kHz (high-frequency noise).

1.4 Low-Dropout Regulators

As we discussed in section 1.3.1, if we can develop a linear regulator with minimized current consumption in the control circuits, and maintain the difference between the input and output voltages at a very low value, the circuit will be very efficient and will have all the valuable specifications of a linear regulator. Using the approximation in Equation (1.26), if we consider a linear regulator circuit with 5 V input and 3.3 V output, the efficiency will be around 66%. If the input is 3.5 for the same output, the theoretical best efficiency can be close to 94%, which is very much better than the efficiency of common switching regulators.

Based on the simple concept discussed above, to power modern portable devices such as cell phones, notebooks, and PDAs, a unique category of linear regulator ICs, low-dropout (LDO) regulators, have emerged during the last two decades. These were used in tandem with switching regulators to power noise-sensitive mixed-signal circuits, RF circuit blocks, and other noise-sensitive circuits. LDOs are available in a wide variety of output voltages and current capacities. Many LDOs are tailored to applications where a good response to a fast-step current transient is important. These devices have captured a large share of power management ICs in the early part of this decade [6]. This kind of a power supply solution is very helpful in low voltage rail based where load current can change rapidly, creating very high current slew rates [7].

1.4.1 Basic Concept of an LDO

Figure 1.11 depicts the simplified block diagram of an LDO regulator IC. The main components are the pass element, precision reference, feedback network, and error amplifier. The input and output capacitors are the only key components of an LDO solution that are not contained within the monolithic LDO. Table 1.1 compares different options available for the pass transistor in a linear regulator and the advantages and disadvantages of the approaches. This is more applicable to modern LDO ICs. In discussing the details and design approaches to LDO-type regulators, let's start from the simple

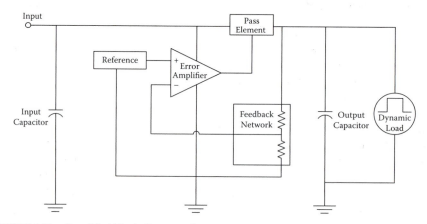

FIGURE 1.11 Simplified block diagram of an LDO.

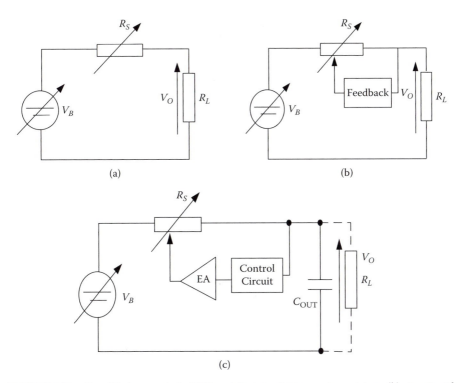

FIGURE 1.12 Simplified concepts in LDO architecture: (a) two series resistors; (b) circuit with feedback; (c) use of control circuit and an error amplifier to implement the control loop.

concept of Figure 1.12(a). If we consider the case of battery input into an LDO where the voltage fluctuates as the load current varies, or as the battery drains the series resistance increases, it can be shown that to keep the output voltage, V_o, constant,

$$R_s = R_L \left(\frac{V_B - V_o}{V_o} \right) \tag{1.31}$$

where V_B and R_S are battery voltage and the resistance due to the LDO series pass element respectively, while R_L represents the load resistance. We can represent the same relationship as

$$R_s = \left(\frac{V_{LDO}}{V_o} \right) R_L \tag{1.32}$$

where V_{LDO} is the voltage across the series pass element. As depicted in Figure 1.12(b), if we have a feedback circuit, the feedback circuit is expected to keep controlling the value of R_S when V_{LDO} fluctuates. As shown in Table 1.1, we can use any configuration of transistors as the series element, and in most new commercial LDOs MOS field effect

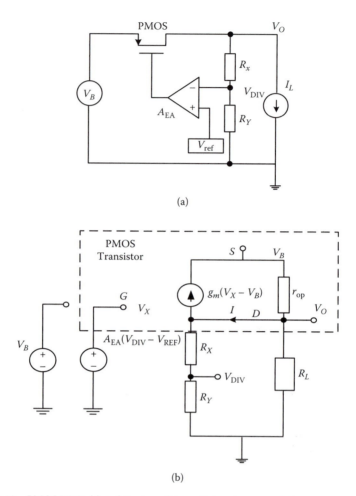

FIGURE 1.13   PMOS LDO: (a) architecture; (b) small signal representation.

transistors are used. Figure 1.12(c) indicates the feedback loop arrangement with an error amplifier, which could be easily implemented by using an op-amp.

The basic PMOS LDO topology shown in Figure 1.13(a) comprises an error amplifier that has the output voltage of V_X and gain of A_{EA}. The power transistor can be represented using the small signal equivalent model of transconductance g_m and output resistance r_{op} as shown in Figure 1.13(b).

According to the above representation in Figure 1.13(b), current I can be written as

$$I = g_m(V_X - V_B) + \frac{(V_o - V_{Div})}{R_X} \tag{1.33}$$

and

$$I = \frac{(V_B - V_o)}{r_{op}} - \frac{V_o}{R_L} \tag{1.34}$$

Also

$$V_X = A_{EA}(V_{DIV} - V_{REF})$$ (1.35)

Using Equations (1.33), (1.34), and (1.35),

$$g_m A_{EA} V_{REF} = V_o\left(\frac{1}{R_L} + \frac{1}{r_{op}} + \frac{1}{R_1}\right) - V_{DIV}\left(g_m A_{EA} + \frac{1}{R_X}\right) - V_B\left(g_m + \frac{1}{r_{op}}\right)$$ (1.36)

and

$$V_{DIV}\left(\frac{1}{R_X} + \frac{1}{R_Y}\right) - \frac{V_o}{R_X} = 0$$ (1.37)

Solving Equations (1.34) and (1.35), V_O becomes

$$V_o = \frac{V_B(1 + A_{PT})\beta + V_{REF} A_{PT} A_{EA}\beta}{\beta[1 + \beta A_{PT} A_{EA} + \frac{r_{op}}{R_L}]}$$ (1.38)

where

$$A_{PT} = g_m r_{op}, \qquad \beta = \frac{R_Y}{R_X + R_Y},$$

and $R_X + R_Y \gg R_L$.
   Therefore

$$V_o = \frac{V_B A_{PT}\beta}{\beta(1 + \beta A_{PT} A_{EA})} + \frac{V_{REF} A_{PT} A_{EA}\beta}{\beta(1 + \beta A_{PT} A_{EA})}.$$ (1.39)

1.4.2 Important Parameters of LDOs

1.4.2.1 Dropout Voltage

This is the minimum voltage difference allowed for the series pass element before the regulator goes out of regulation. Usually a PMOS transistor-based version allows very low value and hence a high efficiency.

1.4.2.2 Input Voltage Range

This is the range of input voltages where the LDO remains in regulation. Lower value depends on the dropout voltage, while the higher end depends on the process capability, the heat sinking requirements, etc.

1.4.2.3 Regulated Output Voltage Range

This is the range of output voltage when the LDO is in regulation under steady-state conditions. However, when the output load current changes fast, transient over- or under-voltage conditions may occur, which will exceed these limits for short durations.

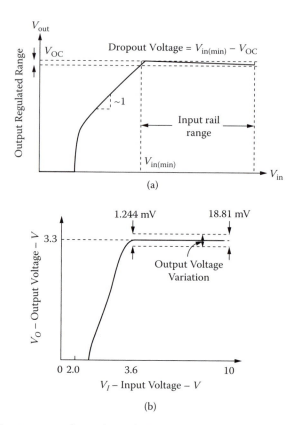

FIGURE 1.14 Input output relationships of LDOs: (a) operation regions and parameters; (b) performance of a typical commercial LDO from Texas Instruments. (From Lee, B. S., Technical Texas Instruments, Application Report SLVA072, 1999. Adapted.)

Figure 1.14(a) indicates the relationship between the input and output voltages, indicating the limits of regulation. Figure 1.14(b) indicates the typical performance of an LDO such as TPS76333 from Texas Instruments, indicating the practical behavior of a 3.3 V output LDO.

There are many practically useful specifications defining the LDO characteristics. These are summarized in Table 1.2.

1.4.3 Application and Design Implications

A common application area of LDOs is the portable products where processors are coupled with many mixed-signal circuitries. In these circumstances two important specifications of a DC power supply become very dominant. These are the output noise and the transient response. Many processors frequently go through sleep and wakeup type sequences where the load currents vary suddenly from very low values to near maximum. These transitions in state-of-the-art products could generate current variations with current slew rates in the range of 10 A/µs to over 250 A/µs. Typical LDO

TABLE 1.2   Important Secondary Specifications of LDOs

Specification	Description	Remarks
Output current range	Output current handling capability of the LDO	Minimum value depends on the stability of the output voltage at low current Maximum value depends on the safe-operating area (SOA) of the series pass device
Load/line transient regulation	A measure of speed of the LDO when the line voltage or load current fluctuates very fast	Usually measured as a margin of allowed variation of the regulated output An important parameter in processor power supplies with high current slew rates
Power supply rejection	This is the ability of the LDO to reject AC ripple on the input side	
Short-circuit current limit	Current drawn from the power supply when the output is short circuited	Lower limit is determined by the maximum regulated output current Upper limit is determined by SOA of the pass transistor
Output capacitor range	Value of the output capacitance to operate within the stability range	Most electrolytic capacitors have wide dependence on the temperature. Stability can be compromised due to this situation There are commercial LDO ICs that can also accept any capacitor value
Overshoot	At startup or during load current transients, output voltage may overshoot. This should be within a maximum limit	

load transient performance for a chip such as TPS763650 from Texas Instruments is shown in Figure 1.15(a), while steady-state performance is shown in Figure 1.15(b). In Figure 1.15(a), ΔV_{LDR} indicates the steady-state response, which is the same value depicted in Figure 1.15(b), but transient changes can always exceed this value as depicted in Figure 1.15(a). In a well-designed LDO-based power supply, this transient fluctuation needs to be minimized.

In order to improve on the transient behavior, there are many improvements incorporated into LDO chips. One such example is the use of a fast transient loop in LDOs such as TPS75433 from Texas Instruments for low and high currents [9].

In the majority of LDOs and quasi-LDOs (where a composite NPN-PNP pair is used as the pass device), the pass device or the driver is a lateral PNP. Even though a PNP is better at providing a lower dropout voltage than an NPN [10], a lateral PNP is a low-frequency cutoff device with a poor transient response. For this reason, proper selection of the external output capacitor is important for the stability of the loop and

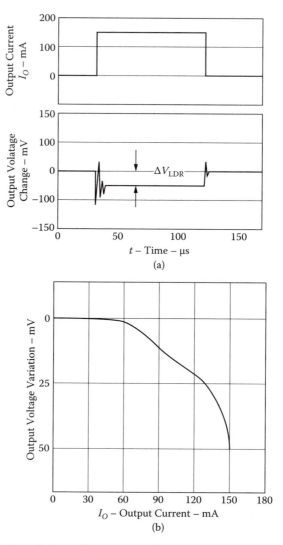

FIGURE 1.15 Load regulation and transient performance of a typical LDO-TPS76350 from Texas Instruments: (a) load transient response; (b) steady-state performance. (Adapted from Lee, B.STexas Instruments, Application Report SLVA072, 1999.)

adequate transient response. The compensation capacitor determines three key characteristics of an LDO: startup delay, load transient response, and loop stability. The startup time is approximately given by

$$T_{startup} = CV_o / I_{limit} \tag{1.40}$$

where C is the value of the output capacitor and I_{limit} is the current limit of the regulator. If C is fully discharged before the regulator is powered up, the regulator will limit current during startup, and the time to reach the nominal V_o will be delayed. Conversely, if C is too small, the output voltage will overshoot the nominal V_o during startup. Because it is impossible to investigate all three characteristics at once, the designer should concentrate on first achieving a stable loop design and then check the startup delay and load transient response. In general if a single pole system can be created, and if the crossover frequency is selected to ensure that the system can quickly respond to load transients without undue ringing at the output, the design will be stable. For stability, the phase margin should be more than 45°. Unfortunately, an LDO has three dominant poles, and two are set by the regulator IC and the third is a function of the load and the output capacitor. The first pole, determined by the error amplifier, generally occurs between 10 and 300 Hz; the second pole, due to the pass device (or the PNP bias device of a compound regulator), is usually between 100 and 300 kHz. The third pole, set by the load and the output capacitor, occurs within the same range as the error amplifier or even slightly lower at light loads. Figure 1.16(a) shows the simplified case of a load and output capacitor combination.

It can be shown that the pole and the zero created by the load are given by

$$f_{pL} = \frac{1}{2\pi(R_L + ESR)C} \tag{1.41a}$$

and

$$f_{zL} = \frac{1}{2\pi(ESR)C} \tag{1.41b}$$

Based on the discussion in chapter 5, section 5.5, where added poles and zeros change the Bode plot, it is apparent that the zero due to capacitor ESR modifies the total response of the circuit. Figure 1.16(b1) shows the case where the output is marginally stable for ESR = 3.0 Ω. As depicted in Figure 3.16(b2), when the ESR is reduced to 1.0 Ω, the system's phase margin increases and the system becomes stable. When the ESR is lowered further, the system can become unstable, as in Figure 1.16(b2). The capacitor used at the output should have some stability within the operational temperature ranges. Figure 5.16 shows typical aluminum electrolytic capacitor characteristics over frequency and temperature. Based on the discussion related to Figure 1.16, it is important for designers to carefully examine the parameter changes of capacitors over frequency and temperature to achieve a stable design. Details can be found in King [9], O'Malley [11], Simpson [12, 13], and Goodenough [14].

1.4.4 LDO Applications and Development Directions

LDOs have gained popularity with the growth of portable battery-powered devices. Many circuit blocks in the portables such as cellular phones, cameras, and laptops have many noise-sensitive mixed-signal components, which may not tolerate the RFI/EMI

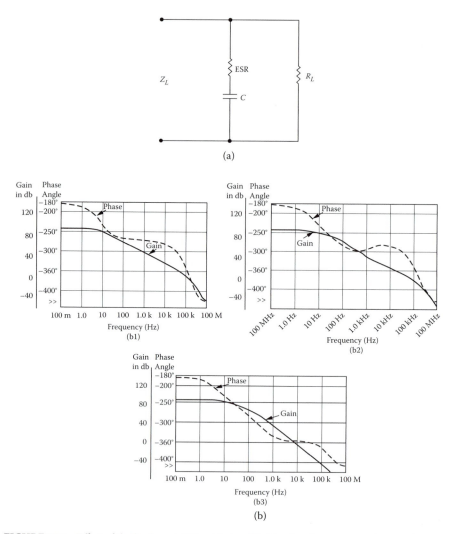

FIGURE 1.16 Effect of the load on stability: (a) simplified load and output capacitor combination; (b1) Bode plot of the output for a practical LDO based on a regulator IC such as CS 8156 from ON Semiconductor for different cases of ESR values: (b1) RO = 120 Ω and C = 22 μF with ESR = 3 Ω; (b2) RO = 120 Ω and C = 22 μF with ESR = 1 Ω; (b3) RO = 120 Ω and C = 22 μF with ESR = 0.01 Ω. (Courtesy of O'Malley, Application Note SR003AN/D, rev. 1, ON Semiconductor, Phoenix, AZ, 2001.)

issues of switching regulators. In these circumstances, LDOs or tandem combinations of LDO and switch-mode regulator are the only practical solution, provided that the efficiency issue can be managed. In most portable devices, LDOs are widely used as they occupy a very small PCB area and do not use any bulky parts such as inductors and the like. LDOs are particularly attractive in systems with noise-sensitive system-on-chip (SoC) applications where battery power is used. LDOs also find applications in automotive environments because of the rapid voltage changes of the 12 V rail during cold startup [15]. Most LDOs are used in powering high-power processors where load current changes in step mode with high current slew rates. Schiff [16] and Rincon-Mora and Allen [17] provide design guidelines to deal with these conditions. With the initiatives to incorporate more commercial-off-the-shelf (COTS) components into military systems, some companies such as Linear Technology and others have developed high-reliability military plastic (MP) packaged LDOs with reverse voltage protection and current limiting over the full range of military operating temperatures [18]. Some devices such as LT3070 from Linear Technology could supply 5 A of load current at digitally programmable voltages from 0.8 V to 1.8 V with dropout voltages as low as 85 mV are examples of these MP-packaged devices. Details on frequency compensation of LDOs are available in Kwok and Mock [19] and Chava and Silva-Martinez [20]. For applications with extra low LDO voltages, ultra-low-dropout (ULDO) linear regulators based on bipolar CMOS-DMOS (BCD) technologies are available [21].

1.4.5 Low-Noise Application Requirements and Noise Measurements for LDO Output

Some of the LDO regulators are specially designed for low-noise requirements [22] within cellular handsets and other portable applications, because most switch-mode power supplies are too noisy for these applications. The noise performance of these components sometimes needs to be quantified, and special measurement setups may be necessary. In this process one should ensure that the LDO meets the system's noise requirement within the entire bandwidth of interest, typically in the range of 10 Hz to 100 kHz. Figure 1.17 indicates a suitable filter structure for testing the noise performance of LDOs in this frequency band. In LDO noise measurement, special consideration should be given to ground loop elimination; hence, all power supplies should be battery based, and thermally responding RMS meters should be used for measurements [23]. General performance verification of LDOs is discussed in Williams and Owen [24].

1.4.6 Adjustable Output LDO Circuits

For applications where nonstandard voltages are required, an adjustable LDO is a good choice, but getting the highest accuracy from such an IC may require a few circuit tricks. Figure 1.18 shows a few examples, including the use of an adjustable reference for improving accuracy [25]. For applications with hot-swap requirements, LDO ICs can be used with special current limiting arrangements [26].

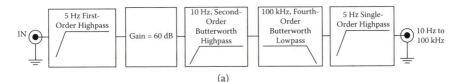

(a)

FIGURE 1.17 Noise measurements for LDOs: (a) block diagram of a filter arrangement; (b) a typical circuit configuration. (Courtesy of Williams, J., and T. Owen, *EDN*, May 11, 2000, 149.)

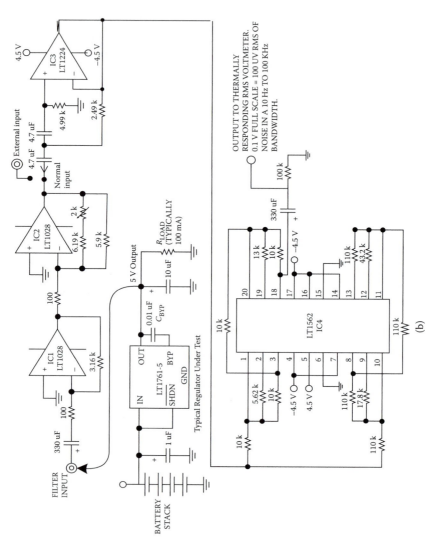

FIGURE 1.17   (Continued)

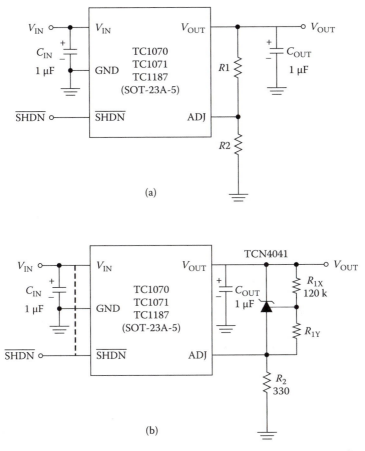

FIGURE 1.18 Adjustable LDO circuits: (a) a simple circuit with two resistors to adjust the output; (b) use of an adjustable reference source for improving accuracy. (Courtesy of Paglia, *EDN*, September 1, 1998.)

1.4.7 Battery-Powered Applications and PMOS-Based LDOs

For battery-powered applications, PMOS-based LDOs provide acceptable solutions. The factors to be considered include dropout voltage, ground current, noise, input voltage, and thermal response. Typical ground current components in an LDO are shown in Figure 1.19(a). Figures 1.19(b) and 1.19(c) show the comparative performance of typical PNP LDOs and PMOS-based LDOs. For details, see Christ [27]. Given the demand from many portable battery-powered applications, there is a considerable research effort on LDOs continuing at universities, and some of these are reflected in References [28–32]. References [33] and [34] provide practical design guidelines for end users, including some thermal design aspects [34].

Another serious possibility is to have very high end-to-end efficiency-based supercapacitor enhancements to LDO-based linear regulator systems [35–37].

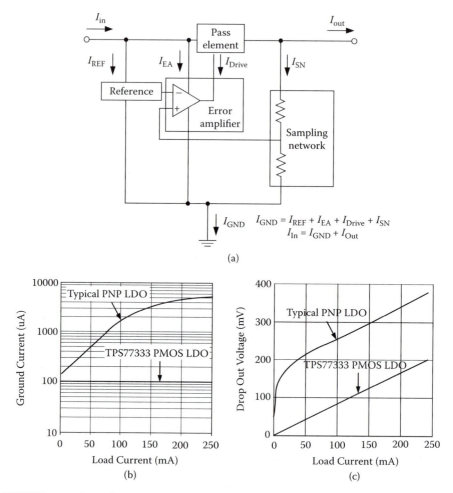

FIGURE 1.19   Ground currents in an LDO and comparison of PNP and PMOS types: (a) ground currents; (b) comparison of ground currents in PNP and PMOS types; (c) comparison of dropout voltage in PNP and PMOS types. (Adapted from Christ, M., *Electronic Engineering*, February 2001, 61.)

References

1. Nowiki, J. R. 1973, Power Supplies for Electronic Equipment London, Leonard Hill books.
2. Goodenough, Frank. 1996. LDO controller handles 250A/μs load transients. *Electronic Design*, November 18, 162–66.
3. O'Malley, K. 1994. Understanding linear regulator compensation. *Electronic Design*, August 22, 123–28.
4. Harrison, L. T. 2005. *Current sources and voltage references*. Burlington, MA: Newnes.

5. Kularatna, N. 2000. *Modern component families and circuit block design*. Burlington, MA: Newnes.

6. Inouye, S., M. Robles-Bruce, and M. Scherer. 2010, August. The 2010 power management—general purpose analog service. Databeans Incorporated.

7. Strik, S., and V. Strik. 2008. LDO design solves load transient problems. *EE Times Asia*, October 1–15, 1–3.

8. Lee, B. S. 1999. Technical review of low dropout regulator operation and performance. Texas Instruments, Application Report SLVA072.

9. King, B. M. 2000. Optimized LDO response to load transients requires the appropriate output capacitor and device performance, *PCIM*, September, 39.

10. Lee, M. 1989. Linear PNP regulator outperforms NPN types. *PCIM*, May, 36.

11. O'Malley, K. 2001. Compensation for linear regulators. Application Note SR003AN/D, rev. 1. ON Semiconductor, Phoenix, AZ.

12. Simpson, C. 2000. Linear regulators: Theory and operation and compensation. Application Note 1148. National Semiconductor, Santa Clara, CA.

13. Simpson, C. 1996. LDO regulators require proper compensation. *Electronic Design*, November 4, 99.

14. Goodenough, F. 1996. LDO controller handles 250A/μs load transients. *Electronic Design*, November 18, 162.

15. Ciscato, S. 1997. Low dropout voltage regulators survive in the automotive environment, *PCIM*, June, 10.

16. Schiff, T. 2001. High-power, high speed processors need stable, fast linear LDO regulators. *PCIM*, January, 57.

17. Rincon-Mora, G. A., and P. E. Allen. 1998. A low-voltage, low quiescent current, low dropout regulator. *IEEE Journal of Solid State Circuits* 33:36.

18. Knoth, S. 2010. Military-grade LDOs battle harsh realms. *Defense Electronics*, June/July, S26–S28.

19. Kwok, K. C., and P. K. T. Mok. 2002. Pole-zero tracking frequency compensation for low dropout regulator. *IEEE International Symposium on Circuits and Systems* 4:IV–735.

20. Chava, C. K., and J. Silva-Martinez. 2002. A robust frequency compensation scheme for LDO regulators. *IEEE International Symposium on Circuits and Systems* 5:V–825.

21. Bontempo, G., T. Signorelli, and F. Pulvirenti. 2001. Low supply voltage, low quiescent current ULDO linear regulator. *8th IEEE International Conference on Electronics, Circuits and Systems* 1:409.

22. Ali, I., and R. Griffith. 2000. A fast response programmable PA regulator subsystem for dual mode CDMA/AMPS handsets. *IEEE MTT-S International Microwave Symposium Digest* 1:139.

23. Williams, J., and T. Owen. 2000. Exacting noise test ensures low-noise performance of low-dropout regulators. *EDN*, May 11, 149.

24. Williams, J., and T. Owen. 2000. Performance verification of low noise low dropout regulators. Application Note AN83-1. Linear Technology Corporation, Milpitas, CA.

25. Paglia, P. 1998. Optimize output-voltage accuracy of adjustable low-dropout regulators, *EDN*, September 1, 105.

26. Wells, E. 1999. LDOs and hot swap power mangers. *PCIM*, January, 46.

27. Christ, M. 2001. Extending the battery life using PMOS LDOs. *Electronic Engineering*, February, 61.

28. Lin, Y. H, K. L. Zheng, and K. Chen. 2008. Smooth hole tracking technique by power MOSFET array in low-dropout regulators. *IEEE Transactions on Power Electronics* 20 (5): 2421–27.

29. Leung, K. N., and P. K. T. Mock. 2003. A CMOS voltage reference based on weighted ΔV_{GS} for CMOS low-dropout linear regulators. *IEEE Journal of Solid-State Circuits* 38 (1, January): 146–50.

30. Lin, C. H, K. Chen, and H. W. Huang. 2009. Low-dropout regulators with adaptive reference control and dynamic push-pull techniques for enhancing transient performance *IEEE Transactions on Power Electronics* 24 (4, April): 1016–22.

31. Sandler, S., and C. E. Hymowitz. 2005. SPICE model supports LDO regulator designs. *Power Electronics Technology*, May, 26–31.

32. Hsieh, C. Y., C. Y. Yang, and K. Chen. 2010. A low-dropout regulator with peak current control topology for overcurrent protection. *IEEE Transactions on Power Electronics* 25 (6): 1386–94.

33. Texas Instruments. 1999. LDO linear regulator design using the universal SOT23 EVM. User's Guide, August, SLUV 019.

34. Chen, H. 2000. Understand LDO and its thermal design. AIC Application Note AN00-LR01EN, May.

35. Kularatna, N., and J. Fernando. 2009. A supercapacitor technique for efficiency improvement in linear regulators. In *IEEE Proceedings of IECON 09*, Portugal, 132–35.

36. Kularatna, N., J. Fernando, K. Kankanamge, and L. Tilakaratna. 2010. Very low frequency supercapacitor techniques to improve the end-to-end efficiency of DC-DC converters based on commercial off the shelf LDOs. In IECON 2010, *36th Annual Conference on IEEE Industrial Electronics Society*, 721–26.

37. Kularatna, N., J. Fernando, K. Kankanamge and X. Zhang 2011. Low frequency supercapacitor circulation technique to improve the efficiency of linear regulators based on LDO ICs, Proceedings of APEC 2011, TX, USA, 1161–65.

2

Switching Power Supply Topologies and Design Fundamentals

2.1 Introduction

In the modern world of electronics, there are three different basic approaches available for DC-to-DC conversion. These are the linear approach, the switching approach, and the charge pump approach. In a practical system, one can mix these three techniques to provide a complex but elegant, overall solution with energy efficiency, effective silicon or PCB area, and noise and transient performance to suit different parts of an electronic system. Switch-mode DC-DC converters, which is a very mature approach to the DC power supply requirements of energy-efficient, compact, and portable systems, was developed mostly around a few fundamental transformer-less topologies such as buck (step-down), boost (step-up), and buck-boost (inverting type), and their transformer-isolated derivatives such as the forward-mode converter, flyback converter, etc. These mature techniques utilize controller chips from various power IC manufacturers, and they have proliferated in most consumer electronic families.

For much higher power density requirements and larger power requirements, switch-mode topologies such as push-pull, half-bridge, and full-bridge versions are used. Another recent development was the use of single-ended primary inductance converter (SEPIC) topology, which has become popular in battery-powered systems.

In this chapter we consider the fundamentals of simple and common topologies, design concepts, and approaches in switch-mode power supplies (SMPSs) with a few design examples of how fundamental design concepts and practices could be applied to develop the power conversion stages based on switch-mode topologies. Due to space constraints, for detailed theoretical aspects and deeper design considerations the detail-minded reader is expected to refer to the many useful references cited.

2.2 Why Switch Modes: An Overall Approach

A low-voltage power supply subsystem must fulfill four essential requirements: (a) isolation from the mains, (b) change of voltage level, (c) conversion to a stable and precise DC value, and (d) energy storage. Given these needs, if we use a basic approach in power supply components for volume or weight reduction, the following essentials can be considered; and in a practical SMPS, designers combine all the following simple concepts to have a lower volume, weight, and efficiency where a higher switching frequency is used for energy conversion.

2.2.1 Transformers and Inductors

Transformers and inductors are the bulkiest components in a power supply. For a transformer, RMS voltage across a winding is given by the relationship

$$V_{rms} = 4.44 B_{max} A N f \qquad (2.1)$$

where B_{max} is the maximum flux density in the core, A is the cross-sectional area of the core, N is the number of turns, and f is the operational frequency. Given this relationship, and for a given core material where B_{max} is a property of the material, for a particular core area if we can operate the transformer at a higher frequency, we can have a reduced number of turns for a required voltage across a winding. Hence if we have a DC power supply, and use a transformer core with a higher maximum flux density (without saturation) to transfer energy to a secondary winding, one can use a higher frequency to have a lesser number of turns. However, for a magnetic core material hysteresis loss (P_h) and Eddie current losses (P_e) can increase with the frequency due to following relationships:

$$P_h \propto f B_{max}^2 \qquad (2.2)$$

and

$$P_e \propto f^2 B_{max}^2 \qquad (2.3)$$

In an overall sense it indicates that if we can use better magnetic materials with a higher operational frequency, we can make the size of a transformer small. Also, if we are to use inductors as filtering elements at higher frequencies, their impedance given by $2\pi f L$ will be higher at a higher frequency.

2.2.2 Capacitors in Power Supplies

Capacitors are used for two basic purposes in a power supply. They are: (1) in the rectification and filtering stage; and (2) used as high-frequency filtering. When we use a capacitor in a full-wave or half-wave rectifier operating at a frequency f, peak-to-peak ripple voltage ($V_{rp\text{-}p}$) is given by

$$V_{rp-p} \propto \frac{I_L}{Cf} \tag{2.4}$$

where I_L is the average DC load current. Given this relationship, if we can rectify and filter the AC input voltage at a higher frequency for the same peak-to-peak ripple voltage, the required capacitor is smaller. Also, in a general high-frequency filtering need where the high frequency needs to be bypassed to ground by a capacitor, we can use a smaller capacitor at a higher frequency, as the impedance of a capacitor is $1/2\pi fC$. In summary, for filtering of ripple superimposed on a DC rail or for general filtering, the required capacitor size is smaller at a higher frequency.

2.2.3 Heat Sinks

In a linear power supply, a series pass element dissipates a higher wattage, due to the higher voltage across the pass element. However, if the transistor can be configured to operate as a switch, dissipation will be lower and required heat sinks can be relatively small. In switch-mode power supplies, this concept is effectively used.

2.3 Basic Switch-Mode Power Supply Topologies

Commonly utilized switch-mode supplies could be subdivided into two basic kinds, nontransformer isolated and transformer isolated. Nontransformer-isolated versions are used in lower power requirements with no necessity to have electrical isolation from the primary energy source. Transformer-isolated topologies are used in higher power requirements, and they come with electrical isolation from the primary supply side. In the following sections we discuss the basic topologies and their fundamental operational relationships.

2.3.1 Nontransformer-Isolated Topologies

Several commonly used nontransformer-isolated topologies are: (a) buck or step-down converter; (b) boost or step-up converter; and (c) buck-boost or inverting type converter. In all three topologies, four basic components (switch, inductor, capacitor, and a diode) are configured in different ways to achieve DC-DC conversion while using a power semiconductor as a switch. Another very commonly used version is called the single-ended primary inductance converter (SEPIC), which is very much used in modern battery-powered portable systems. To provide an overview of the essential concepts used in these, Figure 2.1(a) depicts a simple resistor in series with a switch that operates at a switching frequency of f_c where the total cycle time T_s is comprised of an on time of t_{on} and an off duration of t_{off}. Figure 2.1(b) depicts the output voltage appearing across the resistor R and with an average value of V_o. Figure 2.1(c) illustrates a basic concept to generate the switch control signal using a simple amplifier and a comparator.

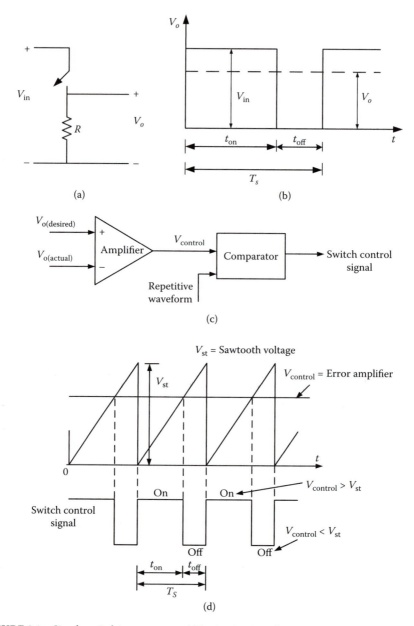

FIGURE 2.1 Simple switching converter: (a) basic circuit without energy storage components; (b) timing diagram; (c) block diagram of a pulse width modulator; (d) comparator signals.

Figure 2.1(d) depicts the comparator signals. In the basic circuit of Figure 2.1(c), a reference signal with a precisely controlled DC voltage is applied to the noninverting input of the amplifier, and the inverting input is supplied with the actual output. Comparator is supplied with a sawtooth waveform of peak voltage V_p and the control signal created by the output of the amplifier, v_c, where the combination generates a pulse width modulated (PWM) signal as in the lower trace of the Figure 2.1(c). In this situation, the duty ratio can be expressed as

$$D = \frac{t_{on}}{T_s} = \frac{v_c}{V_p} \tag{2.5}$$

2.3.1.1 Buck Converter

Figure 2.2(a) depicts the basic arrangement of a buck converter. A controller that has the fundamental characteristics similar to a case described in Figure 2.1(c) can be used to toggle the power switch based on the level of the regulator output. In our analysis here we assume that the four basic components in the power stage, namely the switch, inductor, capacitor, and diode, are ideal. Under these ideal conditions, when the switch is on, if the output capacitor C is very large, the inductor will have a voltage difference of $(V_{in}-V_o)$ and this will cause the inductor current, i_L, to rise steadily until the end of first phase of the switching cycle. Figure 2.1(b) depicts the equivalent circuit when the switch is on. Under this condition,

$$V_{in} - V_o = L \frac{di_L}{dt} \tag{2.6}$$

At the end of the on time, t_{on}, the inductor current's only path to continue is through the diode, and if we consider the diode to be an ideal one, new voltage across the inductor will be $-V_o$ and this will cause the inductor current to gradually decrease until the end of the off time of the switch, t_{off}. See Figure 2.2(c) for an equivalent circuit. Under this condition,

$$-V_o = L \frac{di_L}{dt} \tag{2.7}$$

Under steady-state conditions, neither the inductor nor the capacitor should accumulate energy continuously. Therefore to maintain the volt second balance during the full cycle of time T_s,

$$\int_0^{t_{on}} (V_{in} - V_o)\, dt + \int_{t_{on}}^{T_S} (-V_o)\, dt = 0 \tag{2.8}$$

Given the ideal components, which creates the situation of Figure 2.1(b) and Figure 2.1(c) respectively, we can simplify Equation (2.8) to,

$$(V_{in} - V_o)t_{on} - V_o(T_s - t_{on}) = 0 \qquad (2.9)$$

If D is the duty ratio, this can be reduced to

$$\frac{V_o}{V_{in}} = \frac{t_{on}}{T_S} = D \qquad (2.10)$$

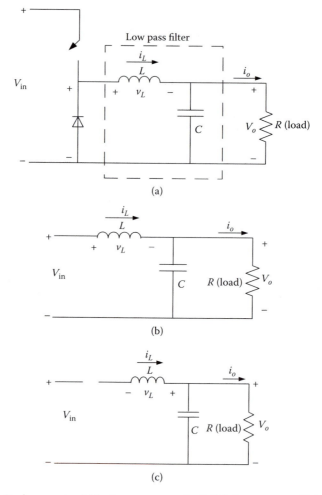

(a)

(b)

(c)

FIGURE 2.2 Buck converter: (a) basic arrangement for DC-DC conversion; (b) first state when the switch is on; (c) second state when the switch is off; (d) waveforms.

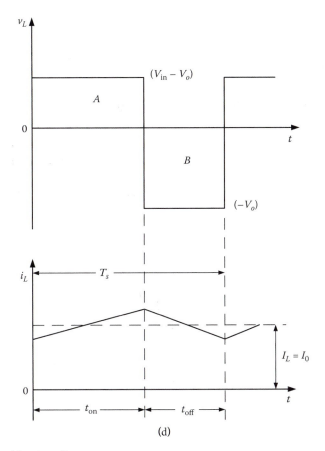

FIGURE 2.2 (Continued).

It is important to note that this condition is applicable only to the case where during the off state of the switch, inductor current does not reach zero, which is termed the continuous-conduction mode (CCM). Figure 2.1(d) indicates the waveforms related to this case.

Under this condition, if all the components are ideal, input power should be equal to output, where

$$V_{in} \cdot I_{in} = V_o I_o \qquad (2.11)$$

From Equations (2.10) and (2.11), we can achieve the following relationship similar to the case of an ideal transformer:

$$\frac{V_o}{V_{in}} = \frac{I_{in}}{I_o} = D \qquad (2.12)$$

Also, it is important to observe that the instantaneous input current i_{in} keeps rising to the peak value of the inductor current, and suddenly drops to zero when switch is off. This situation can create sharp voltage fluctuations at the input, and it is necessary to use a filter capacitor at the input.

2.3.1.2 Buck Converter under Different Modes of Conduction

Given the conditions of the waveforms in Figure 2.2(d), because the inductor average current I_L should be the same as the average output current, if we keep reducing the load current, a condition will be reached where the inductor current will hit zero at the end of the switching cycle as shown in Figure 2.3(a). If we keep reducing it further, there will be a period where the inductor current will be zero, during which period the capacitor will

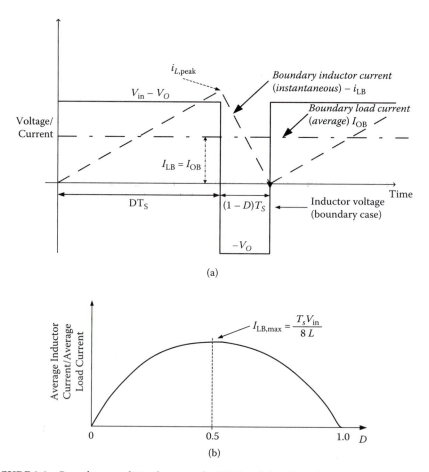

(a)

(b)

FIGURE 2.3 Boundary condition between the CCM and the discontinuous-conduction mode (DCM): (a) current waveform (b)I_{LB} versus D when V_{in} is constant.

feed the load. Under this boundary case where inductor current reaches the zero value at the end of the switching cycle, inductor boundary current is given by

$$I_{LB} = \frac{1}{2} i_{L,peak} = \frac{t_{on}}{2L}(V_d - V_o) = \frac{DT_s}{2L}(V_d - V_o) = I_{OB} \qquad (2.13)$$

Figure 2.3(b) depicts the relationship between the duty ratio D and the boundary value of the output current I_{OB}. As the graph in Figure 2.3(b) indicates, the maximum value of the inductor boundary current occurs at D = 0.5. The practical implication of this relationship is that in a practical buck converter, for a given operating condition of V_d, V_o, D (which are the variable parameters) with selected values of the L and the switching frequency $(1/T_s)$ in the design, when the load current, which is the same as the average inductor current, is below the value from Equation (2.13), the operation becomes discontinuous.

2.3.1.2.1 Analysis of the Case with Constant V_{in}

In a case such as in a motor speed controller, where input drive voltage is constant, V_o will be controlled by adjusting the duty ratio D; we can start our analysis referring to Figure 2.1 (parts b and c). At the edge of continuous-conduction mode, also the relationship of $V_o = DV_{in}$ holds true. Therefore from Equation (2.13),

$$I_{LB} = \frac{T_s V_{in}}{2L} D(1-D) \qquad (2.14)$$

This relationship leads to the graph of Figure 2.3(b) and hence the peak value of boundary load current occurring at D = 0.5. At this maximum, we can achieve that

$$I_{LB,max} = \frac{T_s V_{in}}{8L} \qquad (2.15)$$

Combining Equations (2.14) and (2.15),

$$I_{LB} = I_{LB,max} D(1-D) \qquad (2.16)$$

In Figure 2.4(a) the case of discontinuous mode is indicated where we maintain the switching frequency f_s, L, D, and V_{in} constant while the load current keeps dropping below I_{OB}. This makes the average inductor current drop below I_{LB} and dictates a higher value for V_o than before, resulting in the discontinous conduction mode (DCM).

In analyzing the case of DCM, from graphs in Figure 2.4 we see that the inductor current is zero during the period $\Delta_2 T_s$ and the load current is now supplied by the output capacitor. Inductor voltage is zero during this period. Again considering the volt-second balance for the inductor during the switching cycle,

$$(V_{in} - V_o)DT_s + (-V_o)\Delta_1 T_S = 0 \qquad (2.17a)$$

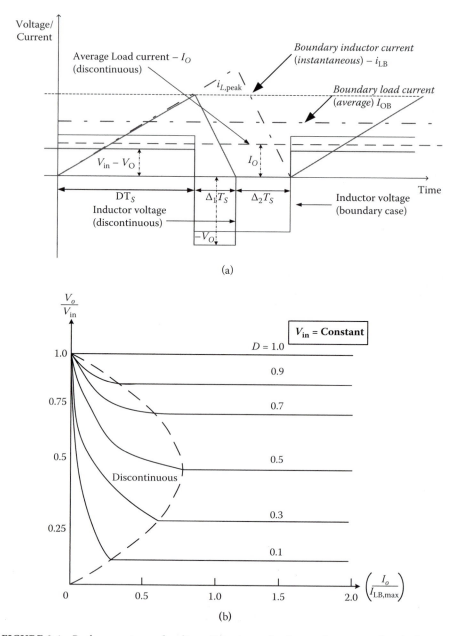

(a)

(b)

FIGURE 2.4 Buck converter under discontinuous-conduction mode compared with the case of boundary condition: (a) waveforms; (b) characteristics under constant V_{in}; (c) characteristics under constant V_o.

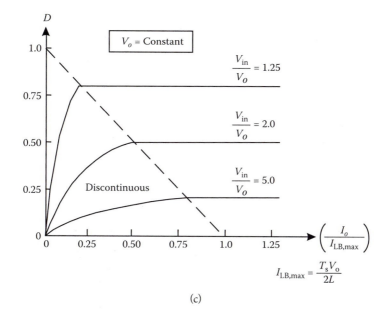

(c)

FIGURE 2.4 (Continued).

while $(D + \Delta_1 < 1)$, this gives,

$$\frac{V_o}{V_{in}} = \frac{D}{D + \Delta_1} \tag{2.17b}$$

From the waveform applicable to DCM in Figure 2.4, and using simple geometry, we can derive the value of peak inductor current, $i_{L,peak}$, as

$$i_{L,peak} = \frac{V_o}{L} \Delta_1 T_s \tag{2.17c}$$

Therefore

$$I_o = i_{L,peak} \frac{D + \Delta_1}{2} \tag{2.17d}$$

Using Equation (2.17c),

$$I_o = \frac{V_o T_s}{2L}(D + \Delta_1)\Delta_1 \tag{2.17e}$$

Using Equation (2.17b),

$$I_o = \frac{V_{in} T_s}{2L} D\Delta_1 \tag{2.17f}$$

Using Equation (2.15),

$$I_o = 4I_{LB,\max} D \Delta_1 \qquad (2.17g)$$

and

$$\Delta_1 = \frac{I_o}{4I_{LB,\max} D} \qquad (2.17h)$$

Combining Equations (2.17b) and (2.17h) we get

$$\frac{V_o}{V_{in}} = \frac{D^2}{D^2 + \frac{1}{4}(I_o/I_{LB,\max})} \qquad (2.18)$$

Figure 2.4(b) indicates the operation of the buck converter characteristics in both modes of operation for a constant input voltage V_{in}. The conversion ratio, V_o/V_{in}, is plotted as a function of $I_o/I_{LB,\max}$ at different duty ratios in this plot, where we can recognize the distinction between the CCM and DCM operations. More details can be found in [1].

2.3.1.2.2 Analysis of the Case with Constant V_o

Now we can analyze the other common DC-DC converter case where we try to maintain the output voltage V_o constant, while V_{in} and I_o vary. Starting from the case of boundary of CCM-DCM, based on Equation (2.13) with $V_{in} = V_o/D$,

$$I_{LB} = \frac{T_s V_o}{2L}(1-D) \qquad (2.19a)$$

With constant V_o the maximum value of I_{LB} occurs at $D = 0$, and hence

$$I_{LB,\max} = \frac{T_s V_o}{2L} \qquad (2.19b)$$

However, this condition of $D = 0$ is hypothetical, since the requirement of V_{in} for that condition is infinite. From Equations (2.19a) and (2.19b),

$$I_{LB} = (1-D)I_{LB,\max} \qquad (2.20)$$

From Equations (2.17b) and (2.17e), we can derive a relationship for the value of D for a given I_o as a ratio of $I_{LB,\max}$ as per the following relationship:

$$D = \frac{V_o}{V_{in}} \left[\frac{I_o/I_{LB,\max}}{1 - V_o/V_{in}} \right]^{1/2} \qquad (2.21)$$

This relationship for the DCM combined with the CCM is plotted in the graph in Figure 2.4(c).

2.3.1.2.3 Output Voltage Ripple

One important output specification in a DC-DC converter is its peak-to-peak ripple, sometimes represented as a percentage of the average output DC voltage. In our previous analysis with ideal components, we assumed that the output capacitor C is very large, in order to assume that the output voltage will be constant. However, in a practical case where we have a finite capacitor value, we should be able to estimate the peak-to peak-ripple voltage, under high load currents, which usually creates the case of CCM. Figure 2.5 depicts this case with the convenient assumption of an ideal capacitor with zero equivalent series resistance (ESR).

Assuming that the average inductor current in the circuit is equal to the average load current, and the average ripple of the inductor current feeds the capacitor in each full cycle, for the CCM case, referring to Figure 2.5,

$$\Delta V_o = \frac{\Delta Q}{C} = \frac{1}{C} \frac{1}{2} \frac{\Delta I_L}{2} \frac{T_s}{2} \qquad (2.21a)$$

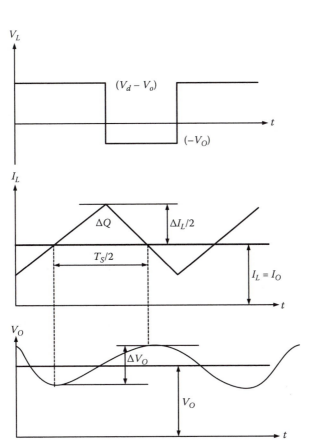

FIGURE 2.5 Output voltage ripple of buck converter under CCM.

From Figure 2.2(b), during t_{off}, which is equal to $(1-D)T_s$,

$$\Delta I_L = \frac{V_o}{L}(1-D)T_s \qquad (2.21\text{b})$$

From Equations (2.21a) and (2.21b),

$$\Delta V_o = \frac{T_s}{8C}\frac{V_o}{L}(1-D)T_s \qquad (2.22\text{a})$$

Therefore

$$\frac{\Delta V_o}{V_o} = \frac{1}{8}\frac{T_s^2(1-D)}{LC} = \frac{\pi^2}{2}(1-D)\left(\frac{f_c}{f_s}\right)^2 \qquad (2.22\text{b})$$

where the switching frequency $f_s = 1/T_S$, and

$$f_c = \frac{1}{2\pi\sqrt{LC}} \qquad (2.22\text{c})$$

In this circuit, the action of the inductor and the capacitor can be considered as a low pass filter with corner frequency f_c, and if these components can be chosen in such a way that $f_c \ll f_s$ percentage ripple can be kept at very low value under CCM conditions. In a similar analysis applied to DCM conditions, one can arrive at the following relationship:

$$\Delta V_o = \frac{[DT_s(V_{in}-V_o)-LI_o][DT_s(V_{in}-V_o)V_o-LI_oV_o+(V_{in}-V_o)(DT_s(V_{in}-V_o)-LI_o)]}{2LCV_o(V_{in}-V_o)}$$

$$(2.23)$$

2.3.1.3 Boost Converter

Boost converter, or the step-up converter, as its name implies, converts the input voltage to a higher value, using the same four components in a power stage as per Figure 2.6(a), controlled by a switching regulator converter IC. Figure 2.6(b) indicates the waveforms under the CCM condition, and Figures 2.6(c) and 2.6(d) indicate the two different states.

Similar to the analysis in the case of buck converter under CCM, by considering the volt-second balance for the inductor for the whole period T_s,

$$\frac{V_o}{V_{in}} = \frac{T_s}{t_{off}} = \frac{1}{1-D} \qquad (2.24\text{a})$$

and, considering a lossless circuit, with input power and output power equal,

$$\frac{I_o}{I_{in}} = (1-D) \qquad (2.24\text{b})$$

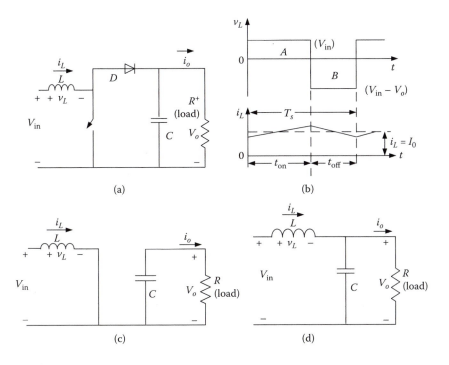

FIGURE 2.6 Step-up converter: (a) basic power stage; (b) CCM waveforms; (c) switch on state; (d) switch off state.

Given the case as in Figure 2.7 indicating the case of boundary condition between CCM and DCM versus DCM condition, we can arrive at the following relationships:

$$I_{oB} = \frac{T_s V_o}{2L} D(1-D)^2 \tag{2.25a}$$

and

$$I_{LB,\text{max}} = \frac{T_s V_O}{8L} \tag{2.25b}$$

From Equation (2.25a), a maximum of I_{oB} occurs at $D = 1/3$, and

$$I_{oB,\text{max}} = \frac{2}{27} \frac{T_s V_o}{L} \tag{2.25c}$$

Given the above relationships, boundary inductor current and boundary load current can be expressed as:

$$I_{LB} = 4D(1-D)I_{LB,\text{max}} \tag{2.26a}$$

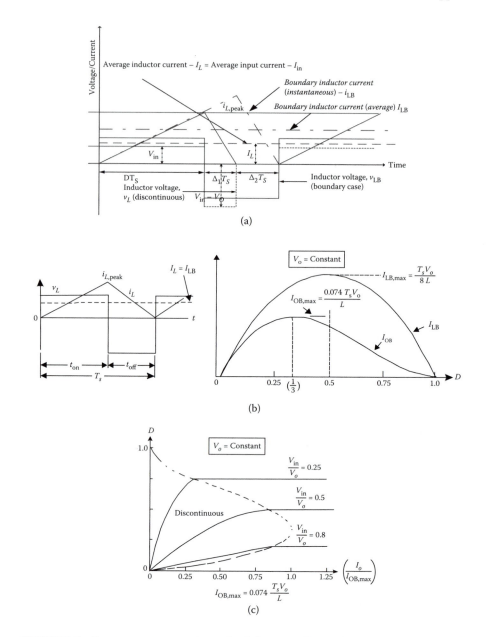

FIGURE 2.7 Boost converter operation under different conditions: (a) waveforms under boundary and DCM conditions; (b) graphs for operation under boundary conditions; (c) graphs for DCM operation.

and

$$I_{oB} = \frac{27}{4} D(1-D)^2 I_{oB,\max} \qquad (2.26b)$$

The analysis is similar to the case of buck converter, and more information is available in [1]. In this case the average diode current will be equal to the average load current, and the instantaneous inductor current (i_L) will be equal to the instantaneous input current (i_{in}). Figure 2.7(b) depicts the graphs for boundary condition for inductor current and the load current in relation to duty ratio D, when the output voltage is constant. Figure 2.7(c) depicts the graphs for D versus output load current, for the ideal case, when the V_o is kept constant. Assuming a constant output current, the peak-to-peak ripple voltage can be expressed as a ratio of average output voltage given by

$$\frac{\Delta V_o}{V_o} = \frac{DT_s}{RC} \qquad (2.27)$$

Figure 2.7(d) indicates the figures related to Equation (2.27), under CCM.
   Though our ideal calculations for CCM give a simple ratio of

$$\frac{1}{1-D}$$

for V_o/V_{in}, parasitic elements in the capacitor, inductor, switch, and diode will lead to a different situation as depicted in the graph of Figure 2.8.

2.3.1.4 Buck-Boost Converter

By reconfiguring the same four elements as in the two previous topologies, we can arrive at another topology, buck-boost converter, which allows us to step up or step down the input DC voltage. Figure 2.9 provides the CCM condition waveforms and two states of switching. An important observation here, as depicted in Figure 2.9(a), is that the output DC voltage has a reverse polarity compared to the input. For this reason this is sometimes called the inverting configuration.
   With a similar analysis to the two previous topologies, we can arrive at the following relationships for CCM:

$$\frac{V_o}{V_{in}} = \frac{D}{1-D} \qquad (2.28a)$$

and

$$\frac{I_o}{I_{in}} = \frac{1-D}{D} \qquad (2.28b)$$

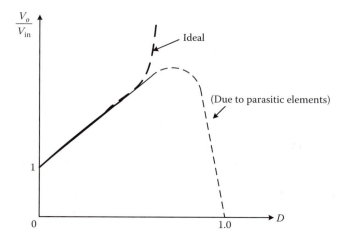

FIGURE 2.8 Effect of parasitic elements in the boost converter.

For this topology under CCM and constant V_o, the boundary inductor current (I_{LB}) and the boundary load current (I_{OB}) can be written as,

$$I_{LB} = I_{LB.\text{max}}(1-D) = \frac{T_s V_o}{2L}(1-D)$$ (2.29a)

and

$$I_{oB} = I_{oB.\text{max}}(1-D)^2 = \frac{T_s V_o}{2L}(1-D)^2$$ (2.29b)

where

$$I_{LB,\text{max}} = I_{oB,\text{max}} = \frac{T_s V_o}{2L}$$ (2.29c)

Peak-to-peak ripple under CCM is given by

$$\frac{\Delta V_o}{V_o} = D\frac{T_s}{\tau}$$ (2.30)

Also, when V_o is kept constant, under DCM duty ratio D can be expressed as

$$D = \frac{V_o}{V_{in}}\sqrt{\frac{I_o}{I_{oB,\text{max}}}}$$ (2.31)

Figure 2.10(a) indicates the waveforms under CCM boundary. Figure 2.10(b) depicts the graphs related to the parameter variations versus variations of D. The effects of parasitic are in Figure 2.10(c).

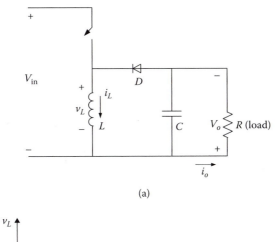

(a)

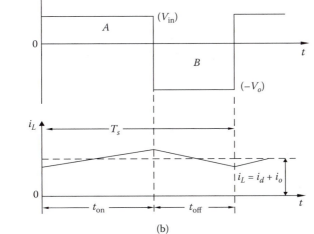

(b)

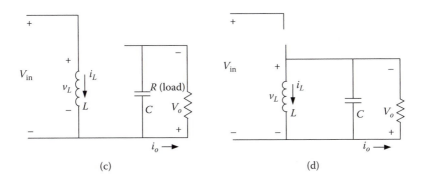

(c)   (d)

FIGURE 2.9 Buck-boost topology: (a) power stage; (b) waveforms under CCM; (c) state under on condition of the switch; (d) off state of the switch.

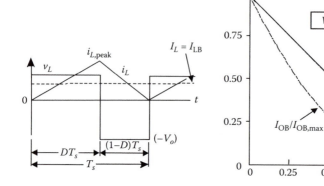

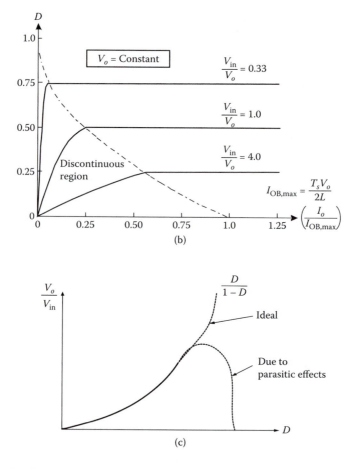

FIGURE 2.10 Buck-boost converter performance: (a) boundary of CCM and DCM; (b) general characteristics under CCM/DCM; (c) effect of parasitic elements in power stage.

2.3.2 Transformer-Isolated Topologies

Nonisolated basic converters (buck, boost, and buck-boost types) are generally used for lower-power PCB-level converter circuits and are not so popular for higher-power applications. Transformer-isolated versions such as forward mode, flyback, and bridge types are generally used for applications where higher power, galvanic isolation, and multiple-output rails are required. In the following paragraphs, a summarized overview on theoretical concepts behind these popular topologies is provided.

It is necessary to use a transformer as a basic building block in these converters. In a quick revision on transformer theory, Figure 2.11(a) indicates the case of an ideal

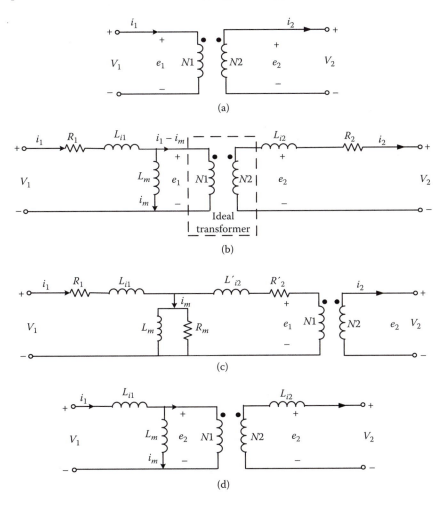

FIGURE 2.11 Transformer equivalent circuits: (a) ideal case: (b) with components indicating winding resistances, leakage inductances, and magnetizing inductance; (c) inclusion of core losses and with further simplifications; (d) equivalent circuit suitable for switching regulator calculations, neglecting all losses in transformer.

transformer, and Figure 2.11(b) indicates the case where practical conditions are taken into account, and in particular, (a) winding resistances, (b) leakage inductances of each winding due to flux not common to both windings (common flux in the core), and (c) a superficial winding (L_m) to represent the magnetizing requirement of the core. Figure 2.11(c) indicates a case where core losses are also indicated. Figure 2.10(d) indicates a simplified case where every secondary item is referred to the primary side, while assuming that the voltage drops across the windings (referred to primary) side are considered small, compared to the input voltage applied to the primary side.

Further discussion on this subject, as applicable to high-frequency switching transformers, is found in [2].

2.3.2.1 Forward-Mode Converters

The forward converter is derived from the buck topology family, generally employing a single switch. The power switch in the forward topology is ground referenced (also called a low-side switch), whereas in buck topology the switch source terminal floats on the switching node. The main advantage of the forward topology is that it provides isolation and the capability to provide step-up or step-down function.

Figure 2.12 shows an idealized forward converter. Initially assuming that the transformer is ideal, when the switch is on, the diode D_1 becomes forward biased, and D_2 becomes reverse biased. Therefore

$$v_L = \frac{N2}{N1}V_{in} - V_o, \quad \text{for } 0 < t < t_{on} \tag{2.32a}$$

When the switch is off,

$$v_L = -V_{out}, \tag{2.32b}$$

for $t_{on} < t < T_s$. Considering the volt-second balance over the cycle,

$$\frac{V_o}{V_{in}} = \frac{N_2}{N_1}D \tag{2.32c}$$

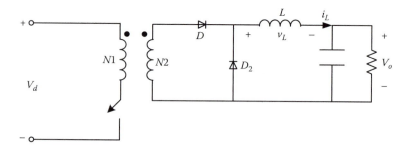

FIGURE 2.12   Simplified forward converter without a demagnetizing winding.

The above relationship indicates that the forward mode converter is similar to the buck converter behavior, however modified by the transformer turns ratio.

One important design consideration in this topology is the magnetizing inductance and the need to reset the transformer core. If the transformer core keeps building up any remaining core flux over each cycle, it will end up in saturation failure. Figure 2.13(a) shows a simplified transformer-isolated forward converter with an additional demagnetizing winding and a diode connected in such a way that when the switch is on, the diode is blocked and vice versa. Therefore during the second part of the switching cycle where the switch is off, diode D_1 gets forward biased and allows current flow in the third winding, to reset the core flux.

When the switch is on,

$$v_1 = V_{in} \tag{2.33a}$$

for $0 < t < t_{on}$. During the period, magnetizing current (i_m) flowing in the winding indicated by L_m keeps increasing toward a peak ($I_{m, peak}$) as in Figure 2.13(c). (We assume that the core does not saturate here.)

When the switch is turned off, i_1, which was flowing in the winding 1, starts entering into the magnetizing winding, and hence during the off period of the switch,

$$i_1 = -i_m \tag{2.33b}$$

With the current directions indicated in Figure 2.13(b), in general,

$$N_1 i_1 + N_3 i_3 = N_2 i_2 \tag{2.33c}$$

However, when the switch is closed D_1 gets reverse biased, and $i_2 = 0$. Therefore from Equation (2.33c),

$$i_3 = \frac{N_1}{N_2} i_m \tag{2.33d}$$

which flows through the diode D_3 into the input supply. During the time interval t_m, this current keeps decaying up to zero value. During this time the primary winding voltage (same as the voltage across L_m) is given by,

$$v_1 = -\frac{N_1}{N_3} V_{in} \tag{2.33e}$$

for $t_{on} < t < t_{on} + t_m$.

Once the transformer is demagnetized, $i_m = -i_1 = 0$ and $v_1 = 0$. The time interval t_m can be calculated by considering the volt-second balance for the winding L_m. Using Equations (2.33a) and (2.33e), combined with volt-second balance,

$$\frac{t_m}{T_s} = \frac{N_3}{N_1} D \tag{2.34}$$

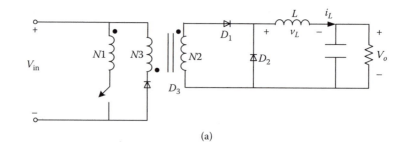

(a)

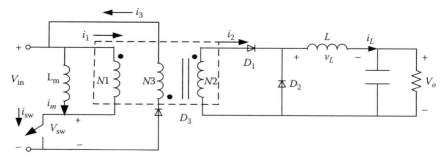

(b)

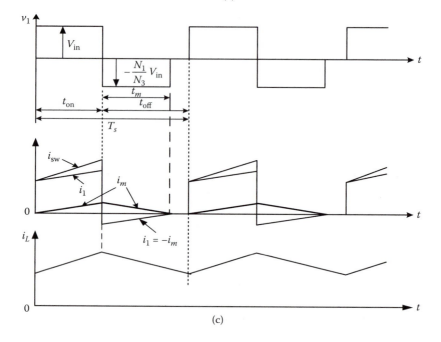

(c)

FIGURE 2.13 Practical forward converter: (a) ideal case; (b) considering the magnetizing inductance; (c) waveforms.

For the transformer to demagnetize before the next cycle begins, the maximum value for t_m/T_s becomes $(1 - D)$. Therefore from Equation (2.34),

$$(1-D_{max}) = \frac{N_3}{N_1} D_{max} \qquad (2.35a)$$

Hence,

$$D_{max} = \frac{1}{1 + \frac{N_3}{N_1}} \qquad (2.35b)$$

In practice, it is quite easy to construct a transformer where N_1 and N_2 are equal, and as N_3 is used only to demagnetize the core, it does not require very much of an isolation requirement. This allows a bifilar winding to be used for N_1 and N_3. Under this case of $N_1 = N_3$, maximum duty cycle ratio, D_{max}, becomes 0.5. Under this situation with a demagnetizing winding also it can be shown that,

$$V_o/V_{in} = \frac{N_2}{N_1} D,$$

which is the same case as per the ideal transformer with no demagnetizing winding.

Given the basic analysis of the idealized forward converter, it is important to indicate that this is one of the very popular topologies in applications such as desktop computer power supplies like the "silver box," where the basic circuit can be modified to have multiple isolated windings, different core demagnetizing (resetting) techniques, or even multiple switches to share the current or the voltage stress across the drain and source terminals of the transistors. One important design consideration in this topology is the magnetizing inductance and the need to reset the transformer core. Figure 2.11(d) shows a simplified transformer model including the magnetizing inductance (L_M) and the leakage inductance (L_L). The value of L_M can be measured at the primary terminals with the secondary winding open-circuited (open-circuit test of a transformer). The peak current in L_M is proportional to the maximum flux density within the core, and a given core can handle only a limited flux density before saturation occurs. At saturation, a rapid reduction of inductance occurs. The other element added to the transformer model is L_L, and this can be measured at the primary terminals with the secondary winding(s) short-circuited (short-circuit test, usually conducted at a lower primary voltage such as 5%–10% of the primary rated value). This term represents the stray value, which does not couple primary to secondary. With careful design, this value can be kept small, and the effect on the converter is limited to voltage spikes on the power switch.

An important consideration in forward-mode converter design is the core-resetting requirement to avoid core saturation. A few techniques are available for this purpose in addition to the common method discussed in our basic analysis. A few advanced techniques used for solving the same problem are active clamp reset and resonant reset forward converters. Figure 2.14 compares these techniques. Mappus [3], King and Gehrke [4], and Hariharan and Schie [5] provide analysis and design aspects of the active clamp reset techniques. Hariharan and Schie [5] and Khasiev [6] detail the important design aspects of resonant reset forward converters.

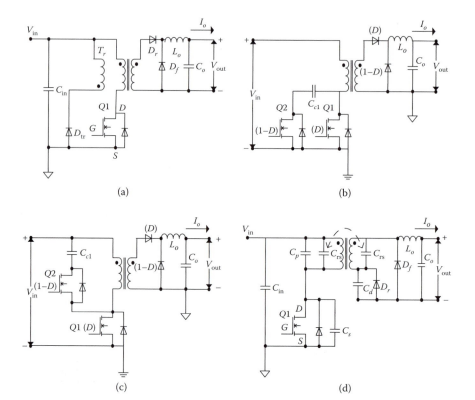

FIGURE 2.14 Core reset techniques: (a) add-on winding and diode; (b) low-side active switch; (c) high-side active clamp technique; (d) single-switch resonant reset technique. (From Mappus, S., Power *Electronics Technology*, July 2004, 24; King, B., and D. Gehrke, *Power Electronics Technology*, June 2003, 52; Hariharan, S., and D. Schie, *Power Electronics Technology*, October 2005, 50. With permission.)

2.3.2.2 Flyback Converters

Flyback converters are derived from the buck-boost topology. Low cost and simplicity are the major advantages of the flyback topology. In multiple-output applications, the addition of a secondary winding, a diode, and an output capacitor is all that is required for an additional output. Flyback converter operation can lead to confusion if the designer approaches the design of its magnetics as if it were a transformer. Except for the case of multiple-output windings, the magnetics in a flyback converter are not a transformer. An easy way to view this is as an energy bucket that is alternately filled (when the switch is on) and dumped (when the switch is off). In other words, a flyback magnetic (sometimes called a transformer choke) is an energy-in, energy-out power transfer device where input and output windings do not conduct current simultaneously. A gapped core is used in general to have adequate leakage inductance at the input side for energy storage during the switch-on period.

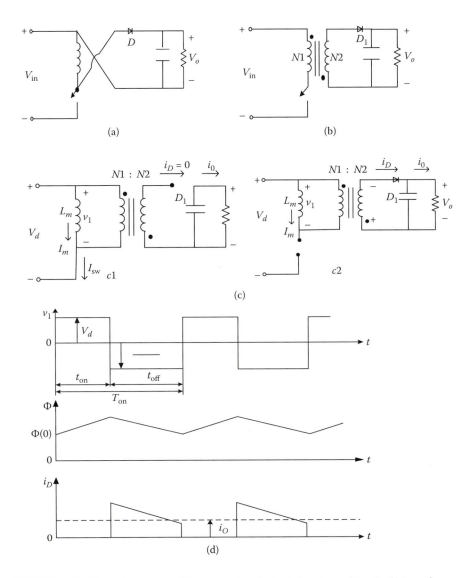

FIGURE 2.15 Flyback converter: (a) rearranging the buck-boost as a flyback; (b) transformer-isolated case; (c) switch-on and switch-off modes; (d) waveforms.

Figure 2.15 depicts the simplified concept of this topology. Figure 2.15(a) depicts the case of buck-boost topology rearranged to show the case of nonisolated flyback, and Figure 2.15(b) depicts the transformer-isolated case. Figure 2.15(c1) depicts the case where a two-winding inductor (which can act as a transformer or an inductor) is represented by its simplified equivalent circuit under on condition of the switch with the diode in reverse-biased condition, while Figure 2.15(c2) depicts the case where the switch is off and the magnetic component is in transformer action, with the diode forward biased.

The CCM condition of the buck-boost converter corresponds to the incomplete demagnetization condition of the flyback converter. If we consider that the flux in the core at the start of the cycle is $\phi(0)$, when the switch is on,

$$\phi(t) = \phi(0) + \frac{V_{in}}{N_1} t \quad \text{for } 0 < t < t_{on} \tag{2.36a}$$

Peak flux at the end of the on period is given by,

$$\phi_{peak} = \phi(t_{on}) = \phi(0) + \frac{V_{in}}{N_1} t_{on} \tag{2.36b}$$

At the end of t_{on} the switch goes off, and the current that was flowing in L_m enters into the primary winding; this causes the voltage across the secondary winding to reverse and make the diode conduct. Under this condition, voltage across the secondary winding $v_2 = -V_o$ (assuming the capacitor at secondary is large), and therefore the flux decreases linearly during t_{off}. This leads to the relationship

$$\phi(t) = \phi_{peak} - \frac{V_o}{N_2}(t - t_{on}) \tag{2.37a}$$

for $t_{on} < t < T_s$. And,

$$\phi(T_s) = \phi_{peak} - \frac{V_o}{N_2}(T_s - t_{on}) \tag{2.37b}$$

Using Equation (2.36b),

$$\phi(T_s) = \phi(0) + \frac{V_{in}}{N_1} t_{on} - \frac{V_o}{N_2}(T_s - t_{on}) \tag{2.37c}$$

Since the net flux through the core should be zero over the whole period,

$$\phi(T_s) = \phi(0) \tag{2.38a}$$

From Equations (2.37c) and (2.38a),

$$\frac{V_0}{V_{in}} = \frac{N_2}{N_1} \frac{D}{1-D} \tag{2.38b}$$

This clearly indicates that the flyback converter's operation under CCM is the same as buck-boost topology, with the transformer ratio modifying the relationship. During the on time of the switch, transformer primary is at the input voltage, while the diode is off and keeping the secondary winding opened. Under this condition, the relationship between the inductor current (i_m) and switch current (i_{sw}) can be written as

$$i_m(t) = i_{sw}(t) = I_m(0) + \frac{V_{in}}{L_m} t \quad \text{for } 0 < t < t_{on} \tag{2.39a}$$

where $I_m(0)$ is the inductor current at the start of the on duration.

Hence,

$$I_{m,peak} = I_{sw,peak} = I_m(0) + \frac{V_{in}}{L_m} t_{on} \qquad (2.39b)$$

During the off interval, switch current is zero, and

$$v_1 = -\frac{N_1}{N_2} V_o.$$

Therefore during the period $t_{on} < t < T_s$, magnetizing and the diode currents can be expressed as

$$i_m(t) = I_{m,peak} - \frac{V_o(N_1/N_2)}{L_m}(t - t_{on}) \qquad (2.40a)$$

and

$$i_D(t) = \frac{N_1}{N_2} i_m(t) = \frac{N_1}{N_2} \left[I_{m,peak} - \frac{V_o(N_1/N_2)}{L_m}(t - t_{on}) \right] \qquad (2.40b)$$

Since the average diode current should be equal to the average output current, I_O, from Equation (2.40b),

$$I_{m,peak} = I_{sw,peak} = \frac{N_2}{N_1} \frac{1}{1-D} I_o + \frac{N_1}{N_2} \frac{(1-D)T_s}{2L_m} V_O \qquad (2.41)$$

Voltage across the switch during the off interval is

$$v_{sw} = V_{in} + \frac{N_1}{N_2} V_o = \frac{V_{in}}{1-D} \qquad (2.42)$$

This indicates that during the off period, the switch has to withstand a much higher voltage than the input maximum voltage. Given this situation, combined with the requirement to design the transformer to operate as an inductor-transformer, the process of design is bit more difficult than other topologies. Also, the transformer core design requirement becomes a bit more involved due to the need of an air gap in most cases, to have the right value for inductance L_m.

2.3.2.3 Push-Pull Converter

This is another popular converter topology with isolation, and Figure 2.16 depicts the basic configuration of the power stage and the waveforms. In this case the center-tapped secondary winding is feeding the inductor alternatively. In the overall switching process, the switches on the primary T_1 and T_2 are alternatively switched with a dead time of Δ in the middle to avoid simultaneous switching of the two switches. When T_1 is switched on, D_1 conducts and D_2 gets reverse biased. This causes a voltage difference across the filter inductor given by

$$v_L = \frac{N_2}{N_1} V_{in} - V_o \qquad (2.43a)$$

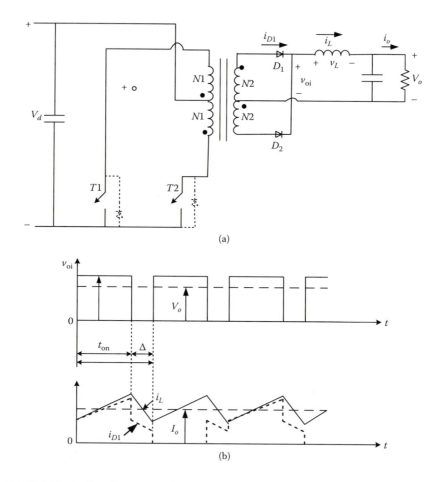

FIGURE 2.16 Push-pull converter: (a) simplified power stage; (b) waveforms.

for $0 < t < t_{on}$. During the dead-time (Δ) when both switches are off, the inductor current splits between the two secondary half-windings, and the voltage across the inductor during the period

$$t_{on} < t < t_{on} + \Delta, \ v_L = -V_O, \tag{2.43b}$$

and

$$i_{D1} = i_{D2} = \frac{1}{2} i_L \tag{2.43c}$$

After a dead time of Δ, the same process happens, and considering the volt-second balance during the period of $\tfrac{1}{2}T_S$ where $(t_{on} + \Delta = \tfrac{1}{2}T_S)$,

$$\frac{V_o}{V_{in}} = 2 \frac{N_2}{N_1} D \quad 0 < D < 0.5 \tag{2.44}$$

Antiparallel diodes on the primary side of the circuit allow a path for the current due to leakage flux of the transformer windings.

2.3.2.4 Half-Bridge and Full-Bridge Converters

Two other commonly used topologies are the half-bridge and full-bridge converters. Figure 2.17 depicts the half-bridge converter and its waveforms. Figure 2.18 depicts the

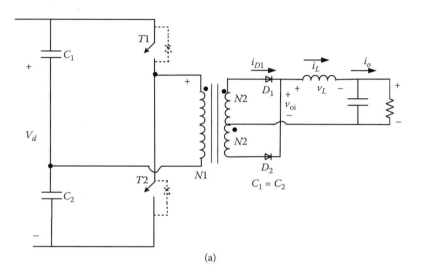

(a)

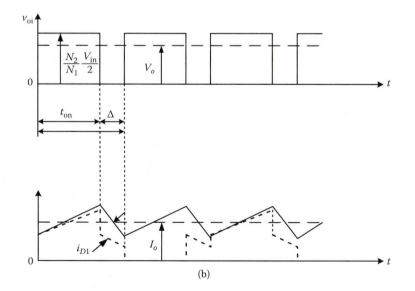

(b)

FIGURE 2.17 Half-bridge converter: (a) simplified block diagram; (b) waveforms.

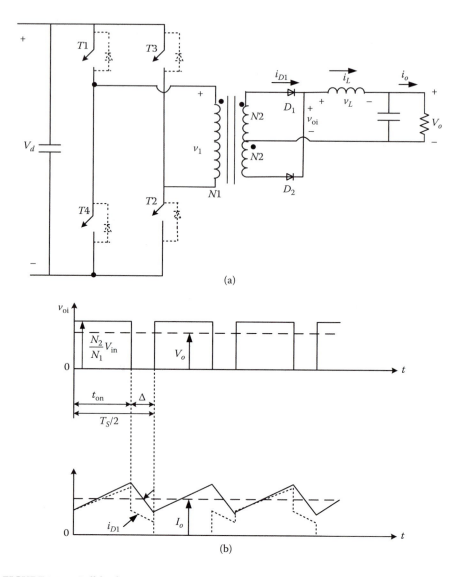

FIGURE 2.18   Full-bridge converter: (a) basic circuit; (b) waveforms.

full-bridge converter. In the half-bridge converter, the two capacitors C_1 and C_2 establish a midpoint voltage of $1/2 V_{in}$, and the two switches are alternatively switching on and off, similar to the case of the push-pull converter. By using similar principles of analysis, for the half-bridge converter we can prove that

$$\frac{V_O}{V_{in}} = \frac{N_2}{N_1} D \qquad\qquad (2.45)$$

In the case of the full-bridge converter, T_1 and T_2 switch on together in one half cycle, and the T_3 and T_4 switch during the second half cycle, maintaining a dead-time of Δ between the two phases. In this case,

$$\frac{V_O}{V_{in}} = 2\frac{N_2}{N_1}D \qquad (2.46)$$

for $0 < D < 0.5$. Antiparallel diodes provide circulation paths to flow the currents due to leakage inductances of the windings.

Comparison of the full-bridge converter and half-bridge converter for identical input and output voltages with the same power output ratings require the following turns ratios:

$$\left(\frac{N_2}{N_1}\right)_{HB} = 2\left(\frac{N_2}{N_1}\right)_{FB} \qquad (2.47)$$

assuming that the transformer magnetizing current is small and neglecting the ripple current in the inductor, switch currents are related by

$$(I_{SW})_{HB} = 2(I_{SW})_{FB}$$

2.4 Applications and Industry-Favorite Configurations

Some relative merits and demerits of the switching converter topologies and typical applications are summarized in subsets of Appendix B, including essential mathematical expressions for important design relationships. The industry has settled on several primary topologies for a majority of applications. Figure 2.19 illustrates the approximate range of usage for these topologies. The boundaries to these areas are determined

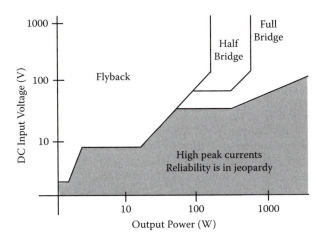

FIGURE 2.19 Industry-favorite configurations.

primarily by the amount of stress the power switches must endure and still provide reliable performance. The boundaries delineated in Figure 2.19 represent approximately 20 A of peak current in power switches.

Nonisolated basic converters (buck, boost, and buck-boost types) are generally used for lower-power PCB-level converter circuits and are not so popular for higher-power applications. Isolated versions such as forward mode, flyback, and bridge types are generally used for applications where higher power, galvanic isolation, and multiple-output rails are required.

As shown in Figure 2.19, bridge converters are generally used for higher-power and higher-voltage converters because there are several power switches to share the dissipation and the voltage stress. In general, full-bridge topology is used for very high power applications, and it is quite important to consider the losses in the circuits and the design complications due to their high-side switches operating with their source terminals (in the case of MOSFETs) or emitters (in IGBTs or power transistors) at floating levels. As indicated in the topology diagrams in Appendices B9 and B10, the transistors on the upper parts of the bridge (high-side transistors) require special circuitry to drive floating gate terminals.

Gate driver ICs help solve this problem. In a high-power DC-DC converter design, to achieve adequate efficiency the designer should develop an "efficiency budget" or a loss calculation. In general, losses are contributed by many different sources, the important ones being:

- Rectification losses (low-frequency rectifiers on the input side and high-frequency rectification circuits on the output side)
- Switching losses in power semis (static and dynamic dissipation)
- Core losses in magnetic components
- Losses due to control and supervisory circuits
- PCB losses associated with high-current tracks of the PCB

Using a simple calculation based on an Excel spreadsheet, the designer can determine where optimization can be achieved. Figure 2.20 indicates the losses associated with a switching power supply with an output capacity of about 10 W. In a larger-capacity power supply, the percentage values may be different.

2.4.1 Use of Gate Driver ICs in High-Power Converters

The essential idea of a gate driver IC is to achieve two important design requirements: to provide correct voltage drive levels required by the MOSFET or IGBT gates where floating voltages are required, and to provide fast charge/discharge gate capacitances for MOSFETs or IGBTs. For example, in half- or full-bridge circuits based on MOSFETs, low-side (n channel) transistors need to be driven by a positive gate voltage with respect to the ground plane, but the high-side transistor gate needs to be driven by a positive voltage with respect to its source terminals, which will be at floating voltage values.

Table 2.1 shows the different techniques used for gate driver circuits and their key features [7]. Gate driver circuits are useful in any switching system topology where two

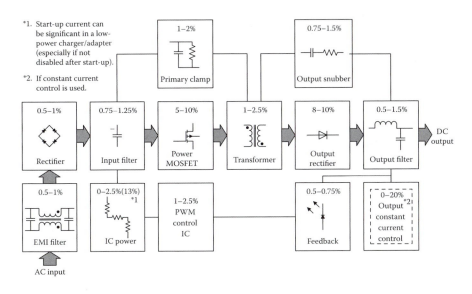

FIGURE 2.20 Losses associated with a switching supply. (From Jovalusky, J., New Energy Standards Banish Linear Supplies, *Power Electronics Technology*, March 2005, 42. With permission.)

switches operate at high and low sides. To justify the use of these for efficient power circuit designs, the designer should understand and pay adequate attention to the parasitic capacitances at the gate input [8]. For IGBT-based bridge topologies, there are hybrid ICs available as gate drivers [9]. In some of these, optoisolators are used for electrical isolation between the drive side and the power stage.

2.4.2 Single-Ended Primary Inductance Converter (SEPIC)

Because of the advantages of operation in buck or boost modes without any voltage inversion, another popular converter used in battery-powered applications is the SEPIC converter. Theoretical concepts related to the SEPIC topology have been of interest since its development in the mid-1970s. However, practical use of the technique was limited until battery-powered applications proliferated, particularly Li-ion types, where the battery pack's useful voltage can range from about 4.2 V to about 2.7 V.

The SEPIC is definitely worth considering for a typical portable system, in which 3 V circuitry is powered by an Li-ion cell. Although SEPIC circuits require more components than buck or boost converters, they allow operation with fewer cells in the battery, where the cost of extra components is usually offset by the savings in the battery. An important use of the SEPIC is in power factor correction (PFC) [11, 12]. SEPIC topologies possess the following advantages:

- They have a single switch.
- They have continuous input current (similar to boost).
- Any output voltage can be used (as in the buck-boost case).

TABLE 2.1   Comparison of Gate Driver Techniques

Comparison of Gate Driver Techniques

Technique	Basic Circuit Configuration	Features
Pulse transformer	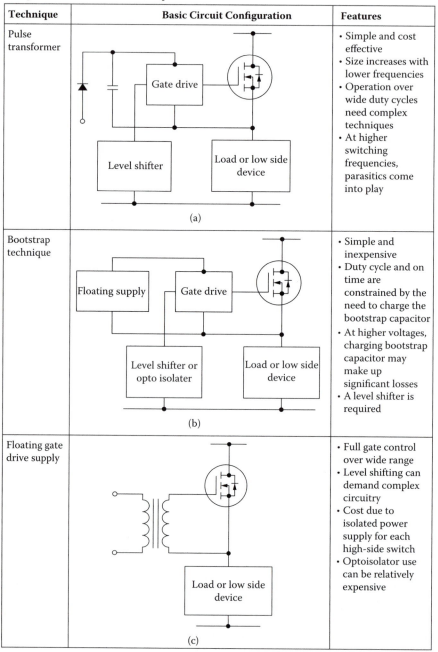 (a)	• Simple and cost effective • Size increases with lower frequencies • Operation over wide duty cycles need complex techniques • At higher switching frequencies, parasitics come into play
Bootstrap technique	(b)	• Simple and inexpensive • Duty cycle and on time are constrained by the need to charge the bootstrap capacitor • At higher voltages, charging bootstrap capacitor may make up significant losses • A level shifter is required
Floating gate drive supply	(c)	• Full gate control over wide range • Level shifting can demand complex circuitry • Cost due to isolated power supply for each high-side switch • Optoisolator use can be relatively expensive

(*continued*)

TABLE 2.1 Comparison of Gate Driver Techniques (Continued)

Technique	Basic Circuit Configuration	Features
Charge pump based	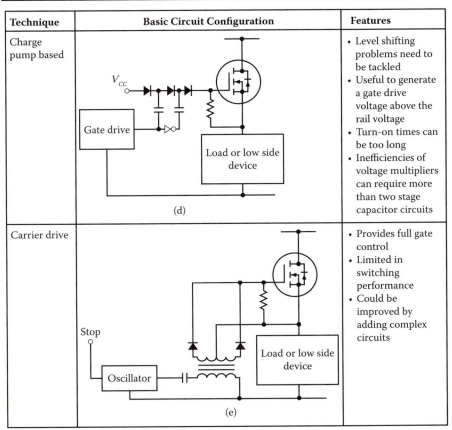(d)	• Level shifting problems need to be tackled • Useful to generate a gate drive voltage above the rail voltage • Turn-on times can be too long • Inefficiencies of voltage multipliers can require more than two stage capacitor circuits
Carrier drive	(e)	• Provides full gate control • Limited in switching performance • Could be improved by adding complex circuits

Source: Adapted from Clemente, S. and Dubhashi A., HV floating MOS gate driver IC, Application Note 978A, International Rectifier, El Segundo, CA, 1990.

- Ripple current can be steered away from the input, reducing the need for input noise filtering.
- They have inrush/overload current limiting capability.
- Switch location is a simple low-side case, hence easier gate drive circuits.
- The outer loop control scheme is similar to a boost converter's case.

Disadvantages of the SEPIC are:

- They have higher switch/diode peak voltages compared to boost topology.
- They have greater bulk capacitor size and cost if operated lower than boost.

Referring to Figure 2.21, SEPIC topology can be considered as an extension of the boost topology. When the switch S_1 is on, current I_1 flows into common rail via the inductor L_1. Similarly, at that time current I_2 flows through L_2 where the capacitor C_1

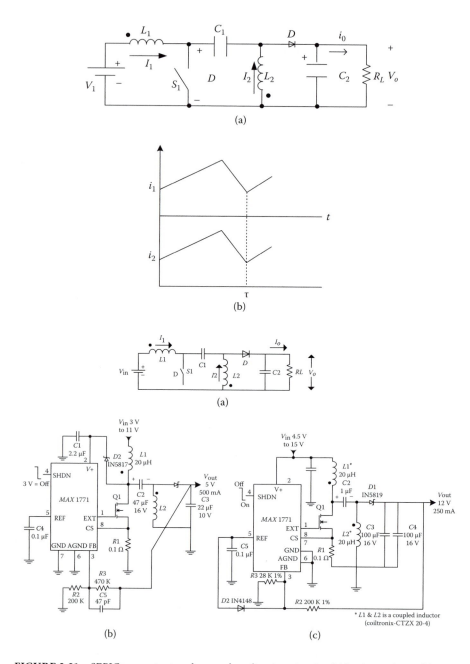

FIGURE 2.21 SEPIC converter topology and application circuits: (a) basic topology; (b) wave-forms for the case when both inductor current are positive and continuous; (c) a circuit for 5 V output from a 3–11 V input; (d) circuit for a 12 V output from a 4.5–15 V input. (Courtesy of Neufled, H., SEPIC Design, *PCIM*, October 1998, 36.)

acts as a voltage source. When S_1 goes off, I_1 flows through C_1 and D into C_2 and R_L (load resistor). I_2 flows through D into C_2. Both currents ramp down as the capacitors are charged, and the load current flows from C_2 to the load. (This is only one of the possible six operation modes of the topology [13].) This is the case where both inductor currents are positive and continuous as in Figure 2.21(b). When this situation occurs, and equating the volt-second balances for each inductor L_1 and L_2 respectively, we can get the following relationships:

$$V_{in}D = (1-D)(V_{C1} + V_o - V_{in}) \qquad (2.48a)$$

and

$$V_{C1}D = (1-D)V_O \qquad (2.48b)$$

These two yield:

$$V_{C1} = V_{in} \qquad (2.48c)$$

From these equations we get the transfer function as

$$\frac{V_o}{V_{in}} = \frac{D}{1-D} \qquad (2.49)$$

This indicates a case similar to buck-boost converter, but without any voltage inversion. More details are available in [13]. A SEPIC converter can have six operating modes. A more detailed analysis [12,13] with design approach can be found in [11–15]. Figures 2.21(c) and 2.21(d) show two SEPIC application circuits. Achievable efficiencies are about 85% [14]. Another useful practical consideration for easy construction and lower cost in SEPIC circuits is to have the two (nearly equal) inductors coupled [13,15]. Some trends of SEPIC applications and advancements are indicated in [16–30].

2.5 A Few Design Examples and Guidelines

There are about 10 different common topologies used in industrial and consumer applications, as summarized in Appendices B1 to B10, which include the topologies discussed in previous sections, and derived versions of flyback and forward converters such as two-transistor flyback and forward-mode converters.

A single-transistor flyback converter is an almost uncontested choice for off-line converters delivering fewer than 150 W. They are inexpensive because the transformer (which really works as a coupled inductor) is part of the output filter, and generating multiple outputs merely requires the addition of another secondary winding along with diodes and output filter capacitors.

However, at power levels greater than 150 W, because of excessive peak currents in the switching transistor and excessive voltages across the switches, this topology reaches its limitations. In these situations the two-transistor forward converter approach is a solution. Design aspects and calculation guidelines for the two-transistor forward converter are available in Gauen [31,32].

The following sections provide a guideline for designing practical DC-DC converters, with some examples of flyback and full-bridge topologies.

2.5.1 Flyback Converter Design Guidelines

A good application example of a flyback converter is an off-the-AC-line power adaptor for a notebook computer or a PDA. Based on the following specifications, one can start developing a flyback converter:

- Nominal AC input voltage (V_{ACnom})
- Minimum and maximum AC input voltage (V_{ACmin} and V_{ACmax})
- Output voltage (V_{out}) (a typical value is about 16 V)
- Maximum output overshoot, full load to no load (ΔV_o)
- Maximum output power (P_o)
- Target efficiency at full load (η)
- Holdup time at nominal AC input voltage and full load at output (T_{hold})

Designing such a power adapter can be a challenge due to recent energy-saving initiatives, such as the European commission Code of Conduct Standby Power requirements, etc. [33]. Figure 2.21 indicates a suggested configuration, as per guidelines in [33].

The above data give the designer the necessary maximum input power,

$$P_{in,\max} = \frac{P_o}{\eta}$$

(2.50a)

To design for low-input line situation with cycle skip hold-up time (when a short duration AC voltage failure) requirements, a minimum DC bus regulation voltage target must be selected and DC bus filter capacitance C_3 in Figure 2.22 must be calculated. Based on an approximate DC bus typical voltage of $V_{DCtyp(pk)} = \sqrt{2}\ V_{ACnom}$, nominal value for DC bus bulk capacitor C_3 can be calculated as

$$C_{Bulk(nom)} = \frac{2P_o \cdot T_{hold}}{(V_{DCtyp(pk)}^2 - V_{DCmin}^2)\eta}$$

(2.50b)

From this DC rail the circuit could operate in two different modes, namely continuous-conduction mode (CCM) with a large primary inductance of the transformer-choke or in discontinuous-conduction mode (DCM) where primary current is shown in

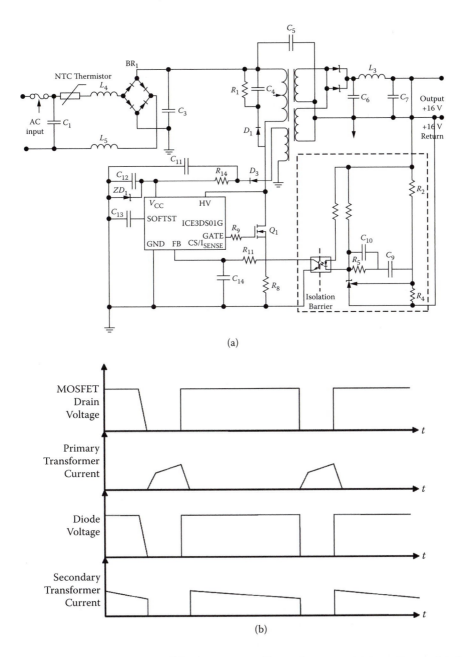

(a)

(b)

FIGURE 2.22 A representative flyback converter and transformer current waveforms related to different modes: (a) basic circuit arrangement; (b) CCM waveforms; (c) DCM waveforms. (Adapted from Hancock, J. M., *Power Electronics Technology*, September 2005, 33–42.)

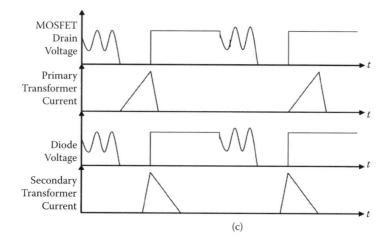

(c)

FIGURE 2.22   (Continued).

Figures 2.22(b) and 2.22(c). For a typical case of DCM with a duty cycle limit $D_{max} = 0.5$, the peak primary current, $I_{PRI(pk)}$, will be

$$I_{PRI(pk)} = \frac{2P_{in\,max}}{V_{DC\,min}D_{max}} \tag{2.50c}$$

Primary RMS current, I_{PRI} (RMS)—which is useful in determining the MOSFET capability and the primary conduction losses—is given by

$$I_{PRI(rms)} = I_{PRI(pk)}\sqrt{\frac{D_{max}}{3}} \tag{2.50d}$$

From the peak primary current and target switching frequency, f_s (determined by the controller in Figure 2.22), the primary inductance of the flyback transformer can be calculated:

$$L_{PRI} = \frac{D_{max}V_{DC\,min}}{I_{PRI(pk)}f_s}. \tag{2.50e}$$

Assuming a maximum flux density value for the core (typically within 0.12 T to 0.3 T for a core such as an ETD30), a core can be selected from a magnetics manufacturer's data sheet. The core set and the gap must be chosen for an A_L product that supports a reasonable number of turns in such a way that it meets the other requirements as well [33]. The number of primary turns can be calculated as

$$N_p = \sqrt{\frac{L_{PRI}}{A_L}}. \tag{2.50f}$$

and rounded down to the nearest integer value. Then the secondary turns are calculated from the following:

$$N_s = \frac{N_P(V_O + V_{Diode})}{V_{R(max)}}. \tag{2.50g}$$

where V_{diode} is the estimated peak diode forward voltage and $V_{R(max)}$ is the reflected voltage on the primary. The reverse voltage for the rectifier diode, $V_{R(Diode)}$, is given by

$$V_{R(Diode)} = V_{OUT} + \left(V_{DCmax} \frac{N_s}{N_p} \right). \tag{2.50h}$$

(In practice the value required may be much higher due to overvoltages related to parasitic inductances and the like.) For more details related to the DCM-type flyback converter, [34,35] are suggested.

Figure 2.22(a) indicates the essential circuit elements of a flyback converter based on a modern SMPS controller chip such as ICE3DSO1 from Infineon Technologies [33] with a power MOSFET driving the primary-side winding. Complete design details for an 80 W, 16 V power supply are given in [33].

2.5.1.1 Flyback Converters Using Power-Integrated Circuits

Another recent approach for flyback converters based on a complete power IC (an SMPS controller and a power MOSFET) is shown in Figure 2.23 from Power Integrations. Figures 2.23(a) to 2.23(d) indicate different levels of feedback circuit arrangements, where output regulation performance can vary from average (lowest cost) to extra-high accuracy. For more details related to these design approaches, see Leman [36] and Power Integrations, Inc. [37].

In this design approach, current waveform parameter K_p simplifies calculations for both continuous and discontinuous modes [36]. For critical mode control-based design approaches, see Basso [38]. Flyback topology design using a MATHCAD-based approach is discussed in Huber and Jovanovic [39, 40]. In this kind of design, for the best performance in charging a battery, the constant voltage mode (CVM), constant power mode (CPM), and constant current mode (CIM) are combined. It is also possible to use either an active clamp or RCD clamp approach for transformer demagnetizing, and critical-mode conduction is used on the boundary of the CVM and CPM regions [40].

2.5.2 Full-Bridge Converter Design Example—240 V DC, 1 kW Output DC-DC Converter with Planar Magnetic and Gate-Driver ICs to Drive the Switches of the Full Bridge

Several years ago the author was asked to develop a DC-DC converter based on the following specifications:

- Input voltage: 20–30 V DC (nominal value of 24 V DC)
- Output voltage: 220 V DC at 1 kW output
- Topology: full bridge

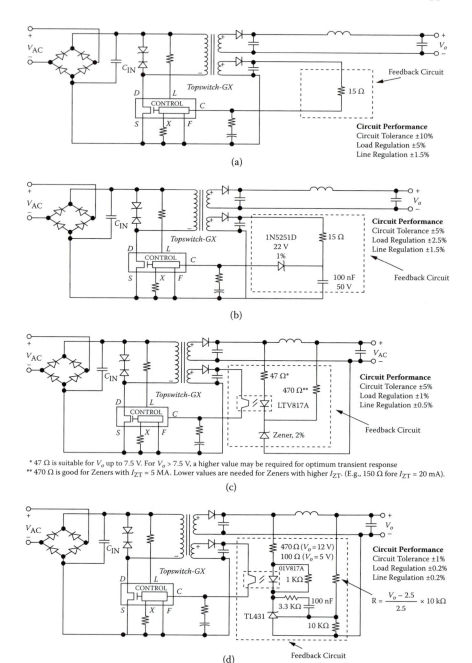

FIGURE 2.23 Reduced component designs using power integrated circuits: (a) low-cost version with simple feedback circuit; (b) with enhanced feedback; (c) opto/zener feedback with tighter regulation; (d) opto/TL431-based feedback with excellent load and line regulation performance. (Courtesy of Power Integrations, Inc., USA.)

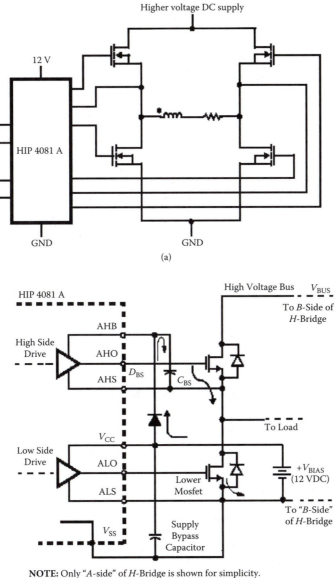

(a)

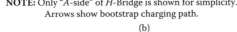

NOTE: Only "*A*-side" of *H*-Bridge is shown for simplicity.
Arrows show bootstrap charging path.

(b)

FIGURE 2.24 HIP4081 MOS gate driver details: (a) basic concept; (b) bootstrap capacitor arrangement. (Adapted from [49])

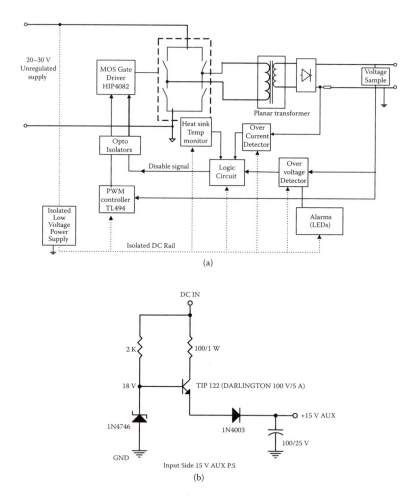

FIGURE 2.25 Design approach to a 1 kW, 24 V input, 220 V DC output full bridge with supervisory circuits and auxiliary power supplies: (a) overall design approach; (b) kick-start power supply based on a simple circuit; (c) load regulation; (d) efficiency.

- Switching frequency range: 150–250 kHz
- Transformer configuration: planar
- Regulation (load and line): ±2%
- Ripple: 0.5 Vpp
- Protection: overload, overvoltage, and over temperature, inhibit control

Several options were available for a project of this nature, and after considering hard-switching PWM to resonant converters, the ultimate decision was for a Harris HIP 408X-based full-bridge configuration, as shown in Figure 2.24(a). In a design of this nature, the gate voltage required to switch on the upper transistors of the bridge should be floating above the source voltage, which will vary as the transistor pairs switch on alternatively. HIP 4082 or a similar device is designed to achieve this requirement, and also provide

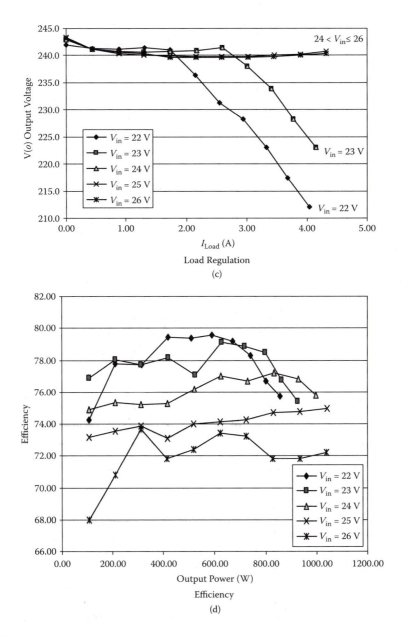

FIGURE 2.25 (Continued).

a high charging current (in the order of 2.5 A max) to quickly charge the gate-source capacitance of the MOSFET. Figure 2.24(a) provides the basic concept to achieve this condition. Figure 2.24(b) provides the details of the MOS gate driver internals indicating the bootstrap capacitors, which provide the floating drive requirements to the upper MOSFETs.

With the requirement for an extremely compact version, with a percentage-efficiency target in the high 70s, an estimate of losses was done based on a set of MOSFETs with $V_{DS} = 70V$, $I_{D, max} = 180A$, and $R_{DS(on)} = 6m\Omega$ (or better) as the four switches. The overall unit was expected to have an isolated low-voltage power supply for the control and supervisory circuits, as shown in the bottom left-hand corner of the Figure 2.25(a). For initial startup requirements, a simple auxiliary power supply of 15 V was proposed based on a simple open-loop linear regulator. Once the 220 V DC output appeared, an auxiliary winding in the planar transformer would handle overpowering of the control and supervisory circuits. Figure 2.25(b) shows this auxiliary (kick-start) supply. With the decision to hard-switch the PWM, a TL494 was chosen as the PWM controller. As indicated in the power stage of Figure 2.25(a), a MOSgate driver of the type HIP 4081 [41] was used to simplify the design and to achieve a smaller PCB area after carefully considering the simplicity achievable by the pulse transformers. To achieve an extremely flat profile with a very small PCB area, a planar transformer from Payton America, Inc. was used [43] as similar to Figure 7.11. In achieving a low component count and associated high reliability, it was necessary to drop the temptation to use standard logic IC-based supervisory circuits and use (a component count optimized) simple comparator circuit-based subcircuit. A kick-start supply was used to power up the TL494 PWM controller [42], gate drivers, etc., during startup, and once the system started running, a single-turn auxiliary winding in the transformer would take over, increasing efficiency. The efficiency achieved was about 75% at full load. The performance of the circuit is shown in Figures 2.25(c) and 2.25(d).

References

1. Mohan, N., T. M. Undeland, and W. P. Robins. 2003. *Power electronics, converters, applications and design.* New York: Wiley.
2. den Bossche, A. V., and V. C. Valchev. 2005. *Inductors and transformers for power electronics.* Boca Raton, FL: CRC Press.
3. Mappus, S. 2004. Active clamp transformer reset: High or low side. *Power Electronics Technology*, July, 24.
4. King, B., and Gehrke, D. 2003. Active clamp control boosts forward converter efficiency. *Power Electronics Technology*, June, 52.
5. Hariharan, S., and D. Schie. 2005. Designing single switch forward converters. *Power Electronics Technology*, October, 50.
6. Khasiev, V. 2003. Moving forward converters to higher efficiency. *Power Electronics Technology*, May, 38.
7. Jovalusky, J. 2005. New energy standards banish linear supplies. *Power Electronics Technology*, March, 42.
8. Clemente, S., and A. Dubhashi. 1990. HV floating MOS gate driver IC. Application Note 978A. International Rectifier, El Segundo, CA.
9. McGinty, J. 1998. Gate drivers. *PCIM*, May, 28.
10. Motto, E., and J. Donlon. 2005. Hybrid ICs drive high-power IGBT modules. *Power Electronics Technology*, March, 24.

11. Dixon, L. 1993. High power factor preregulator using the SEPIC converter. Unitrode Corporation, Merrimack, NH, 6-1-6-11. Available at http://focus.ti.com/lit/ml/slup103/slup103.pdf.

12. Dixon, L. 1993. High power factor preregulator using the SEPIC converter. Unitrode Corporation, Merrimack, NH, 7-1-7-7.

13. Nuefeld, H. 1998. SEPIC design. *PCIM*, October, 36.

14. Rahban, T. 2002. Consider SEPIC topology for new designs. *Power Electronics Technology*, November, 18.

15. Dixon, L. 1993. High power factor preregulator using the SEPIC converter. Unitrode Corporation, Merrimack, NH, 8-1-8-4.

16. Oishi, H., H. Okada, K. Ishizaka, and R. Itoh. 1995. SEPIC-derived three-phase sinusoidal rectifier operating in discontinuous current conduction mode. *IEE Proceedings on Electric Power Applications* 142 (4): 239–45.

17. Buso, S., G. Spiazzi, and D. Tagliavia. 2000. Simplified control technique for high-power-factor flyback Cuk and Sepic rectifiers operating in CCM. *IEEE Transactions on Industry Applications* 36 (5): 1413–18.

18. Simonetti, D. S. L., J. Sebastian, and J. Uceda. 1997. The discontinuous conduction mode Sepic and Cuk power factor preregulators: Analysis and design. *IEEE Transactions on Industrial Electronics* 44 (5): 630–37.

19. Veerachary, M. 2005. Power tracking for nonlinear PV sources with coupled inductor SEPIC converter. *IEEE Transactions on Aerospace and Electronic Systems* 41 (3): 1019–29.

20. de Melo, P. F., R. Gules, E. F. R. Romaneli, and R. C. Annunziato. 2010. A modified SEPIC converter for high-power-factor rectifier and universal input voltage applications. *IEEE Transactions on Power Electronics* 25 (2): 310–21.

21. Jozwik, J. J., and M. K. Kazimierczuk. 1989. Dual SEPIC PWM switching-mode DC/DC power converter. *IEEE Transactions on Industrial Electronics* 36: (1): 64–70.

22. Zhu, M., and F. L. Luo. 2008. Series SEPIC implementing voltage-lift technique for DC-DC power conversion. *IEEE Transactions on Power Electronics* 1 (1): 109–21.

23. Chung, H. S. -H., K. K. Tse, S. Y. R. C. M. Hui C. M. Mok, and M. T. Ho. 2003. A novel maximum power point tracking technique for solar panels using a SEPIC or Cuk converter. *IEEE Transactions on Power Electronics* 18 (3): 717–24.

24. Adar, D., G. Rahav, and S. Ben-Yaakov. 1996. Behavioural average model of SEPIC converters with coupled inductors. *Electronics Letters* 32 (17): 1525–26.

25. Ismail, E. H. 2009. Bridgeless SEPIC rectifier with unity power factor and reduced conduction losses. *IEEE Transactions on Industrial Electronics* 56 (4): 1147–57.

26. Lin, B.-R. and C. L. Huang. 2009. Analysis and implementation of an integrated SEPIC-forward converter for photovoltaic based light emitting diode lighting. *IEEE Transactions on Power Electronics* 2 (6): 635–45.

27. Shieh, J.-J. 2000. SEPIC derived three-phase switching mode rectifier with sinusoidal input current. *IEEE Proceedings on Electric Power Applications* 147 (4): 286–94.

28. Chiang, S. J., H.-J. Shieh, and M.-C. Chen. 2009. Modeling and control of PV charger system with SEPIC converter. *IEEE Transactions on Industrial Electronics* 56 (11): 4344–53.

29. Al-Saffar, M. A., E. H. Ismail, A. J. Sabzali, and A. A. Fardoun. 2008. An improved topology of SEPIC converter with reduced output voltage ripple. *IEEE Transactions on Power Electronics* 23 (5): 2377–86.

30. Kwon, J.-M., W.-Y. Choi, J.-L. Lee, E.-H. Kim, and B.-H. Kwon. 2006. Continuous-conduction-mode SEPIC converter with low reverse-recovery loss for power factor correction. *IEEE Proceedings on Electric Power Applications* 153 (5): 673–81.

31. Gauen, K. 1991. Considerations for a two transistor, current mode forward converter—part I: Overall design concepts. *PCIM*, April, 38.

32. Gauen, K. 1991. Considerations for a two transistor, current mode forward converter—part II: Power circuit design. *PCIM*, May, 38.

33. Hancock, J. M. 2005. Improving the performance of flyback power supplies. *Power Electronics Technology*, September, 33–42.

34. Ruble, R., and R. Clarke. 1994. Designing flyback converters—part I, design basics. *PCIM*, January, 43.

35. Ruble, R., and R. Clarke. 1994. Designing flyback converters—part II, 48V dual output converter. *PCIM*, April, 23.

36. Leman, B. R. 1995. Finding the keys to flyback power supplies produces efficient design. *EDN*, April 13, 101.

37. TOPSwitch-GX® flyback design methodology. 2004. Application Note 32. Power Integrations, Inc., San Jose, CA. Available at http://www.powerint.com/PDFFiles/an32.pdf.

38. Basso, C. 1998. Critical mode control stabilizes switch-mode power supplies. *EDN*, April 23, 171.

39. Huber, L., and M. M. Jovanovic. 1996. Optimizing flyback topologies for portable AC/DC adapter/charger applications: Part I—adapter charger requirements. *PCIM*, August, 68.

40. Huber, L., and M. M. Jovanovic. 1996. Optimizing flyback topologies for portable AC/DC adapter/charger applications: Part II—DC/DC converter design. *PCIM*, September, 34.

41. Danz, J. E. 2003. HIP4081, 80V high frequency H-bridge driver. Application Note 9325.3. Intersil Corporation, Milpitas, CA.

42. TL494 data sheet. 2002. Texas Instruments, Dallas, TX. Available at http://www.datasheetcatalog.com/datasheets_pdf/T/L/4/9/TL494.shtml.

43. Product catalog. Payton America, Boca Raton, FL. Available at http://www.payton-group.com/catalogue/20-21.pdf.

44. Hancock, J. M. 2005. Improving the performance of flyback power supplies. *Power Electronics Technology*, September, 33.

45. Ruble, R., and R. Clarke. 1994. Designing flyback converters—part I, design basics. *PCIM*, January, 43.

46. Ruble, R., and R. Clarke, 1994. Designing flyback converters—part II, 48V dual output converter. *PCIM*, April, 23.

47. Leman, B. R. 1995. Finding the keys to flyback power supplies produces efficient design. *EDN*, April 13, 101.
48. TOPSwitch-GX® flyback design methodology. 2004. Application Note 32. Power Integrations, Inc., San Jose, CA. Available at http://www.powerint.com/PDFFiles/an32.pdf.
49. Intersil, Inc. 2007. HIP4081A, 80 V high frequency H-bridge driver. Intersil Application Note AN 9405-05, December.

3

Power Semiconductors

3.1 Introduction

During the elapsed six and a half decades since the invention of the transistor, the power electronics world has been able to enjoy the benefit of many different types of power semiconductor devices. These devices are able to handle voltages from a few volts to several kilovolts and switching currents from a few milliamperes to kiloamperes. Within a decade from the invention of the transistor, the thyristor was commercialized.

Around 1968 power transistors began replacing the thyristors in switch-mode power systems. The power MOSFET, as a practical commercial device, has been available since 1976. When "smart power" devices appeared in the market, designers were able to make use of the insulated gate bipolar transistor (IGBT) from the early 1990s.

In 1992 MOS-controlled thyristors were commercially introduced, and around 1995 semiconductor materials such as GaAs and silicon carbide (SiC) opened new vistas for better-performing power diodes for high-frequency switching systems. Presently, the spectrum of what are referred to as "power devices" spans a very wide range of devices and technology. Discrete power semiconductors continued to be the leading edge for power electronics in the 1990s. Improvements on the fabrication processes for basic components such as diodes, thyristors, and bipolar power transistors have paved the way to high-voltage, high-current, and high-speed devices. Some major players in the industry have invested in manufacturing capabilities to transfer the best and newest power semiconductor technologies from research areas to production.

Commercially available power semiconductor devices could be categorized into several basic groups, such as diodes, thyristors, bipolar junction power transistors (BJTs), power metal oxide semiconductor field effect transistors (power MOSFETs), insulated gate bipolar transistors (IGBTs), MOS-controlled thyristors (MCTs), gate turn-off thyristors (GTOs), and insulated gate commutated thyristors (IGCTs). This chapter provides an overview of the characteristics, performance factors, and limitations of several of these device families.

3.2 Power Diodes and Thyristors

3.2.1 Power Diodes

The diode is the simplest semiconductor device, comprising a P-N junction. In attempts to improve its static and dynamic properties, numerous diode types have evolved. In power applications, diodes are used principally to rectify, that is, to convert alternating current to direct current. However, a diode is used also to allow current freewheeling. That is, if the supply to an inductive load is interrupted, a diode across the load provides a path for the inductive current and prevents high-voltage Ldi/dt from damaging sensitive components of the circuit.

The basic parameters characterizing the diodes are the maximum forward average current $I_{F(Ave)}$ and the peak inverse voltage (PIV). This parameter is sometimes termed *blocking voltage* (V_{rrm}). There are two main categories of diodes, namely general-purpose P-N junction rectifiers and fast-recovery P-N junction rectifiers. General-purpose types are used in circuits operating at line frequencies such as 50 or 60 Hz. Fast-recovery (or fast-turn-off) types are used in conjunction with other power electronics systems with fast-switching circuits.

Classic examples of the second type are switch-mode power supplies (SMPSs), inverters, etc. Figure 3.1(a) indicates the capabilities of a power device manufacturer catering to very high power systems, and Figure 3.1(b) indicates the capabilities of a manufacturer catering to a wide range of applications.

In high-frequency situations such as inverters and SMPSs, two other important phenomena dominate the selection of rectifiers. Those are the "forward recovery" and the "reverse recovery."

3.2.1.1 Forward Recovery

The turn-on transient can be explained with Figure 3.2. When the load time constant L/R is compared to the time for turn-on t_{fr} (forward recovery time), load current will hardly change during this period. For the time $t < 0$, the switch S_w is closed. Steady conditions prevail and the diode D is reverse biased at $-V_S$. It is in the off state, and $i_D = 0$.

At $t = 0$, the switch S_w is opened. The diode becomes forward biased, providing a path for the load current in R and L, so that the diode current i_D rises to I_F ($\approx I_l$) after a short time t_r (rise time) and the diode voltage drop falls to its steady value after a further time t_f (fall time). This is shown in Figure 3.2(b). The diode turn-on time is the time t_{fr}, which comprises $t_r + t_f$. It takes this time t_{fr} for charge to change from one equilibrium state (off) to the other (on).

The total drop V_D reaches a peak forward value V_{FR} that may be from 5 to 20 V, a value much greater than the steady value V_{DF} generally between 0.6 and around 1.2 V. The time t_r for the voltage to reach V_{FR} is usually about 0.1 μs. At a time $t > t_r$, the current i_D becomes constant at I_l (which will be the forward diode current I_F).

Further, conductivity modulation takes place due to the growth of excess carriers in the semiconductor accompanied by a reduction of resistance. Consequently, the $i_D R_D$ voltage drop reduces. In the equilibrium state, which may take a time of t_f, with a uniform distribution of excess carriers, the voltage drop v_D reaches its minimum steady-state value V_{DF}.

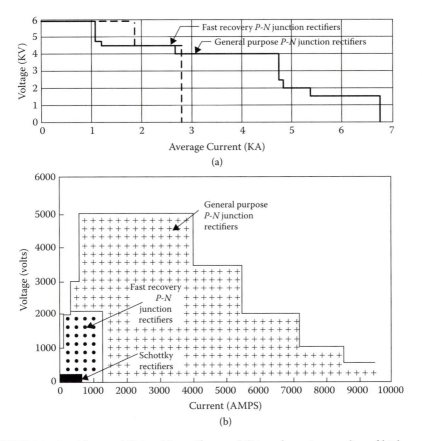

FIGURE 3.1 Rectifier capabilities: (a) rectifier capabilities of a major supplier of high-power semiconductors (reproduced by permission of GEC Plessey Semiconductors, UK); (b) rectifier capabilities of a manufacturer catering to a wide range of applications. (Reproduced from International Rectifier Inc., USA. With permission.)

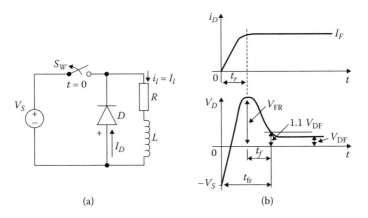

FIGURE 3.2 Turn-on characteristics: (a) circuit; (b) waveforms.

During the turn-on interval t_p the current is not uniformly distributed so the current density can be high enough in some parts to cause hot spots and possible failure. Accordingly, the rate of rise of current di_D/dt should be limited until the conduction spreads uniformly and the current density decreases. Associated with the high-voltage V_{FR} at turn-on, there is high current, so there is extra power dissipation that is not evident from the steady-state model. The turn-on time varies from a few nanoseconds to about 1 ms, depending on the device type.

3.2.1.2 Reverse Recovery

The turn-off phenomenon can be explained using Figure 3.3. In Figure 3.3(a) except for the diode, the circuit elements of this simple chopper are considered to be ideal. Switching S_w at a regular frequency, the source of constant voltage, V_s, maintains a constant current I_1 in the RL load, because it is assumed that the load time constant (L/R) is long compared with the period of the switching.

While the switch S_w is closed, the load is being charged and the diode should be reverse biased. While the switch S_w is open, the diode D provides a freewheeling path for

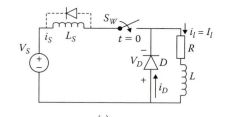

(a)

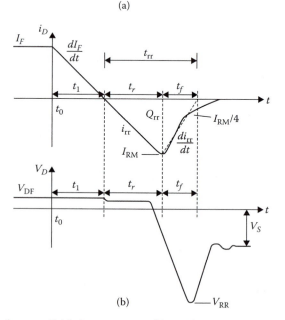

(b)

FIGURE 3.3 Diode turn-off: (a) chopper circuit; (b) waveforms.

the load current I_l. The inductance L_S is included for practical reasons and may be the lumped source inductance and snubber inductance, which should have a freewheeling diode to suppress high voltages when the switch is opened.

Let us consider that steady conditions prevail. At the time $t = 0^-$ the switch S_w is open, the load current is $i_1 = I_l$, the diode current is $i_D = I_1 = I_F$, and the voltage drop V_D across the diode is small (about 1 V).

The important concern is what happens after the switch is closed at $t = 0$. Figure 3.3(b) depicts the waveforms of the diode current i_D and voltage v_D. At $t = 0^-$ there was the excess charge carrier distribution of conduction in the diode. This distribution cannot change instantaneously so at $t = 0^+$ the diode still looks like a virtual short circuit, with $v_D = 1$ V. Kirchhoff's current law provides us with the relation:

$$i_D = I_1 - i_s \tag{3.1}$$

and Kirchhoff's voltage law yields:

$$V_s = L_s \frac{di_s}{di_t} = L_s \frac{d(I_1 - i_D)}{dt} = -L_s \frac{di_D}{dt} \tag{3.2}$$

Accordingly, the diode current changes at the rate,

$$\frac{di_D}{dt} = -\frac{V_s}{L_s} = \text{Constant} \tag{3.3}$$

This means that it takes a time $t_1 = L_s(I_l/V_s)$ seconds for the diode current to fall to zero. At the time $t = t_1$ the current i_D is zero, but up to this point the majority carriers have been crossing the junction to become minority carriers, so the P-N junction cannot assume a blocking condition until these carriers have been removed. At zero current, the diode is still a short circuit to the source voltage. Equations (3.1) to (3.3) still apply, and the current i_D rises above I_l at the same rate. The diode voltage V_D changes little while the excess carriers remain. The diode reverse current rises over a time t_r during which the excess charge carriers are swept out of the region.

At the end of the interval t_r, the reverse current i_D can have risen to a substantial value I_{RR} (peak **re**verse **re**covery current), but by this time, sufficient carriers have been swept out and recombined so that current cannot be supported. Therefore over a fall-time interval t_f, the diode current i_D reduces to almost zero very rapidly while the remaining excess carriers are swept out or recombined.

It is during the interval t_f that the potential barrier begins to increase both to block the reverse-bias voltage applied by the source voltage as i_D reduces, and to suppress the diffusion of majority carriers because the excess carrier density at the junction is zero. The reverse voltage creates the electric field that allows the depletion layer to acquire space charge and widen.

That is, the electric field causes electrons in the n region to be forced away from the junction toward the cathode and causes holes to be forced away from the junction toward the anode. The blocking voltage v_D can rise above the voltage V_s transiently because of the additional voltage $L_s(di_D/dt)$ as i_D falls to zero over the time t_f.

The sum of the intervals $t_r + t_f = t_{rr}$ is known as the reverse recovery time, and it varies generally (from 10 ns to over 1 microsecond) for different diodes. This time is also known as the storage time because it is the time that is taken to sweep out the excess charge Q_{RR} from the silicon by the reverse current. Q_{RR} is a function of $I_D = I_1$, di_D/dt, and the junction temperature. It has an effect on the reverse recovery current I_{RR} and the reverse recovery time t_{rr}, so it is usually quoted in the data sheets. The fall time t_f can be influenced by the design of the diode. It would seem reasonable to make it short to decrease the turn-off time, but the process is expensive. The bulk of the silicon can be doped with gold or platinum to reduce carrier lifetimes and hence to reduce t_f. The advantage is an increased frequency of switching.

There are two disadvantages associated with this gain in performance. One is an increased on-state voltage drop and the other is an increased voltage recovery overshoot V_{RR}, which is caused by the increased $L_s(di_s/dt)$ as i_D falls more quickly.

Waveforms associated with the reverse recovery behavior of diodes are slightly different in the two specific common cases, the freewheel mode and the rectifier mode. Compared to the case of freewheel mode in Figure 3.3, the rectifier case mode could have more overshooting in the negative direction with high-frequency ringing. This is discussed in [28].

Of the two effects, reverse recovery usually results in the greater power loss and can also generate significant EMI. However, these phenomena were considered to be no big deal at 50 or 60 Hz. With the advent of semiconductor power switches, power conversion began to move into the multikilohertz range, and faster rectifiers were needed.

The relatively long minority carrier lifetime in silicon (tens of microseconds) causes a lot more charge to be stored than is necessary for effective conductivity modulation. In order to speed up reverse recovery, early "fast" rectifiers used various lifetime-killing techniques to reduce the stored minority charge in the lightly doped region. The reverse recovery times of these rectifiers were dramatically reduced, down to about 200 ns, although forward recovery and forward voltage were moderately increased as a side effect of the lifetime-killing process. As power conversion frequencies increased to 20 kHz and beyond, there eventually became a growing need for even faster rectifiers, which caused the "epitaxial" rectifier to be developed.

References [29–32] provide more information on the recovery process of high-frequency rectifiers. Reference [33] provides details on approaches to the model recovery process.

3.2.1.3 Fast and Ultrafast Rectifiers

The foregoing discussion reveals the importance of the switching parameters such as (a) forward recovery time (t_{fr}), (b) forward recovery voltage (V_{FR}), (c) reverse recovery time (t_{rr}), (d) reverse recovery charge (Q_{rr}), and (e) reverse recovery current I_{RM}, during the transition from forward to reverse and vice versa. With various process improvements, fast and ultrafast rectifiers have been achieved within the voltage and current limitations shown in Figure 3.1.

The figure shows that technology is available for devices up to 2000 V ratings and over 1000 A current ratings, which are mutually exclusive. In these diodes although cold t_{rr}

values are good, at high junction temperature, t_{rr} is three to four times higher, increasing switching losses and, in many cases, causing thermal runaway.

Several methods exist to control the switching characteristics of diodes, and each leads to a different interdependency of forward voltage drop V_F, blocking voltage V_{RRM}, and t_{rr} values. It is these interdependencies (or compromises) that differentiate the ultrafast diodes available on the market today. The important parameters for the turn-on and turn-off behavior of a diode are V_{FR}, V_F, t_{fr}, I_{RM}, and t_{rr}, and the values vary depending on the manufacturing processes.

Several manufacturers, such as IXYS Semiconductors and International Rectifier, manufacture a series of ultrafast diodes, termed fast-recovery epitaxial diodes (FREDs), which gained wide acceptance during the 1990s. For an excellent description of these components, see [21].

3.2.1.4 Schottky Rectifiers

Schottky rectifiers occupy a small corner of the total spectrum of available rectifier voltage and current ratings illustrated in Figure 3.1(b). They are, nonetheless, the rectifier of choice for low-voltage switching power supply applications, with output voltages up to a few tens of volts, particularly at high switching frequency. For this reason, Schottkys account for a major segment of today's total rectifier usage. The Schottkys' unique electrical characteristics set them apart from conventional P-N junction rectifiers, in the following important respects:

- Lower forward voltage drop
- Lower blocking voltage
- Higher leakage current
- Virtual absence of reverse recovery charge

The two fundamental characteristics of the Schottky that make it a winner over the P-N junction rectifier in low-voltage switching power supplies are its lower forward voltage drop and virtual absence of minority carrier reverse recovery.

The absence of minority carrier reverse recovery means virtual absence of switching losses within the Schottky itself. Perhaps more significantly, the problem of switching voltage transients and attendant oscillations is less severe for Schottkys than for P-N junction rectifiers. Snubbers are therefore smaller and less dissipative.

The lower forward voltage drop of the Schottky means lower rectification losses, better efficiency, and smaller heat sinks. Forward voltage drop is a function of the Schottky's reverse voltage rating. The maximum voltage rating of today's Schottky rectifiers is about 150 V. At this voltage, the Schottky's forward voltage drop is lower than that of a fast-recovery epitaxial P-N junction rectifier by 150 to 200 mV.

At lower voltage ratings, the lower forward voltage drop of the Schottky becomes progressively more pronounced, and more of an advantage. A 45 V Schottky, for example, has a forward voltage drop of 0.4 to 0.6 V, versus 0.85 to 1.0 V for a fast-recovery epitaxial P-N junction rectifier. A 15 V Schottky has a mere 0.3 to 0.4 V forward voltage drop.

A conventional fast-recovery epitaxial P-N junction rectifier, with a forward voltage drop of 0.9 V, would dissipate about 18% of the output power of a 5 V supply. A Schottky, by contrast, reduces rectification losses to the range of 8% to 12%. These are the simple reasons why Schottkys are virtually always preferred in low-voltage, high-frequency switching power supplies. For any given current density, the Schottky's forward voltage drop increases as its reverse repetitive maximum voltage (V_{RRM}) increases. The basic hallmarks of any process are its maximum-rated junction temperature—the T_{jmax} class and the "prime" rated voltage, the V_{rrm} class. These two basic hallmarks are set by the process; they in turn determine the forward voltage drop and reverse leakage current characteristics. Figure 3.4 indicates this condition for T_{jmax} of 150°C.

3.2.1.4.1 Leakage Current and Junction Capacitance of Schottky Diodes

Figure 3.5 shows the dependence of leakage current on the operating voltage and junction temperature within any given process. Reverse leakage current increases with applied reverse voltage and with junction temperature. Figure 3.6 shows typical relationship between operating temperature and leakage current, at rated V_{RRM}, for the 150°C/45 V and 175°C/45 V Schottky processes.

An important circuit characteristic of the Schottky is its junction capacitance. This is a function of the area and thickness of the Schottky die, and of the applied voltage. The higher the V_{RRM} class, the greater the die thickness and the lower the junction capacitance. This is illustrated in Figure 3.7. Junction capacitance is essentially independent of the Schottky's T_{jmax} class and of operating temperature.

Around early 2000, manufacturers such as International Rectifier have introduced better Schottky diodes, based on trench technology, which have lower forward voltage drop and lower reverse leakage current than the common planar Schottky diodes. The new optimized trench Schottky diodes (an example is 80CPT015) offer 15% reduction in forward drop and a fourfold improvement in reverse leakage current [34]. Another

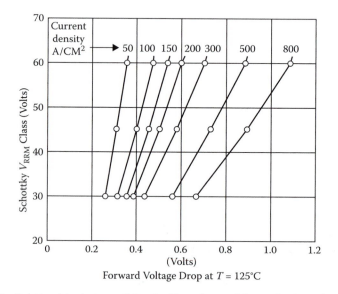

FIGURE 3.4 Relationships between Schottky V_{RRM} class and forward voltage drop, for 150°C T_{jmax} class devices. (Reproduced from International Rectifier, USA. With permission.)

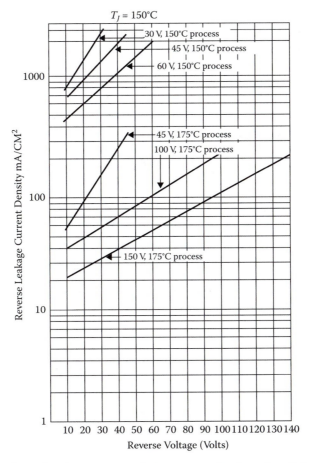

FIGURE 3.5 Relationships between reverse leakage current density and applied reverse voltage. (Reproduced from International Rectifier, USA. With permission.)

such advancement reported was the trench MOS barrier Schottky diodes, introduced by Vishay Semicondcutors, which offers low forward voltage together with higher blocking voltages such as 120 V to 150 V, for example parts such as Vishay VTS40100CT with 40 A, 100 V ratings [35].

3.2.1.5 GaAs Power Diodes

Efficient power conversion circuitry requires rectifiers that exhibit low forward voltage drop, low reverse recovery current, and fast recovery time. Silicon has been the material of choice for fast, efficient rectification in switched power applications. However, technology is nearing the theoretical limit for optimizing reverse recovery in silicon devices.

To increase speed, materials with faster carrier mobility are needed. Gallium arsenide (GaAs) has a carrier mobility that is five times that of silicon [24]. Since Schottky

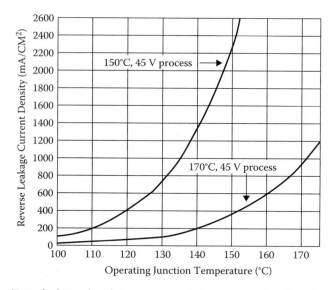

FIGURE 3.6 Typical relationships between reverse leakage current density and operating junction temperature. (Reproduced from International Rectifier, USA. With permission.)

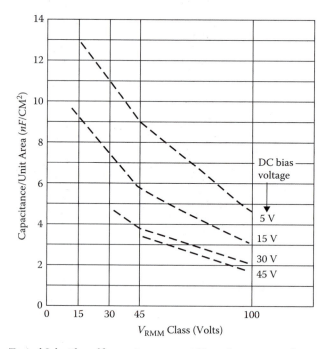

FIGURE 3.7 Typical Schottky self-capacitance versus V_{RRM} class, measured at various bias voltages. (Reproduced from International Rectifier, USA. With permission.)

technology for silicon devices is difficult to produce at voltages above 200 V, development has focused on GaAs devices with ratings of 180 V and higher. The advantages realized by using GaAs rectifiers include fast switching and reduced reverse recovery-related parameters. An additional benefit is that the variation of parameters with temperature is much less than silicon rectifiers.

For example, Motorola's 180 V and 250 V GaAs rectifiers are being used in power converters that produce 24, 36, and 48 V DC outputs. Converters producing 48 V DC, especially popular in telecommunications and mainframe computer applications, could gain the advantage of GaAs parts compared to similar silicon-based parts at switching frequencies around 1 MHz [25].

The 180 V devices offered by Motorola can increase power density in 48 V DC applications up to 90 W/in^3 [21]. These devices allow designers to switch converters at 1 MHz without generating large amounts of EMI.

Figures 3.8(a) and 3.8(b) indicate typical forward voltage and typical reverse current for 20 A, 180 V GaAs parts from Motorola.

For further details, the reader is directed to [1,24,25].

3.2.1.6 Silicon Carbide (SiC) Schottky Diodes

A new option for high-speed diode requirements is the SiC Schottky diodes, which started appearing on the market around 2002. These devices have a much bigger blocking voltage capability up to about 3.5 kV, a bandgap 3 times higher than silicon, a breakdown field 10 times higher than silicon, and a thermal conductivity comparable with copper. Figure 3.9(a) depicts blocking voltage capabilities of silicon, GaAs, and SiC devices. Figure 3.9(b) compares the reverse recovery characteristics of 600 V capability ultrafast diodes and an SiC device. More details on SiC power semiconductors can be found in [36–40].

3.2.2 Thyristors

The thyristor is a four-layer, three-terminal device as depicted in Figure 3.10. The complex interactions between three internal P-N junctions are then responsible for the device characteristics. However, the operation of the thyristor and the effect of the gate in controlling turn-on can be illustrated and followed by reference to the two-transistor model of Figure 3.11. Here, the p_1-n_1-p_2 layers are seen to make up a PNP transistor, and the n_2-p_2-n_1 layers create an NPN transistor with the collector of each transistor connected to the base of the other.

With a reverse voltage, cathode positive with respect to the anode, applied to the thyristor, the p_1-n_1 and p_2-n_2 junctions are reverse biased and the resulting characteristic is similar to that of the diode with a small reverse leakage current flowing up to the point of reverse breakdown as shown by Figure 3.12(a). With a forward voltage applied and no gate current supported, the thyristor is in the forward-blocking mode. The emitters of the two transistors are now forward biased and no conduction occurs. As the applied voltage is increased, the leakage current through the transistors increases to the point at which the positive feedback resulting from the base/collector connections drives both transistors into saturation, turning them, and hence the thyristor, on. The thyristor is

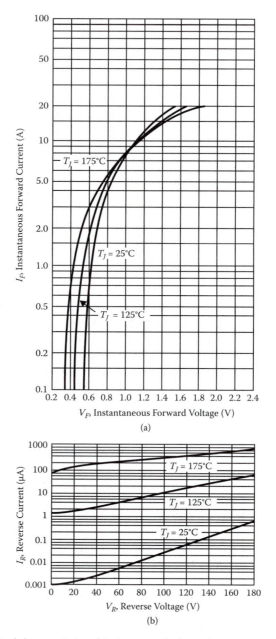

FIGURE 3.8   Typical characteristics of GaAs power diodes with 20 A, 180 V ratings: (a) forward voltage; (b) reverse current. (Copyright © Motorola, Inc. With permission.)

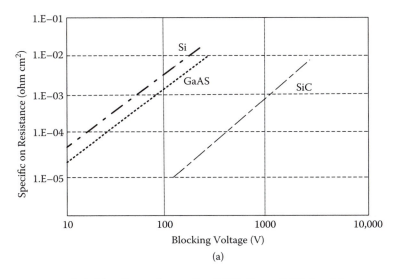

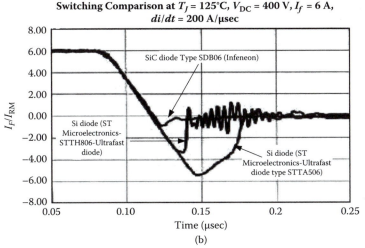

FIGURE 3.9 Comparison SiC diodes with other devices: (a) blocking voltage capability; (b) reverse recovery characteristics. (Adapted from Kapels, H., M. Krach, et al., *Power Electronics Technology*, January 2002, 15–21.)

now conducting, and the forward voltage drop across it falls to a value of the order of 1 to 2 V. This condition is also shown in the thyristor static characteristic of Figure 3.12(a).

If a current is injected into the gate at a voltage below the breakover voltage, this will cause the NPN transistor to turn on. The positive feedback loop will then initiate the turn-on of the PNP transistor. Once both transistors are on, the gate current can be removed because the action of the positive feedback loop will be to hold both transistors, and hence the thyristor, in the on state.

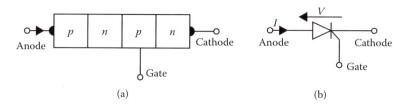

FIGURE 3.10   The thyristor: (a) construction; (b) circuit symbol.

The effect of the gate current is therefore to reduce the effective voltage at which forward breakover occurs, as illustrated by the Figure 3.12(b). After the thyristor has been turned on, it will continue to conduct as long as the forward current remains above the holding current level, irrespective of gate current or circuit conditions.

3.2.2.1 Ratings and Different Types of Devices

The operation of all power semiconductors is limited by a series of ratings that define the operating boundaries of the device. These ratings include limits on the peak, average, and RMS currents, the peak forward and reverse voltages for the devices, maximum rates of change of device current and voltage, device junction temperature, and, in the case of the thyristor, the gate current limits.

The current ratings of a power semiconductor are related to the energy dissipation in the device and hence the device junction temperature. The maximum value of on-state current $(I_{av(max)})$ is the maximum continuous current the device can sustain under defined conditions of voltage and current waveform without exceeding the permitted temperature rise in the device. The RMS current rating (I_{RMS}) is similarly related to the permitted temperature rise when operating into a regular-duty cycle load.

In the case of transient loads, as the internal losses and hence the temperature rise in a power semiconductor are related to the square of the device forward current, the relationship between the current and the permitted temperature rise can be defined in

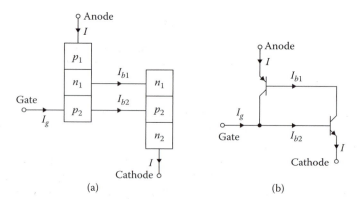

FIGURE 3.11   The two-transistor model of a thyristor: (a) structure; (b) the PNP and NPN transistor combination.

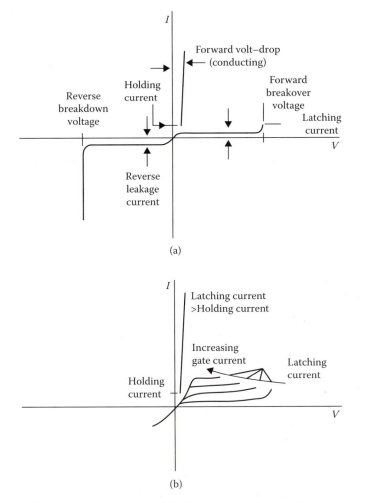

FIGURE 3.12 Thyristor characteristics: (a) thyristor characteristics with zero gate current; (b) switching characteristics.

terms of an i^2dt rating for the device. On turn-on, current is initially concentrated into a very small area of the device cross section and the device is therefore subject to a di/dt rating, which sets a limit to the permitted rate of rise of forward current.

The voltage ratings of a power semiconductor device are primarily related to the maximum forward and reverse voltages that the device can sustain. Typically, values will be given for the maximum continuous reverse voltage ($V_{RC(max)}$), the maximum repetitive reverse voltage ($V_{RR(max)}$), and the maximum transient reverse voltage ($V_{RT(max)}$). Similar values exist for the forward-voltage ratings.

The presence of a fast transient of forward voltage can cause a thyristor to turn on, and a dv/dt rating is therefore specified for the device. The magnitude of the imposed dv/dt can be controlled by the use of a snubber circuit connected in parallel with the thyristor. Data sheets for thyristors always quote a figure for the maximum surge current I_{TSM} that the device can survive.

This figure assumes a half sine pulse with a width of either 8.3 or 10 ms, which are the conditions applicable for 60 or 50 Hz mains, respectively. This limit is not absolute; narrow pulses with much higher peaks can be handled without damage, but little information is available to enable the designer to determine a current rating for short pulses. See [2] for guidelines in this area.

Ever since its introduction, circuit design engineers have been subjecting the thyristor to increasing levels of operating stress and demanding that these devices perform satisfactorily there. The different stress demands that the thyristor must be able to meet are:

Higher blocking voltages
More current-carrying capability
Higher *di/dt*'s
Higher *dv/dt*'s
Shorter turn-off times
Lower gate drive
Higher operating frequencies.

There are many different thyristors available today that can meet one or more of these requirements, but, as always, an improvement in one characteristic is usually gained only at the expense of another. As a result, different thyristors have been optimized for different applications. Modern thyristors can be classified into several general types, namely:

Phase control thyristors
Inverter thyristors
Asymmetrical thyristors
Reverse-conducting thyristors (RCTs)
Light-triggered thyristors.

The voltage and current capabilities of phase control thyristors and inverter thyristors from a power device manufacturer are summarized in Figure 3.13.

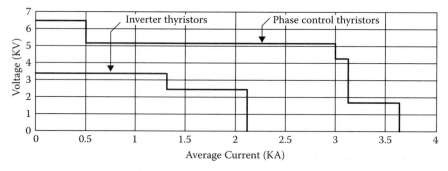

FIGURE 3.13 Thyristor rating capabilities. (Reproduced from GEC Plessey Semiconductors, UK. With permission.)

3.2.2.1.1 Phase Control Thyristors

Phase control or *converter* thyristors generally operate at line frequency. They are turned off by natural commutation and do not have special fast-switching characteristics.

Current ratings of phase control thyristors cover the range from a few amperes to about 3500 A, and voltage ratings from 50 to over 6500 V. To simplify the gate drive requirement and increase sensitivity, the use of an amplifying gate, which was originally developed for fast-switching "inverter" thyristors, is widely adopted in phase control SCR.

3.2.2.1.2 Inverter Thyristors

The most common feature of an inverter thyristor that distinguishes it from a standard phase control type is that it has fast turn-off time, generally in the range of 5 to 50 μs, depending upon voltage rating. Maximum average current ratings of over 2000 and 1300 A have been achieved with 2000 V- and 3000 V-rated inverter thyristors, respectively.

Inverter thyristors are generally used in circuits that operate from DC supplies, where current in the thyristor is turned off either through the use of auxiliary comutating circuitry, by circuit resonance, or by "load" commutation. Whatever the circuit turn-off mechanism, fast turn-off is important because it minimizes sizes and weight of comutating and/or reactive circuit components.

3.2.2.1.3 Asymmetrical Thyristors

One of the main salient characteristics of asymmetrical thyristors (ASCRs) is that they do not block significant reverse voltage. They are typically designed to have a reverse-blocking capability in the range of 400 to 2000 V.

The ASCR finds applications in many voltage-fed inverter circuits that require anti-parallel feedback rectifiers that keep the reverse voltage to less than 20 V. The fact that ASCR needs to block voltage only in the forward direction provides an extra degree of freedom in optimizing turn-off time, turn-on time, and forward-voltage drop.

3.2.2.1.4 Reverse-Conducting Thyristors

The reverse-conducting thyristor (RCT) represents the monolithic integration of an asymmetrical thyristor with an antiparallel rectifier. Beyond obvious advantages of the parts count reduction, the RCT eliminates the inductively induced voltage within the thyristor-diode loop (virtually unavoidable to some extent with separate discrete components). Also, it essentially limits the reverse voltage seen by the thyristor to only the conduction voltage of the diode.

3.2.2.1.5 Light-Triggered Thyristors

Many developments have taken place in the area of light-triggered thyristors. Direct irradiation of silicon with light-created electron-hole pairs, which, under the influence of an electric field, produce a current that triggers the thyristors.

The turn-on of a thyristor by optical means is an especially attractive approach for devices that are to be used in extremely high-voltage circuits. A typical application area is in switches for DC transmission lines operating in the hundreds of kilovolts range, which use series connections of many devices, each of which must be triggered on

TABLE 3.1 Thyristor Types and Popular Names

JEDEC Titles	Popular Names, Types
Reverse-blocking diode thyristor	[a]Four-layer diode, silicon unilateral switch (SUS)[a]
Reverse-blocking triode thyristor	Silicon-controlled rectifier (SCR)
Reverse-conducting diode thyristor	[a]Reverse-conducting four-layer diode
Reverse-conducting triode thyristor	Reverse-conducting SCR
Bidirectional triode thyristor	Triac
Turn-off thyristor	Gate turn-off switch (GTO)

[a] Not generally available.

command. Optical firing in this application is ideal for providing the electrical isolation between trigger circuits and the thyristor, which floats at a potential as high as hundreds of kilovolts above ground.

The main requirement for an optically triggered thyristor is high sensitivity while maintaining high dv/dt and di/dt capabilities. Because of the small and limited quantity of photo energy available for triggering the thyristor from practical light sources, very high gate sensitivity of the order of 100 times that of the electrically triggered device is needed.

3.2.2.1.6 JEDEC Titles and Popular Names

Table 3.1 compares the Joint Electronic Device Engineering Council (JEDEC) titles for commercially available thyristors types with popular names. JEDEC is an industry standardization activity cosponsored by the Electronic Industries Association (EIA) and the National Manufacturers Association (NEMA). Silicon-controlled rectifiers (SCRs) are the most widely used power control elements. Triacs are quite popular in lower-current (<40 A) AC power applications.

3.3 Gate Turn-Off Thyristors

A gate turn-off thyristor (GTO) is a thyristorlike latching device that can be turned off by application of a negative pulse of current to its gate. This gate turn-off capability is advantageous because it provides increased flexibility in circuit application. It now becomes possible to control power in DC circuits without the use of elaborate commutation circuitry.

Prime design objectives for GTO devices are to achieve fast turn-off time and high current turn-off capability and to enhance the safe-operating area during turnoff. Significant progress has been made in both areas during the last few years, largely due to a better understanding of the turn-off mechanisms. The GTO's turn-off occurs by removal of excess holes in the cathode-base region by reversing the current through the gate terminal.

The GTO has gained popularity in switching circuits, especially in equipment that operates directly from European mains. The GTO offers the following advantages over a bipolar transistor: high blocking voltage capabilities, in excess of 1500 V, and also high overcurrent capabilities. It also exhibits low gate currents, fast and efficient turn-off, as well as outstanding static and dynamic dv/dt capabilities.

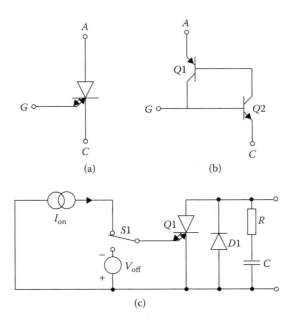

FIGURE 3.14 GTO symbol, equivalent circuit, and basic drive circuit: (a) symbol of GTO; (b) two-transistor equivalent of GTO; (c) basic drive circuit.

Figure 3.14(a) depicts the symbol of GTO, and Figure 3.14(b) shows its two-transistor equivalent circuit. Figure 3.14(c) shows a basic drive circuit. The GTO is turned on by a positive gate current, and it is turned off by applying a negative gate cathode voltage.

A practical implementation of a GTO gate drive circuit is shown in Figure 3.15. In this circuit when transistor Q_2 is off, emitter follower transistor Q_1 acts as a current source pumping current into the gate of the GTO through a 12 V zener Z_1 and polarized

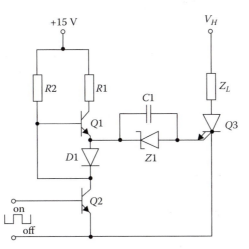

FIGURE 3.15 Practical realization of a GTO gate drive circuit.

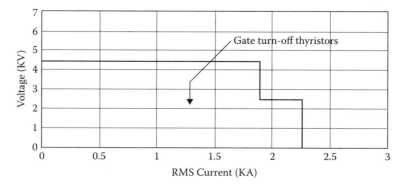

FIGURE 3.16 Ratings covered by available GTOs. (Reproduced from GEC Plessey Semiconductors, UK. With permission.)

capacitor C_1. When the control voltage at the base of Q_2 goes positive, transistor Q_2 turns on, while transistor Q_1 turns off since its base now is one diode drop more negative than its emitter. At this stage the positive side of capacitor C1 is essentially grounded, and C1 will act as a voltage source of approximately 10 V, turning the GTO off. Isolated gate drive circuits may also be easily implemented to drive the GTO.

With improved cathode emitter geometries and better optimized vertical structures, today's GTOs have made significant progress in turn-off performance (the prime weakness of earlier GTOs). Figure 3.16 shows the available GTO ratings and, as can be seen, they cover quite a wide spectrum. However, the main applications lie in the higher voltage end (>1200 V) where bipolar transistors and power MOSFETs are unable to compete effectively. In the present-day market there are GTOs with current ratings over 3000 A and voltage ratings over 4500 V. For further details on GTOs, see [3–5].

3.4 Bipolar Power Transistors

During the last two decades, attention has been focused on high-power transistors as switching devices in inverters, SMPSs, and similar switching applications. New devices with faster switching speeds and lower switching losses are being developed that offer performance beyond that of thyristors. With their faster speed, they can be used in an inverter circuit operating at frequencies over 200 KHz. In addition, these devices can be readily turned off with a low-cost reverse base drive without the costly commutation circuits required by thyristors.

3.4.1 Bipolar Transistor as a Switch

The bipolar transistor is essentially a current-driven device. That is, by injecting a current into the base terminal a flow of current is produced in the collector. There are essentially two modes of operation in a bipolar transistor: the linear and saturating modes. The linear mode is used when amplification is needed, while the saturating mode is used to switch the transistor either on or off.

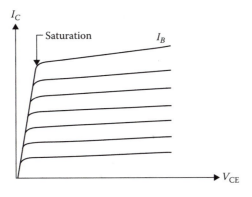

FIGURE 3.17 Typical output characteristics of BJT.

Figure 3.17 shows the V-I characteristic of a typical bipolar transistor. Close examination of these curves shows that the saturation region of the V-I curve is of interest when the transistor is used in a switching mode. At that region a certain base current can switch the transistor on, allowing a large amount of collector current to flow, while the collector-to-emitter voltage remains relatively small.

In actual switching applications a base drive current is needed to turn the transistor on, while a base current of reverse polarity is needed to switch the transistor back off. In practical switching operations certain delays and storage times are associated with transistors. In the following section are some parameter definitions for a discrete bipolar transistor driven by a step function into a resistive load.

Figure 3.18 illustrates the base-to-emitter and collector-to-emitter waveforms of a bipolar NPN transistor driven into a resistive load by a base current pulse I_B. The

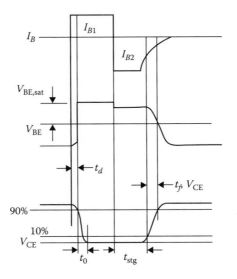

FIGURE 3.18 Bipolar transistor switching waveforms.

following are the definitions associated with these waveforms:

Delay Time, t_d: Delay time is defined as the interval of time from the application of the base drive current I_{B1} to the point at which the collector-emitter voltage V_{CE} has dropped to 90% of its initial off value.

Rise Time, t_r: Rise time is defined as the interval of time it takes the collector-emitter voltage V_{CE} to drop to 10% from its 90% off value.

Storage Time, t_{stg}: Storage time is the interval of time from the moment reverse base drive I_{B2} is applied to the point where the collector-emitter voltage V_{CE} has reached 10% of its final off value.

Fall Time, $t_{f,VCE}$: Fall time is the time interval required for the collector-emitter voltage to increase from 10% to 90% of its off value.

3.4.2 Inductive Load Switching

In the previous section, the definitions for the switching times of the bipolar transistor were made in terms of collector-emitter voltage. Since the load was defined to be a resistive one, the same definitions hold true for the collector current. However, when the transistor drives an inductive load, the collector voltage and current waveforms will differ. Since current through an inductor does not flow instantaneously with applied voltage, during turn-off one expects to see the collector-emitter voltage of a transistor rise to the supply voltage before the current begins to fall. Thus two fall time components may be defined, one for the collector-emitter voltage $t_{f,VCE}$ and the other for the collector current $t_{f,Ic}$. Figure 3.19 shows the actual waveforms.

Observing the waveforms in Figure 3.19 we can define the collector-to-emitter fall time $t_{f,VCE}$ in the same manner as in the resistive case, while the collector fall time $t_{f,Ic}$ may be defined as the interval in which collector current drops from 90% to 10% of its initial value. Normally, the load inductance L behaves as a current source, and therefore it charges the base-collector transition capacitance faster than the resistive load. Thus for the same base and collector currents, the collector-emitter voltage fall time $t_{f,VCE}$ is shorter for the inductive circuit.

3.4.3 Safe Operating Area and V-I Characteristics

The output characteristics (I_C versus V_{CE}) of a typical NPN power transistor are shown in Figure 3.20(a). The various curves are distinguished from each other by the value of the base current.

Several features of the characteristics should be noted. First, there is a maximum collector-emitter voltage that can be sustained across the transistor when it is carrying substantial collector current. The voltage is usually labeled BV_{SUS}. In the limit of zero base current, the maximum voltage between collector and emitter that can be sustained increases somewhat to a value labeled BV_{CEO}, the collector-emitter breakdown voltage when the base is open circuited. This latter voltage is often used as the measure of the transistor's voltage standoff capability because usually the only time the transistor will see large voltages is when the base current is zero and the BJT is in cutoff.

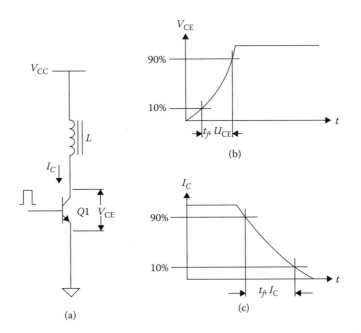

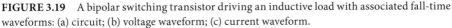

FIGURE 3.19 A bipolar switching transistor driving an inductive load with associated fall-time waveforms: (a) circuit; (b) voltage waveform; (c) current waveform.

The voltage BV_{CBO} is the collector-base breakdown voltage when the emitter is open circuited. The fact that this voltage is larger than BV_{CEO} is used to advantage in so-called open-emitter transistor turn-off circuits.

The region labeled *primary breakdown* is due to conventional avalanche breakdown of the collector-base junction and the attendant large flow of current. This region of the characteristics is to be avoided because of the large power dissipation that clearly accompanies such breakdown.

The region labeled *second breakdown* must also be avoided because large power dissipation also accompanies it, particularly at localized sites within the semiconductor. The origin of second breakdown is different from that of avalanche breakdown and will be considered in detail later in this chapter. BJT failure is often associated with second breakdown.

The major observable difference between the I-V characteristics of a power transistor and those of a logic-level transistor is the region labeled *quasi-saturation* on the power transistor characteristics of Figure 3.20(a). Quasi-saturation is a consequence of the lightly doped collector drift region found in the power transistor.

Logic-level transistors do not have this drift region and so do not exhibit quasi-saturation. Otherwise all of the major features of the power transistor characteristic are also found on those of logic-level devices.

Figure 3.20(b) indicates the relative magnitudes of NPN transistor collector breakdown characteristics, showing primary and secondary breakdown with different base bias conditions. With low-gain devices V_a is close to V_b in value, but with high-gain devices V_b may be two to three times that of V_a. Notice that negative resistance characteristics occur

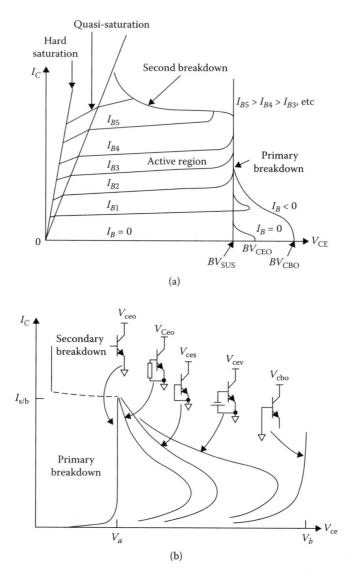

FIGURE 3.20 Current-voltage characteristics of an NPN power transistor showing breakdown phenomenon: (a) indication of quasi saturation; (b) relative primary and secondary breakdown conditions for different bias levels.

after breakdown, as is the case with all the circuit-dependent breakdown characteristics. Reference [22] provides a detailed explanation on the behavior.

3.4.3.1 Forward-Bias Secondary Breakdown

In the switching process BJTs are subjected to great stress, during both turn-on and turn-off. It is imperative that the engineer clearly understand how the power bipolar

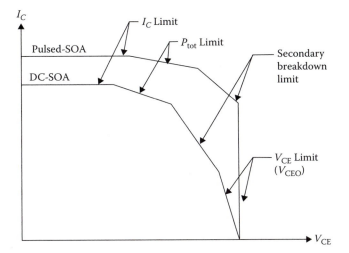

FIGURE 3.21   DC and pulse SOA for BJT.

transistor behaves during forward- and reverse-bias periods in order to design reliable and trouble-free circuits.

The first problem is to avoid secondary breakdown of the switching transistor at turn-on, when the transistor is forward biased. Normally the manufacturer's specifications will provide a safe-operating area (SOA) curve, such as the typical one shown in Figure 3.21. In this figure collector current is plotted against collector-emitter voltage. The curve locus represents the maximum limits at which the transistor may be operated. Load lines that fall within the pulsed forward-bias SOA curve during turn-on are considered safe, provided that the device thermal limitations and the SOA turn-on time are not exceeded.

The phenomenon of forward-biased secondary breakdown is caused by hot spots that are developed at random points over the working area of a power transistor, caused by unequal current conduction under high-voltage stress. Since the temperature coefficient of the base-to-emitter junction is negative, hot spots increase local current flow. More current means more power generation, which in turn raises the temperature of the hot spot even more.

Since the temperature coefficient of the collector-to-emitter breakdown voltage is also negative, the same rules apply. Thus the voltage stress is not removed, ending the current flow, the collector-emitter junction breaks down, and the transistor fails because of thermal runaway.

3.4.3.2 Reverse-Bias Secondary Breakdown

It was mentioned in previous paragraphs that when a power transistor is used in switching applications, the storage time and switching losses are the two most important parameters with which the designer has to deal extensively.

On the other hand the switching losses must also be controlled since they affect the overall efficiency of the system. Figure 3.22 shows turn-off characteristics of a high-voltage power transistor in resistive and inductive loads. Inspecting the curves we can

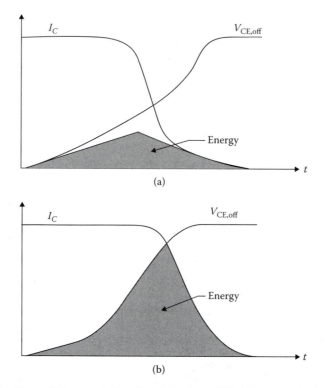

I_C $V_{CE,off}$

Energy

t

(a)

I_C $V_{CE,off}$

Energy

t

(b)

FIGURE 3.22   Turn-off characteristics of a high-voltage BJT: (a) resistive load; (b) inductive load.

see that the inductive load generates a much higher peak energy at turn-off than its resistive counterpart. It is then possible, under these conditions, to have a secondary breakdown failure if the reverse-bias safe-operating area (RBSOA) is exceeded.

The RBSOA curve (Figure 3.23) shows that for voltages below V_{CEO} the safe area is independent of reverse-bias voltage V_{EB} and is limited only by the device collector current IC. Above V_{CEO} the collector current must be derated depending upon the applied reverse-bias voltage.

It is then apparent that the reverse-bias voltage V_{EB} is of great importance and its effect on RBSOA very interesting. It is also important to remember that avalanching the base-emitter junction at turn-off must be avoided, since turn-off switching times may be decreased under such conditions. In any case, avalanching the base-emitter junction may not be considered relevant, since normally designers protect the switching transistors with either clamp diodes or snubber networks to avoid such encounters.

3.4.4 Darlington Transistors

The on-state voltage $V_{CE(sat)}$ of the power transistors is usually in the 1–2 V range, so that the conduction power loss in the BJT is quite small. BJTs are current-controlled devices, and base current must be supplied continuously to keep them in the on state. The DC

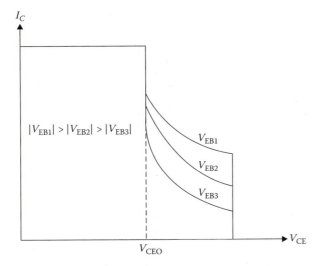

FIGURE 3.23 RBSOA plot for a high-voltage BJT as a function of reverse-bias voltage V_{EB}.

current gain h_{FE} is usually only 5–10 in high-power transistors and so these devices are sometimes connected in a Darlington or triple Darlington configuration as is shown in Figure 3.24 to achieve a larger current gain. However, some disadvantages accrue in this configuration including slightly higher overall $V_{CE(sat)}$ values and slower switching speeds. The current gain of the pair h_{FE} is

$$h_{FE} = h_{FE1} \cdot h_{FE2} + h_{FE1} + h_{FE2} \tag{3.4}$$

Darlington configurations using discrete BJTs or several transistors on a single chip (a monolithic Darlington [MD]) have significant storage time during the turn-off transition. Typical switching times are in the range of a few hundred nanoseconds to a few microseconds.

BJTs including MDs are available in voltage ratings up to 1400 V and current ratings of a few hundred amperes. In spite of a negative temperature coefficient of on-state

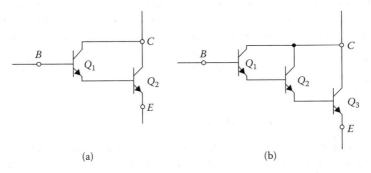

FIGURE 3.24 Darlington configurations: (a) Darlington; (b) triple Darlington.

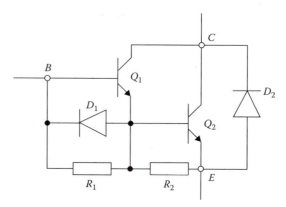

FIGURE 3.25   A practical monolithic Darlington pair.

resistance, modern BJTs fabricated with good quality control can be paralleled provided that care is taken in the circuit layout and that some extra current margin is provided.

Figure 3.25 shows a practical monolithic Darlington pair with diode D_1 added to speed up the turn-off time of Q_1 and D_2 added for half- and full-bridge circuit applications. Resistors R_1 and R_2 are low-value resistors and provide a leakage path for Q_1 and Q_7.

3.5 Power MOSFETs

3.5.1 Introduction

Compared to BJTs, which are current-controlled devices, field-effect transistors (FETs) are voltage-controlled devices. There are two basic FETs: the junction FET (JFET) and the metal-oxide semiconductor FET (MOSFET). Both have played important roles in modern electronics. The JFET has found wide application in such cases as high-impedance transducers (scope probes, smoke detectors, etc.) and the MOSFET in an ever-expanding role in integrated circuits, where CMOS (complementary MOS) is perhaps the most well known.

Power MOSFETs differ from bipolar transistors in operating principles, specifications, and performance. In fact, the performance characteristics of MOSFETs are generally superior to those of bipolar transistors: significantly faster switching time, simpler drive circuitry, the absence of a second breakdown failure mechanism, the ability to be paralleled, and stable gain and response time over a wide temperature range. The MOSFET was developed out of the need for a power device that could work beyond the 20 kHz frequency spectrum, anywhere from 100 kHz to above 1 MHz, without experiencing the limitations of the bipolar power transistor.

3.5.2 General Characteristics

Bipolar transistors are described as minority-carrier devices in which injected minority carriers recombine with majority carriers. A drawback of recombination is that it limits the device's operating speed. Current-driven base-emitter input of a bipolar transistor presents a low-impedance load to its driving circuit. In most power circuits, this low-impedance input requires somewhat complex drive circuitry.

By contrast, a power MOSFET is a voltage-driven device whose gate terminal is electrically isolated from its silicon body by a thin layer of silicon dioxide (SiO_2) as shown in the simplified schematic of a MOSFET shown in Figure 3.26(a). The n-channel device formed on a p-substrate carries two n+ regions for the drain and the source areas between which a channel of length l is formed. As a majority-carrier semiconductor, the MOSFET operates at much higher speed than its bipolar counterpart because there is no charge-storage mechanism. A positive voltage applied to the gate of an n-type MOSFET

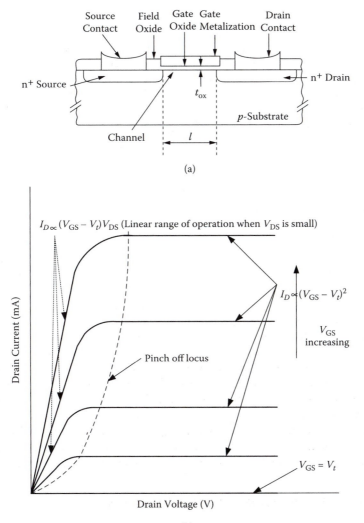

(a)

(b)

FIGURE 3.26 Structure of a MOSFET, characteristics, and symbol: (a) basic concept; (b) characteristics; (c) symbol for n-channel device.

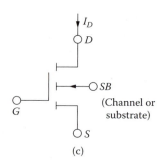

FIGURE 3.26   (Continued).

creates an electric field in the channel region just beneath the gate oxide region with a thickness of t_{ox}; that is, the electric charge on the gate causes the p-region beneath the gate to convert to an n-type region, as shown in Figure 3.26(a).

This conversion, called the surface-inversion phenomenon, allows current to flow between the drain and source through an n-type material. In effect, the MOSFET ceases to be an NPN device when in this state. The region between the drain and source can be represented as a resistor, although it does not behave linearly, as a conventional resistor would. Because of this surface-inversion phenomenon, the operation of a MOSFET is entirely different from that of a bipolar transistor.

There are different commercial variations of power MOSFETS, usually referred to as *planar* and *trench* designs. Figure 3.27(a) indicates the details of planar MOSFET, and Figures 3.27(b) and 3.27(c) indicate trench MOSFET versions known as current-crowding V-groove Trench MOSFET and the truncated V-groove MOSFET.

By virtue of its electrically isolated gate, a MOSFET is described as a high-input impedance, voltage-controlled device, compared to a bipolar transistor. As a majority-carrier semiconductor, a MOSFET stores no charge and so can switch faster than a bipolar device. Majority-carrier semiconductors also tend to slow down as temperature increases. This effect brought about by another phenomenon called carrier mobility makes a MOSFET more resistive at elevated temperatures, and much more immune to the thermal runaway problem experienced by bipolar devices. *Mobility* is a term that defines the average velocity of a carrier in terms of the electrical field imposed on it.

A useful by-product of the commercially produced MOSFETs is the internal parasitic diode formed between source and drain, as shown in right-hand side of Figure 3.27(b). (There is no equivalent for this diode in a bipolar transistor other than in a bipolar Darlington transistor.) Its characteristics make it useful as a clamp diode in inductive-load switching.

Different manufacturers use different techniques for constructing a power FET, and names like *VMOS*, *TMOS*, *Cool MOS*, etc., have become trademarks of specific companies.

3.5.3 MOSFET Structures and On Resistance

Most power MOSFETs are manufactured using various proprietary processes by various manufacturers on a single silicon chip structured with a large number of closely packed identical cells. For example, Intersil (formerly Harris Semiconductor) power MOSFETS

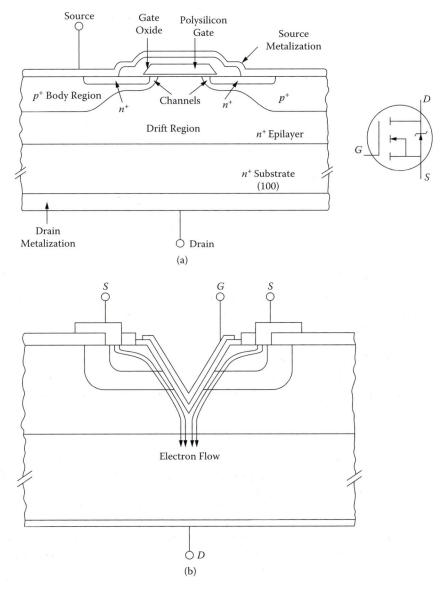

FIGURE 3.27 Different commercial implementations of power MOSFET: (a) planar MOSFET; (b) current crowding in V-Groove Trench MOSFET; (c) truncated V-groove MOSFET. (Reproduced from Barkhordarian, V., *PCIM*, June 1997, 28–36. With permission.)

are manufactured using a vertical double-diffused process, called VDMOS or simply DMOS. In these cases, a 120 mil² chip contains about 5000 cells and a 240 mil² chip has more than 25,000 cells.

One of the aims of multiple-cells construction is to minimize the MOSFET parameter $R_{DS(ON)}$ when the device is in the on-state. When $R_{DS(ON)}$ is minimized, the device

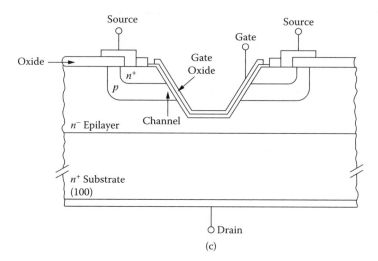

FIGURE 3.27   (Continued)

provides superior power-switching performance because the voltage drop from drain to source is also minimized for a given value of drain-source current. Reference [6] provides more details.

Figure 3.28(a) indicates different components of resistances (contributing to $R_{DS(ON)}$) and stray capacitances contributing to delays in operation of the devices. In addition a power MOSFET comes with a few parasitic devices such as diodes, BJT, and a JFET in general. The on resistance of a power MOSFET is made up of several components as shown in Figure 3.28(b) where

$$R_{DS(on)} = R_{Source} + R_{Ch} + R_A + R_J + R_D + R_{sub} + R_{wcml}$$

where R_{source} is the source diffusion resistance, R_{Ch} is the channel resistance, R_A is the accumulation resistance, R_J is the JFET component, R_D is the drift region resistance, and R_{Sub} is the substrate resistance.

R_{wcml} is the sum of many components contributed by the bond wire, contact resistance between the source and drain, metallization, the silicon and any lead frame contributions. Following the relationship in [42] for the specific on resistance, $R_{on, sp}$, neglecting secondary effects, indicates that the higher-voltage devices usually have a higher on resistance

$$R_{on,sp} = 8.06 * 10^{-6} \frac{(BV_{DSS})^{2.5}}{\mu}$$

where $R_{on, sp}$ is the $R_{DS\,(on)}$ per unit area, and μ is the mobility of electrons (for n-channel devices), and BV_{DSS} is the maximum drain-source voltage. Wafers with substrate resistivity of up to 20 mΩ·cm are used for high-voltage devices and less than 50 mΩ.cm are used for low-voltage devices in general. R_{wcml} component's contribution in high-voltage devices is negligible in general. At high voltages the $R_{DS(on)}$ is dominated by the contribution from the epi-layer components [41].

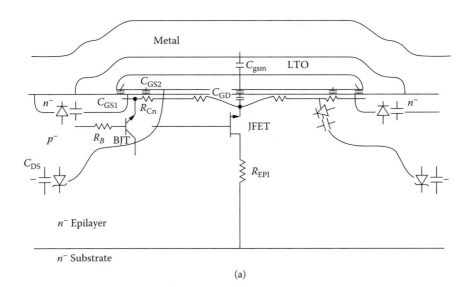

(a)

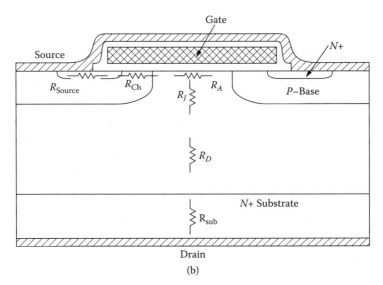

(b)

FIGURE 3.28 Power MOSFET details related to fabrication: (a) internal resistances, stray capacitances, and parasitic devices; (b) components contributing to $R_{DS(on)}$. (Reproduced from Barkhordarian, V., *PCIM*, June 1997, 28–36. With permission.)

3.5.4 I-V Characteristics

Figure 3.26(b) shows the drain-to-source operating characteristics of the power MOSFET. Although the curve is similar to the case of the bipolar power transistor (Figure 3.17), there are some fundamental differences.

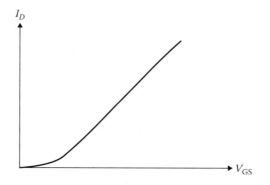

FIGURE 3.29(a)   Transfer characteristics of a power MOSFET.

The MOSFET output characteristic curves reveal two distinct operating regions, namely a "constant resistance" and a "constant current." Thus as the drain-to-source voltage is increased, the drain current increases proportionally, until a certain drain-to-source voltage called *pinch-off* is reached. After pinch-off, an increase in drain-to-source voltage produces a constant drain current.

When the power MOSFET is used as a switch, the voltage drop between the drain and source terminals is proportional to the drain current; that is, the power MOSFET is working in the constant resistance region, and therefore it behaves essentially as a resistive element. Consequently, the on-resistance $R_{DS(ON)}$ of the power MOSFET is an important figure of merit because it determines the power loss for a given drain current, just as $V_{CE,sat}$ is of importance for the bipolar power transistor.

By examining Figure 3.29(a), we note that the drain current does not increase appreciably when a relatively low gate-to-source voltage is applied; in fact, drain current starts to flow after a threshold gate voltage has been applied, in practice somewhere between 2 and 4 V. Beyond the threshold voltage, the relationship between drain current and gate voltage is approximately linear. Thus the transconductance g_{fs}, which is defined as the rate of change of drain current to gate voltage, is practically constant at

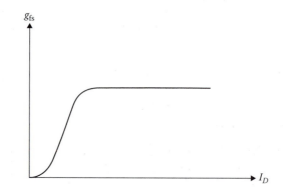

FIGURE 3.29(b)   Relationship of transconductance (gfs) to ID of a power MOSFET.

higher values of drain current. Figure 3.29(a) illustrates the transfer characteristics of I_D versus V_{GS}, while Figure 3.29(b) shows the relationship of transconductance g_{fs} to drain current.

It is now apparent that a rise in transconductance results in a proportional rise in the transistor gain, i.e., larger drain current flow, but unfortunately this condition swells the MOSFET input capacitance. Therefore carefully designed gate drivers must be used to deliver the current required to charge the input capacitance in order to enhance the switching speed of the MOSFET.

3.5.5 Gate Drive Considerations

The MOSFET is a voltage-controlled device; that is, a voltage of specified limits must be applied between gate and source in order to produce a current flow in the drain.

Since the gate terminal of the MOSFET is electrically isolated from the source by a silicon oxide layer, only a small leakage current flows from the applied voltage source into the gate. Thus the MOSFET has an extremely high gain and high impedance.

Threshold voltage, V_t, is defined as the minimum gate electrode bias required to strongly invert the surface under the poly-silicon gate region and form a conducting channel between the source and the drain regions. V_t is usually measured at a drain current of 250 μA in general. Common voltages are 2–4 V for high-voltage devices and 1–2 V for lower-voltage logic-compatible devices. With these devices finding increasing use in portable systems with battery power at a premium, the trend is toward lower values of $R_{DS(on)}$ and V_t.

The switching performance of a device is determined by the time required to establish voltage changes across capacitances and current changes in inductances. In the equivalent circuit shown in Figure 3.29(c), R_G is the distributed resistances of the gate and is approximately inversely proportional to the active area. L_s and L_D are source and drain lead inductances that are around few 10s of nH. Typical values of input (C_{iss}), output (C_{oss}), and

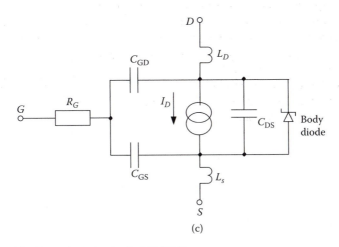

FIGURE 3.29(c)   Equivalent circuit of a MOSFET.

reverse transfer (C_{rss}) capacitances given in data sheets are related to capacitances shown in Figure 3.29(c). The following relationships are used to estimate the typical values.

$$C_{iss} = C_{GS} + C_{GD} + C_{DS}(shorted),$$

$$C_{rss} = C_{GD},$$

and

$$C_{OSS} = C_{DS} + C_{GD}$$

C_{GD}, gate-to-drain capacitance, is a nonlinear function and is the most important parameter because it provides a feedback loop between the output and the input of the circuit. This parameter is also called the Miller capacitance because it causes the total dynamic input capacitance to become greater than the sum of the static capacitances.

In designing gate drive circuits, more useful set of parameters are gate charges than the gate capacitances. Figure 3.30(b) depicts typical charging process and the waveforms. When the gate is connected to a voltage source, gate source voltage starts to increase until it reaches V_t at which point the drain current starts to flow. With the charging of C_{GS} gate voltage continues to rise (during the period t_1–t_2), and the drain voltage rises proportionately. At time t_2, C_{GS} is completely charged, and the drain current reaches the predetermined value (based on external circuit) and stays constant. Referring to the test circuit in Figure 3.30(a) when the C_{GS} is fully charged at t_2, V_{GS} becomes constant and the drive current starts to charge the Miller capacitance C_{DG}. This continues until time t_3. From t_2 to t_3 with the drain voltage falling linearly, the gate drive is charging the C_{DG}. Once both capacitors are fully charged, gate voltage increases until it reaches the maximum gate drive voltage. Good circuit design practice dictates the use of a higher gate voltage than the bare minimum required for switching, and therefore the gate charged used in calculations should be the charge Q_G corresponding to t_4. This maximum-charge approach simplifies the case of estimating any current source values (such as gate driver capabilities) where basic capacitor charge calculations can be used to estimate the charging times.

In order to turn a MOSFET on, a gate-to-source voltage pulse is needed to deliver sufficient current to charge the input capacitor in the desired time. The MOSFET input capacitance C_{iss} is the sum of the capacitors formed by the metal-oxide gate structure, from gate to drain (C_{GD}) and gate to source (C_{GS}). Thus the driving voltage source impedance R_g must be very low in order to achieve high transistor speeds. A way of estimating the approximate driving generator impedance, plus the required driving current, is given in the following equations:

$$R_g = \frac{t_r(or\ t_f)}{2 \cdot 2C_{iss}} \tag{3.5}$$

and

$$I_g = C_{iss} \cdot \frac{dv}{dt} \tag{3.6}$$

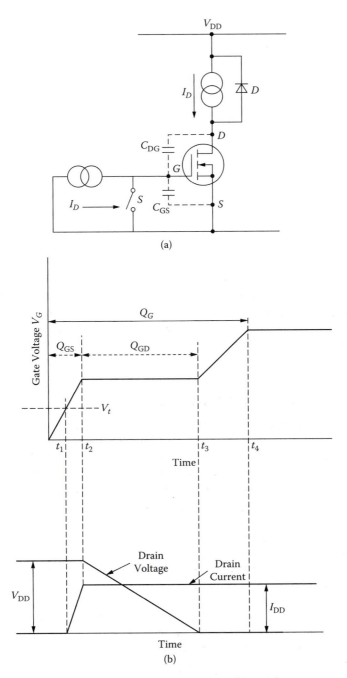

FIGURE 3.30 MOSFET switching behavior: (a) test circuit; (b) waveforms.

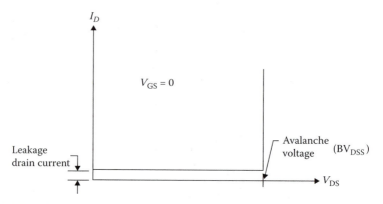

FIGURE 3.31a   Drain-to-source blocking characteristics of the MOSFET.

where R_g = generator impedance; C_{iss} = MOSFET input capacitance, pF; and dv/dt = generator voltage rate of change, V/ns.

Figure 3.31(a) illustrates the relationship of drain current, I_D, versus V_{DS} when there is zero V_{GS} applied to gate, which is the blocking characteristic of the power MOSFET. Leakage drain current (I_{DSS}) continues to flow until the device V_{DS} reaches the breakdown voltage, BV_{DSS}, at which point the body diode breaks down. BV_{DSS} is usually measured at 250 μA drain current.

3.5.6 Temperature Characteristics

The high operating temperatures of bipolar transistors are a frequent cause of failure. The high temperatures are caused by hot spotting, the tendency of current in a bipolar device to concentrate in areas around the emitter. Unchecked, this hot spotting results in the mechanism of thermal runaway, and eventual destruction of the device. MOSFETs do not suffer this disadvantage because their current flow is in the form of majority carriers. The mobility of majority carriers in silicon decreases with increasing temperature.

This inverse relationship dictates that the carriers slow down as the chip gets hotter. In effect, the resistance of the silicon path is increased, which prevents the concentrations of current that lead to hot spots. In fact, if hot spots do attempt to form in a MOSFET, the local resistance increases and defocuses or spreads out the current, rerouting it to cooler portions of the chip.

Because of the character of its current flow, a MOSFET has a positive temperature coefficient of resistance, as shown by the curves of Figure 3.31(b).

The positive temperature coefficient of resistance means that a MOSFET is inherently stable with temperature fluctuation, and provides its own protection against thermal runaway and second breakdown. Another benefit of this characteristic is that MOSFETs can be operated in parallel without fear that one device will rob current from the others. If any device begins to overheat, its resistance will increase, and its current will be directed away to cooler chips.

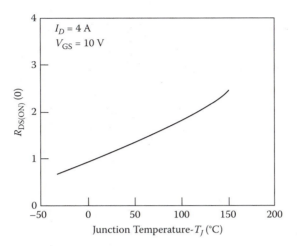

FIGURE 3.31(b)   Positive temperature coefficient of a MOSFET.

3.5.7 Safe Operating Area and Avalanche Rating

In the discussion of the bipolar power transistor, it was mentioned that, in order to avoid secondary breakdown, the power dissipation of the device must be kept within the operating limits specified by the forward-bias SOA curve. Thus at high collector voltages, the power dissipation of the bipolar transistor is limited by its secondary breakdown to a very small percentage of full-rated power. Even at very short switching periods, the SOA capability is still restricted, and the use of snubber networks is incorporated to relieve transistor-switching stress and avoid secondary breakdown.

In contrast, the MOSFET offers an exceptionally stable SOA, since it does not suffer from the effects of secondary breakdown during forward bias. Thus both the DC and pulsed SOA are superior to that of the bipolar transistor. In fact, with a power MOSFET it is quite possible to switch rated current at rated voltage without the need of snubber networks. Of course, during the design of practical circuits, it is advisable that certain derating must be observed.

Figure 3.31(c) shows typical MOSFET and equivalent bipolar transistor curves superimposed in order to compare their SOA capabilities. Secondary breakdown during reverse bias is also nonexistent in the power MOSFET, since the harsh reverse-bias schemes used during bipolar transistor turn-off are not applicable to MOSFETs. Here, for the MOSFET to turn off, the only requirement is that the gate is returned to 0 V. For more details on SOA, refer to [43].

3.5.7.1 Dynamic Stresses and Avalanche Failures

MOSFET manufacturers usually specify the SOA based on static operating conditions. However, most power electronic circuits carry inductive and capacitive components, which can cause significant transients due to

$$\frac{di}{dt}$$

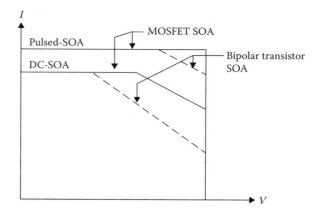

FIGURE 3.31(c)   SOA curves for power MOSFET.

and

$$\frac{dv}{dt}$$

effects, and these effects require quite complex approaches to quantify accurately. Any transient has the potential to drive the drain to source of a MOSFET into avalanche. To help cope with the transients there are avalanche-resistant devices that are rated over a wide range of current, junction temperature, and time.

In a power MOSFET used in a switching circuit where inductance is switched by the device, when the device is switched off (similar to the case of a boost converter with load suddenly switching off simultaneously) inductive current could finally end up flowing through the body diode, which enters into avalanche. The maximum value of this diode current, I_{AS}, which is given in data sheets as single-pulse drain source avalanche current, is dependent on the starting junction temperature, the time in avalanche, etc. Avalanche survivability of a power MOSFET centers on the condition that the parasitic bipolar transistor inside the device never turns on [44]. This rating, known as the unclamped inductive switching (UIS) rating, which becomes application specific, is discussed in [44–46] and is also applicable to modern high-speed power rectifiers [46]. A test method to measure these is discussed in [46]. A practically valuable discussion with application considerations in various circuits including resonant converters is in [47]. References [48–52] are useful in estimating the reliability and modeling of power MOSFETs.

3.5.8 Practical Components

3.5.8.1 High-Voltage and Low-On-Resistance Devices

Denser geometries, processing innovations, and packaging improvements in the 1990s were resulting in power MOSFETS that have ever-higher voltage ratings and current-handling capabilities. See [7–9] for more information. Bipolar transistors have always been available with very high voltage ratings, and those ratings do not carry onerous price penalties. Achieving good high-voltage performance in power MOSFETs, however, has been problematic, for several reasons.

First, the $R_{DS(ON)}$ of devices of equal silicon area increases exponentially with the voltage rating. To get the on-resistance down, manufacturers would usually pack more parallel cells onto a die. But this denser packing causes problems in high-voltage performance. Propagation delays across a chip, as well as silicon defects, can lead to unequal voltage stresses and even to localized breakdown.

Manufacturers resort to a variety of techniques to produce high-voltage (>1000 V), low-$R_{DS(ON)}$ power MOSFETs that offer reasonable yields. Advanced Power Technology (APT), for example, deviates from the trend toward smaller and smaller feature sizes in its quest for low on-resistance. Instead, the company uses large dies to get $R_{DS(ON)}$ down. APT manufactures power MOSFETs using dies as large as 585 × 738 mil and reaching voltage ratings as high as 1000 V. APT10026JN, a device from their product range, has a current rating of 1000 V with 690 W power rating. On-resistance of the device is 0.26 Ω.

When the devices were aimed out to low power applications such as laptop and notebook computers, cellular phones, etc., extremely low $R_{DS(on)}$ values from practical devices were necessary.

Specific on-resistance of double-diffused MOSFETs (DMOSFETs), more commonly known as power MOSFETs, has continually shrunk over the past two decades. In other words, the $R_{DS(ON)}$ per unit area has dropped. The reduced size with regard to low-voltage devices (those rated for a maximum drain-to-source voltage V_{DS} of under 100 V) was achieved by increasing the cell density.

Most power-MOSFET suppliers now offer low-voltage FETs from processes that pack 4 million to 8 million cells/in^2, in which each cell is an individual MOSFET. The drain, gate, and source terminals of all the cells are connected in parallel. Manufacturers such as International Rectifier (IOR) have developed many generations of MOSFETs based on DMOS technology. For example, the HEXFET family from IOR has gone through five generations, gradually increasing the number of cells per in^2 with almost 10-fold decrease in the $R_{DS(ON)}$ parameter as per Figure 3.32. $R_{DS(ON)}$ times the device die area in this figure is a long-used figure of merit (FOM) for power semiconductors. This is called *specified on resistance*. For details, [23] is suggested.

Because Generation 5 die are smaller than the previous generation, there is room within the same package to accommodate additional devices such as a Schottky diode. The FETKY family from IOR [26], which uses this concept of integrating a MOSFET with a Schottky diode, is aimed at power converter applications such as synchronous regulators and the like.

In designing their DMOSFETs, Siliconix borrowed a DRAM process technique called the trench gate. They then developed a low-voltage DMOSFET process that provides 12 million cells/in^2 and offers lower specific on-resistance—$R_{DS(ON)}$—than present planar processes.

The first device from the process, the n-channel Si4410DY, comes in Siliconix's data book "Little Foot" eight-pin DIP. The Si4410DY sports a maximum on-resistance of 13.5 mΩ, enhanced with 10 V of gate-to-source voltage (V_{GS}). At a V_{GS} of 4.5 V, $R_{DS(ON)}$ nearly doubles, reaching a maximum value of 20 mΩ.

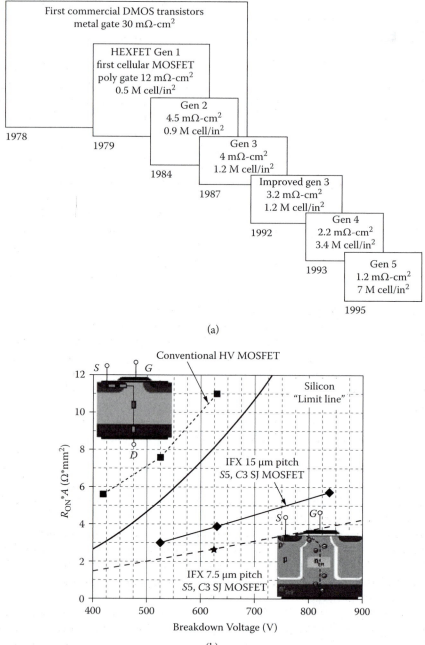

(a)

(b)

FIGURE 3.32 The power MOSFET generations related to DMOS and the impact of superjunction (SJ) technology: (a) DMOS generations up to 1995; (b) performance of SJ technology with conventional high-voltage MOSFETs. (Part [a] courtesy of International Rectifier. Part [b] reproduced from Davis, S., *Power Electronics Technology*, March 2010, 10–13. With permission.)

During the year 1997 Temic Semiconductors (formerly Siliconix) has further improved their devices to carry 32 million cells/in^2 using their TrenchFET technology. These devices come in two basic families, namely (a) low-on-resistance devices and (b) low-threshold devices. Maximum on-resistance was reduced to 9 mΩ and 13 mΩ, compared to the case of Si 4410DY devices with corresponding values of 13.5 mΩ (for V_{GS} of 10 V) and 20 mΩ (for V_{GS} of 4.5 V), for these low-on-resistance devices. In the case of the low-threshold family, these values were 10 mΩ and 14 mΩ for gate source threshold values of 4.5 V and 7.5 V, respectively. For details [27] is suggested.

By the early part of 2000, high-density DC-DC converters were mandatory to power high-power processors and processor-based portables where battery run-time optimization was a key factor in any portable design. With this power MOSFET suppliers were pressured to develop better-performing devices. Driving factors for this were the following:

- Higher efficiency with reduced switching losses
- Low R$_{DS(on)}$
- Lower power dissipation
- Improved reliability
- Improved UIS
- Eliminatation of the need for paralleling devices for higher power
- Lower gate charge and capacitance
- Faster switching speeds
- Increased power package density

Trench-gated vertical DMOS silicon became common in the mid-2000s, providing documented and established advantages over their planar counterparts. Typical Generation 2 devices from companies such as Vishay Siliconix (an example is Si4842DY) came out around 2003 for both n-channel and p-channel devices with 32 million to 50 million cells/in^2 with on-resistances in the range of 4–5mΩ and with gate charge levels of less than 25 nC [52]. These devices were used in DC-DC converters with percentage efficiencies in the high 80s. Superjunction (SJ) MOSFETS, initially developed and introduced by Seimens, followed later by ST Microelectronics, employ a novel drain structure [54]. The first generation of these devices, known as Cool MOS, provides a better efficiency compared to common DMOS technology as the chip area increases. For details on this technology, which appears to have beaten the silicon limit line, is discussed in [54]. Figure 3.32(b) indicates this superjunction MOSFET technology performance in terms of $R_{DS(on)} \cdot A$ versus the breakdown voltage.

3.5.8.2 P-Channel MOSFETs

Historically, p-channel FETs were not considered as useful as their n-channel counterparts. The higher resistivity of p-type silicon, resulting from its lower carrier mobility, put it at a disadvantage compared to n-type silicon.

Due to the approximately 2:1 superior mobility on n-type devices, n-channel power FETs dominate the available devices, because they need about half the area of silicon for a given current or voltage rating. However, as the technology matures, and with the demands of power-management applications, p-channel devices are starting to become available.

They make possible power CMOS designs and eliminate the need for special high-side drive circuits. When a typical n-channel FET is employed as a high-side switch running off a plus supply rail with its source driving the load, the gate must be pulled at least 10 V above the drain. A p-channel FET has no such requirement. A high-side p-channel MOSFET and a low-side n-channel MOSFET tied with common drains make a superb high-current "CMOS equivalent" switch.

Because on-resistance rises rapidly with device voltage rating, it was only recently (1994/1995) that high-voltage p-channel power MOSFETs were introduced commercially. One such device is IXTH11P50 from IXYS Semiconductors with a voltage rating of 500 V, a current rating of 11 A, and an on-resistance of 900 mΩ.

Such high-current devices eliminate the need to parallel many lower-current FETs. These devices make possible complementary high-voltage push-pull circuits and simplified half-bridge and H-bridge motor drives.

Recently introduced low-voltage p MOSFETs from the TrenchFET family of Temic Semiconductors [27] have typical $R_{DS(ON)}$ values between 14 mΩ and 25 mΩ.

3.5.8.3 More Advanced Power MOSFETs

With the advancement of processing capabilities, industry benefits with more advanced power MOSFETs such as:

Current-sensing MOSFETs (Figure 3.33(a))
Logic-level MOSFETs (Figure 3.33(b))
Current-limiting MOSFETs (Figure 3.33(c))
Voltage-clamping, current-limiting MOSFETs (Figure 3.33(d))

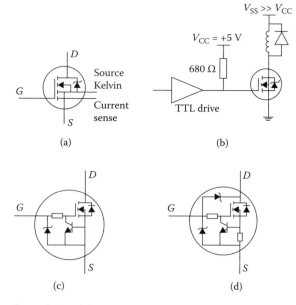

FIGURE 3.33 Advanced Monolithic MOSFETs: (a) current-sensing MOSFET; (b) application of a logic-level MOSFET; (c) current-limiting MOSFET; (d) voltage-clamping, current-limiting MOSFET.

The technique of current mirroring for source current-sensing purposes involves connecting a small fraction of the cells in a power MOSFET to a separate sense terminal. The current in this terminal (see Figure 3.33(a)) is a fixed fraction of the source current feeding the load. Current sense lead provides an accurate fraction of the drain current that can be used as a feedback signal for control and/or protection.

It's also valuable if you must squeeze the maximum switching speed from a MOSFET. For example, you can use the sense terminal to eliminate the effects of source-lead inductance in high-speed switching applications. Several manufacturers such as Harris, IXYS, and Phillips, manufacture these components.

Another subdivision of the rapidly diversifying power-MOSFET market is a class of devices called logic-level FETs. Before the advent of these units, drive circuitry had to supply gate-source turn-on levels of 10 V or more. The logic-level MOSFETs accept drive signals from CMOS or TTL ICs that operate from a 5 V supply. Suppliers of these types include International Rectifier, Harris, IXYS, Phillips-Amperex, and Motorola. Similarly, other types shown in Figure 3.33(a) and (b) above are also available in monolithic form, and some of these devices are categorized under "intelligent discretes."

3.5.8.4 GaN MOSFETs

Around March 2010, Efficient Power Corporation (EPC) started producing and sampling GaN MOSFETS employing a CMOS foundry, with switching speeds in the orders of GHz [55–57]. EPC produces an enhancement-mode GaN transistor using a proprietary process with a GaN-on silicon structure. Figure 3.34 provides some details of the device structure, a plot of theoretical resistance times die area (specific on resistance) versus breakdown voltage and transfer characteristic, and on-resistance versus V_{GS}.

Reference [56] compares the figure of merit (FOM), which is $R_{DS(on)} \cdot Q_G$ *(gate charge)*, of first-generation EPC devices EPC1001, EPC 1010, etc., with other silicon devices.

3.6 Insulated Gate Bipolar Transistor (IGBT)

MOSFETs have become increasingly important in discrete power device applications due primarily to their high input impedance, rapid switching times, and low on-resistance. However, the on-resistance of such devices increases with increasing drain-source voltage capability, thereby limiting the practical value of power MOSFETs to application below a few hundred volts.

To make use of the advantages of power MOSFETs and BJTs together, a newer device, the insulated gate bipolar transistor (IGBT), was introduced in the 1990s.

With the voltage-controlled gate and high-speed switching of a MOSFET and the low saturation voltage of a bipolar transistor, the IGBT is better than either device in many high-power applications. It is a composite of a transistor with an n-channel MOSFET connected to the base of the PNP transistor.

Figure 3.35(a) shows the symbol, and Figure 3.35(b) shows the equivalent circuit. Typical IGBT characteristics are shown in Figure 3.35(c). Physical operation of the IGBT is closer to that of a bipolar transistor than to that of a power MOSFET. The IGBT consists of a PNP transistor driven by an n-channel MOSFET in a pseudo-Darlington configuration.

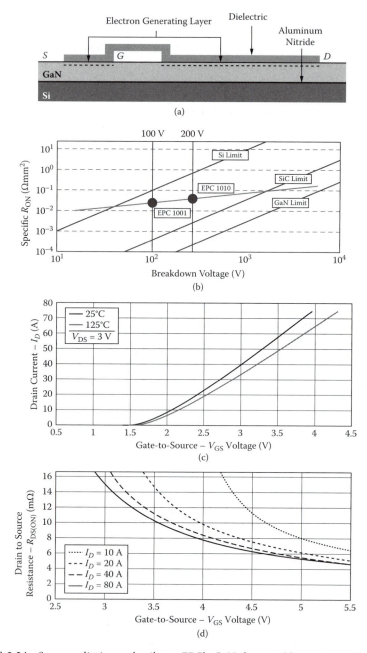

FIGURE 3.34 Some preliminary details on EPC's GaN devices: (a) structure; (b) theoretical resistance * die area for Si/SiC/GaN, etc.; (c) transfer characteristics; (d) RDS(on) versus VGS. (Reproduced from Davis, S., *Power Electronics Technology*, March 2010, 10–13. With permission.)

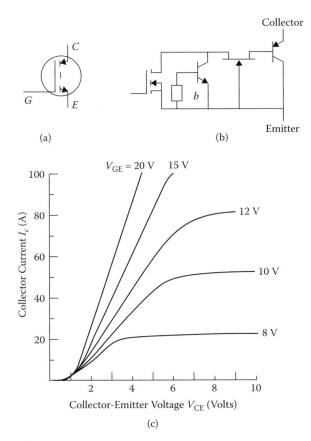

FIGURE 3.35   IGBT: (a) symbol; (b) equivalent circuit; (c) typical output characteristics.

The JFET supports most of the voltage and allows the MOSFET to be a low-voltage type, and consequently have a low $R_{DS(ON)}$ value. The absence of the integral reverse diode gives the user the flexibility of choosing an external fast-recovery diode to match a specific requirement. This feature can be an advantage or a disadvantage, depending on the frequency of operation, cost of diodes, current requirement, etc.

In IGBTs, on-resistance values have been reduced by a factor of about 10 compared with those of conventional n-channel power MOSFETs of comparable size and voltage rating.

IGBT power modules are rapidly gaining applications in systems such as inverters, UPS systems, and automotive environments. The device ratings are reaching beyond 1800 V and 600 A. The frequency limits from early values of 5 kHz are now reaching beyond 20 kHz while intelligent IGBT modules that include diagnostic and control logic along with gate drive circuits are gradually entering the market.

TABLE 3.2 Characteristics Comparison of IGBTs, Power MOSFETs, Bipolars, and Darlingtons

	Power MOSFETs	IGBTs	Bipolars	Darlingtons
Type of drive	Voltage	Voltage	Current	Current
Drive power	Minimal	Minimal	Large	Medium
Drive complexity	Simple	Simple	High (large positive and negative currents required)	Medium
Current density for given voltage drop	High at low voltage Low at high voltages	Very high (small trade-off with switching speed)	Medium (severe trade-off with switching speed)	Low
Switching losses	Very low	Low to medium (depending on trade-off with conduction losses)	Medium to high (depending on trade-off with conduction losses)	High

A characteristics comparison of IGBTs, power MOSFETs, bipolars, and Darlingtons is contained in Table 3.2. References [10–15] provide more details on IGBTs and their applications.

Following is a brief guideline for applying IGBT in practical systems, together with some details from the Advanced Power Technology (APT) application note [58].

State-of-the-art IGBT devices are efficient and less costly, and they are used not only in motor drivelike applications but also in SMPS systems running into a few 100 kHz. Figure 3.36(a) depicts the comparison of similar size MOSFETs versus IGBTs from APT. From Figure 3.36(a) it is clear that above currents of 15 A, IGBT on-state voltage is much less than a similar-size MOSFET. Also, we can see that temperature has much less impact on the on-state voltage of an IGBT. This kind of performance encourages designers to use IGBTs in high-power SMPS applications.

There are two different types of IGBTs, namely punch-through (PT) devices and non-punch-through (NPT) devices. Figure 3.36(b) indicates the cross section of a PT-IGBT where the additional n+ buffer layer is present, which does not exist in an NPT device. Usually an NPT device is preferred in an application such as a motor drive where short-circuit capability of the device is preferred, while in higher-speed applications such as SMPSs a PT device is attractive. For a given switching speed, an NPT technology generally provides a higher $V_{CE(on)}$ voltage than a PT device. NPT devices are typically short-circuit rated while PT devices are not.

Given the case that there is a parasitic NPN transistor within an IGBT, this could create the equivalent of thyristor, which is an NPNP device. If this parasitic transistor ever turns on, latch-up can occur. For this reason careful adherence to data sheet values is recommended. Figure 3.36(c) indicates a simplified equivalent circuit, and Figure 3.36(d) indicates the IGBT model including the parasitic thyristor. In general an IGBT has a longer switching-off period due to long tail current, and this needs to be considered in a design. References [59] and [60] give very useful application information from a designer's viewpoint.

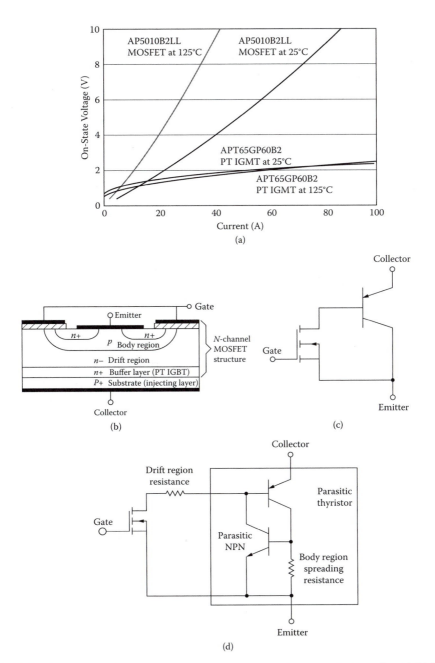

FIGURE 3.36 Practical IGBT devices (a) comparison of similar size MOSFET and an IGBT (b) cross section of a PT device (c) simplified equivalent circuit (d) IGBT model showing parasitic thyristor (Sources: Ref[58–59], parts (a)/(d) reproduced with permission from Power Electronics Technology magazine).

3.7 MOS-Controlled Thyristor (MCT)

MOS-controlled thyristors are a new class of power semiconductor devices that combine thyristor current and voltage capability with MOS-gated turn-on and turn-off. Various subclasses of MCTs can be made: p-type or n-type, symmetric or asymmetric blocking, one or two-sided Off-FET gate control, and various turn-on alternatives including direct turn-on with light.

All of these subclasses have one thing in common: turn-off is accomplished by turning on a highly interdigitated Off-FET to short out one or both of the thyristor's emitter-base junctions. The device, first announced a few years ago by General Electric's power semiconductor operation (later part of Harris Semiconductor, USA), was developed by Vic Temple. Harris was the only supplier of MCTs; however, ABB has introduced a new device called the insulated gate commutated thyristor (IGCT), which is in the same family of devices.

Figure 3.37 depicts the MCT equivalent circuit. Most of the characteristics of an MCT can be understood easily by reference to the equivalent circuit shown here. MCT closely approximates a bipolar thyristor (the two-transistor model is shown) with two opposite-polarity MOSFET transistors connected between its anode and the proper layers to turn it on and off. Since MCT is an NPNP device rather than a PNPN device, an output terminal or cathode must be negatively biased.

Driving the gate terminal negative with respect to the common terminal or anode turns the P channel FET on, firing the bipolar SCR. Driving the gate terminal positive with respect to the anode turns on the N-channel FET, shunting the base drive to the PNP bipolar transistor making up part of the SCR, causing the SCR to turn off. It is obvious from the equivalent circuit that when no gate to anode voltage is applied to the

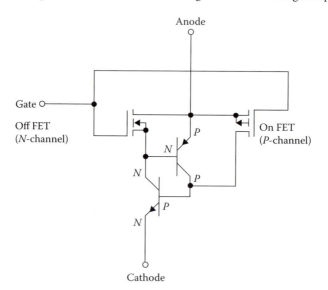

FIGURE 3.37   MCT-equivalent circuit.

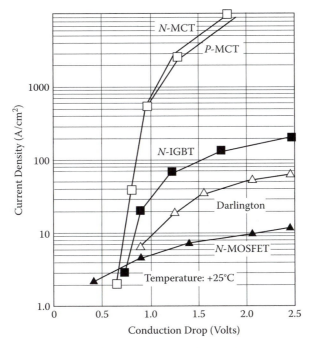

FIGURE 3.38 Comparison of 600 V devices. (Copyright © by Harris Corporation. Reprinted from Harris Semiconductor Sector. With permission.)

gate terminal of the device, the input terminals of the bipolar SCR are unterminated. Operation without gate bias is not recommended.

In the P-MCT, a P-channel On-FET is turned on with a negative voltage, which charges up the base of the lower transistor to latch on the MCT. The MCT turns on simultaneously over the entire device area, giving the MCT excellent *di/dt* capability. Figure 3.38 compares different 600 V power-switching devices. Figure 3.39 compares the characteristics of a 1000 V P-MCT device with an N-IGBT device of the same voltage rating. Note that the MCT typically has 10 to 15 times the current capability at the same voltage drop.

The MCT will remain in the on state until current is reversed (like a normal thyristor) or until the off-FET is activated by a positive gate voltage. Just as the IGBT looks like a MOSFET driving a BJT, the MCT looks like a MOSFET driving a thyristor (an SCR). SCRs and other thyristors turn on easily, but their turn-off requires stopping, or diverting virtually all of the current flowing through them for a short period of time. On the other hand, the MCT is turned off with voltage control on the high-impedance gate. The MCT offers a lower specific on-resistance at high voltage than any other gate-driven technology.

That is, just as the IGBT operates at a higher current density than the DMOSFET, the MCT (like all thyristors) operates at even higher current densities. In the future, the ultimate power switch may well be the MOS-controlled thyristor (MCT). References [16–20] provide details for designers.

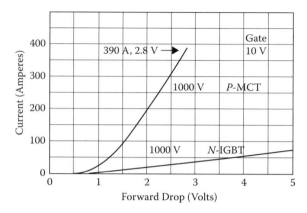

FIGURE 3.39 Comparison of forward voltage drop of 1000 V P-MCT and N-IGBT at 150°C. (Copyright © by Harris Corporation. Reprinted from Harris Semiconductor Sector. With permission.)

MCT as a commercial device was not so popular due to competition from other switching devices. Another new device introduced in early 2000s was the integrated gate commutated thyristor (IGCT) [61]. This is an off-shoot from the GTO technology introduced by ABB.

References

1. Ashkianazi, G., J. Lorch, and M. Nathan. 1995. Ultrafast GaAs power diodes provide dynamic characteristics with better temperature stability than silicon diodes." *PCIM*, April, 10–16.
2. Hammerton, C. J. 1989. Peak current capability of thyristors. *PCIM*, November, 52–55.
3. Coulbeck, L., W. J. Findlay, and A. D. Millington. 1994. Electrical trade-offs for GTO thyristors. *Power Engineering Journal*, February, 18–26.
4. Bassett, Roger J., and Colin Smith. 1989. A GTO tutorial: Part I. *PCIM*, July, 35–39.
5. Bassett, Roger J., and Colin Smith. 1989. A GTO tutorial: Part II—gate drive. *PCIM*, August, 21–28.
6. McNulty, Tom. 1993. Understanding power MOSFETs. Application Note AN 7244.2, Harris Semiconductor, September.
7. Travis, Bill. 1989. Power MOSFETs & IGBTS. *EDN*, January, 128–47.
8. Goodenough, Frank. 1995. Trench-gate DMOSFETs in S0-8 switch 10A at 30 V. *Electronic Design*, March, 65–77.
9. Goodenough, Frank. 1994. DMOSFETs switch milliwatts to megawatts. *Electronic Design*, September, 57–65.
10. Furuhata, Sooichi, and Tadashi Miyasaka. 1990. IGBT power modules challenge bipolars: MOSFETs in invertor applications. *PCIM*, January, 24–28.
11. Russel, J. P., et al. 1997. The IGBTs—a new high conductance MOS-gated device. Application Note AN 8607.1. Harris Semiconductor, May.

12. Wojslawowicz, J. E. 1995. Third generation IGBTS approach ideal switch capability. *PCIM*, January, 28–37.
13. Frank, Randy, and John Wertz. 1994. IGBTS integrate protection for distributorless ignition systems. *PCIM*, February, 42–49.
14. Dierberger, K. 1992. IGBT Do's and Don'ts. *PCIM*, August, 50–55.
15. Clemente, S., A. Dubhashi, and B. Pelly. 1990. Improved IGBT process eliminates latch-up, yields higher switching speed—part I. *PCIM*, October, 8–16.
16. Temple, V., D. Watrous, S. Arthur, and P. Kendle. 1992. MOS-controlled thyristor (MCT) power switches—part I: MCT basics. *PCIM*, November, 9–16.
17. Temple, V., D. Watrous, S. Arthur, and P. Kendle. 1993. MOS-controlled thyristor (MCT) power switches—part II: Gate drive and applications. *PCIM*, January, 24–33.
18. Temple, V., D. Watrous, S. Arthur, and P. Kendle. 1993. MOS-controlled thyristor (MCT) power switches—part III: Switching, applications and the future. *PCIM*, February, 24–33.
19. Temple, V. A. K. 1986. MOS-controlled thyristors—a new class of power devices. *IEEE Transactions on Electronic Devices* ED–33 (10): 1609–18.
20. Temple, V. A. K. 1989. Advances in MOS-controlled thyristor technology. *PCIM*, November, 12–15.
21. Burkel, R., and T. Schneider. 1994. Fast recovery epitaxial diodes characteristics—applications—examples. *IXYS Technical Information* 33 (Publication No. D94004E).
22. Williams, B. W. 1992. *Power electronics: Devices, drivers, applications and passive components*. New York: Macmillan.
23. Kinzer, Dan. 1995. Fifth-generation MOSFETs set new benchmarks for low on-resistance. *PCIM*, August, 59.
24. Delaney, S., A. Salih, and C. Lee. 1995. GaAs diodes improve efficiency of 500 kHz DC-DC converter. *PCIM*, August, 10–11.
25. Deuty, S. 1996. GaAs rectifiers offer high efficiency in a 1 MHz 400 Vdc to 48 Vdc converter. *HFPC Conference Proceedings*, September, 24–35.
26. Davis, C. 1997. Integrated power MOSFET and Schottky diode improves power supply designs. *PCIM*, January, 10–14.
27. Goodenough, Frank. 1997. Dense MOSFET enables portable power control. *Electronic Design*, April 14, 45–50.
28. Peter, J. M. 1989. High frequency power rectification. *PCIM*, April, 24–38.
29. Peter, J. M. 1984. Switching behavior of fast diodes in the converter circuits. *Power Conversion International*, September, 64–70.
30. Boreneman, E. H. 1977. Choosing the correct recovery rectifier. *Solid State Power Conversion*, November/December, 14–16.
31. Anderson, S., K. Gauen, and W. C. Roman. 1991. Low loss, low noise diodes improve high frequency power supplies, part II. *PCIM*, March, 39–42.
32. Carsten, B. 1986. Reverse recovery characteristics of high speed rectifiers. *PCIM*, February, 42–48.
33. Lauritzen, P. O., and C. L. Ma. 1991. A simple diode model with reverse recovery. *IEEE Transcations on Power Electronics* 6 (2, April): 188–91.
34. Chiola, D. 2002. Trench technology boosts Schottky rectifier. *Power Electronics Technology*, December, 32–34.

35. Chen, M., H. Kuo, and K. Sweetman. 2006. High-voltage TMBS diodes challenge planar Schottkys. *Power Electronics Technology*, October, 22–32.
36. Kapels, H., M. Krach, et al. 2002. SiC Schottky diodes ready for blast off. *Power Electronics Technology*, January, 15–21.
37. Zverev, I. 2003. SiC Schottky diodes improve boost converter performance. *Power Electronics Technology*, March, 38–49.
38. O'Neill, M. 2005. SiC puts new spin on motor drives. *Power Electronics Technology*, January, 14–22.
39. Hodge, S. 2004. SiC Schottky diodes in power factor correction. *Power Electronics Technology*, August, 14–22.
40. Cooper, J. A. 1998. Advances in silicon carbide power switching devices. *Proceedings of Power Electronics*, November, 235–45.
41. Barkhordarian, V. 1996. Power MOSFET basics. *PCIM*, June, 28–36.
42. Giovanni, F. D. 1998. Mesh overlay MOSFET. *PCIM*, May, 50–56.
43. Ronan, H. R. 1998. MOSFET SOA—thermal current focussing affects power MOSFET operation. *PCIM*, August, 51–55.
44. Ronan, H. R. 1997. One equation quantifies a power MOSFET's UIS rating. *PCIM*, August, 26–35.
45. Hammerton, C. J. 1996. Avalanche overshot poses a hazard for MOSFETs. *PCIM*, January, 52–56.
46. Ronan, H. R. 1998. Power rectifier unclamped inductive switching. *PCIM*, October, 16–24.
47. Wu, T., H. Tran, and C. Nguyen. 2000. Dynamic stresse scan cause power MOSFET failure. *PCIM*, April, 28–44.
48. Blake, C., T. McDonald, D. Kinzer, et al. 2005. Evaluating the reliability of power MOSFETS. *Power Electronics Technology*, November, 40–44.
49. Kazmirski, T. 2006. Protecting MOSFETS against overcurrent events. *Power Electronics Technology*, January, 14–21.
50. Benczkowski, S., and R. Mancini. 1998. Improved MOSFET model. *PCIM*, September, 64–68.
51. Pandya, K. 2007. Tool eases modelling of power MOSFETs. *Power Electronics Technology*, March, 30–33.
52. Pearson, S. S. Tran, and S. Sapp. 2007. Improved MOSFET model achieves higher accuracy. *Power Electronics Technology*, January, 14–19.
53. Moxey, G. 2003. TrenchFETs enhance DC-DC converter performance. *Power Electronics Technology*, February, 36–38.
54. Hancock, M. 2005. Superjunction FETS boost efficiency in PWMs. *Power Electronics Technology*, July, 20–29.
55. Davis, S. 2010. Enhancement mode GaN MOSFET delivers impressive performance. *Power Electronics Technology*, March, 10–13.
56. Davis, S. 2010. GaN powers MOSFET figure of Merit past silicon-based devices. *Power Electronics Technology*, March,14–17.
57. Alderman, A. N. 2010. Where are the high voltage GaN products? *Power Electronics Technology*, June, 34–38.

58. Dodge, J., and J. Hess. 2002. IGBT tutorial. Application Note APT0201, rev. B. Advanced Power Technology, July 1.

59. Dodge, J. 2003. Why opt for IGBTs in SMPS applications? *Power Electronics Technology*, July, 34–39.

60. McArthur, R. 2001. Making use of gate charge information in MOSFET and IGBT data sheets. Application Note APT0103, rev. B. Advanced Power Technology, October.

61. Carroll, E., and J. Siefgken. 2002. IGCTs: Moving on the right track. *Power Electronics Technology*, August, 16–22.

Bibliography

Anderson, S., K. Gauen, and C. W. Roman. "Low Loss, Low Noise Diodes Improve High Frequency Power Supplies." *PCIM*, February 1991, 6–13.

Adler, Michael, et al. "The Evolution of Power Device Technology." *IEEE Transactions on Electronic Devices*, ED-31, no. 11 (November 1984): 1570–91.

Arthur, S. D., and V. A. K. Temple. "Special 1400 Volt N—MCT Designed for Surge Applications." *Proceedings of EPE '93* 2 (1993): 266–71.

Barkhordarian, V. "Power MOSFET Basics." *PCIM*, June 1996, 28–39.

Bird, B. M., K. G. King, and D. A. G. Pedder. *An Introduction to Power Electronics*, 2nd ed. (New York: Wiley, 1993).

Borras, R., P. Aloisi, and D. Shumate. "Avalanche Capability of Today's Power Semiconductors." *Proceedings of EPE '93* 2 (1993): 167–71.

Bose B. K. *Modern Power Electronics* (New York: IEEE Press, 1997).

Bradley, D. A. *Power Electronics* (London: Chapman & Hall, 1995).

Consoli, A., et al. "On the Selection of IGBT Devices in Soft Switching Applications." *Proceedings of EPE '93* (1993): 337–43.

Deuty, Scott, Emory Carter, and Ali Salih. 1995. "GaAs Diodes Improve Power Factor Correction Boost Converter Performance." *PCIM*, January, 8–19.

Driscoll, J. 1990. "Bipolar Transistors and High Side Switches in High Voltage, High Frequency Power Supplies." *Proceedings of Power Conversion Conference*, October.

Driscoll, J. C. 1990. "High Current Fast Turn-On Pulse Generation Using Power Tech PG–5xxx Series of 'Pulser' Gate Assisted Turn-Off Thyristors (GATO's)." Power Tech App Note.

Eckel, H. G., and L. Sack. 1993. "Optimization of the Turn-Off Performance of IGBT at Overcurrent and Short-Circuit Current." *Proceedings of EPE '93*, 317–21.

Frank, Randy, and Richard Valentine. 1990. "Power FETS Cope with the Automotive Environment." *PCIM*, February, 33–39.

Gauen, K., and W. Chavez. 1993. "High Cell Density MOSFETs: Low on Resistance Affords New Design Options." *Proceedings of PCI*, October, 254–64.

Goodenough, Frank. 1994. "DMOSFETs, IGBTS Switch High Voltage." *Electronic Design* 7 (November): 95–105.

Heumann, K., and M. Quenum. 1993. "Second Breakdown and Latch-Up Behavior of IGBTs." *Proceedings of EPE '93*, 301–5.

International Rectifier. 1997. *Schottky Diode Designer's Manual.*

Lynch, Fernando. 1994. "Two terminal power semiconductor technology breaks current/voltage/power barrier." *PCIM*, October, 10–14.

Locher, R. E. 1996. "1600V BIMOSFET™ Transistors Expand High Voltage Applications." *PCIM*, August, 8–21.

Mitlehner, H., and H. J. Schulze. 1994. "Current Developments in High Power Thyristors." *EPE Journal* 4 (1, March): 36–47.

Mitter, C. S. 1995. "Introduction to IGBTs." *PCIM*, December, 32–39.

Mohan, N., T. M. Undeland, and W. P. Robbins. 1989. *Power Electronics: Converter, Applications, and Design* (New York: Wiley).

Nilsson, T. 1986. "The Insulated Gate Bipolar Transistor Response in Different Short Circuit Situations." *Proceedings of EPE '93*, 328–31.

Peter, Jean Marie. 1986. "State of the Art and Development in the Field of Medium Power Devices." *PCIM*, May, 14–27.

Polner, Alex. 1995. "Characteristics of ultra High Power Transistors." *Proceedings of First National Solid State Power Conversion Conference*, March.

Ramshaw, R. S. 1993. *Power Electronics Semiconductor Switches* (London: Chapman & Hall).

Rippel, Wally E. 1989. "MCT/FET Composite Switch: Big Performance with Small Silicon." *PCIM*, November, 16–27.

Roehr, Bill. 1989. "Power Semiconductor Mounting Considerations." *PCIM*, September, 8–18.

Sasada, Yorimichi, Shigeki Morita, and Makato Hideshima. 1986. "High Voltage, High Speed IGBT Transistor Modules." *Toshiba Review* 157 (Autumn): 34–38.

Schultz, Warren. 1985. "Ultrafast-Recovery Diodes Extend the SOA of Bipolar Transistors." *Electronic Design* 14 (March): 167–74.

Serverns, Rudy, and Jack Armijos. 1984. *MOSPOWER Applications Handbook*. Siliconix Inc.

Smith, Colin, and Roger Bassett. 1989. "GTO Tutorial Part III—Power Loss in Switching Applications." *PCIM*, September, 99–105.

Travis, B. 1996. "MOSFETs and IGBTs Differ in Drive Methods and Protections Needs." *EDN*, March 1, 123–37.

4

Resonant Converters and Wireless Power Supplies

Aiguo Patric Hu

4.1 Introduction

The concept of resonance based on an LC tank circuit has long been used in high-power systems for generating sinusoidal voltage and current waveforms for testing and easy switching purposes. However, due to its circuit complexity it had not been widely employed in low-power DC-DC converter design until about the early 1990s. The thrust toward resonant power supplies has been fueled by the industry's demand for miniaturization, high-power efficiency, as well as low EMI.

Resonant converters are based on soft-switching techniques, which turn on and turn off semiconductor switches in a switch power supply at either zero-voltage or zero-current crossings, namely zero-voltage switching (ZVS) and zero-current switching (ZCS) respectively. During the switching period, if either the voltage or current is zero, the switching power loss would be zero. Thus the converter can operate at a high frequency without causing high switching losses, and therefore the heat sinks may be reduced or even eliminated for low-power applications. Moreover, high-frequency operation would reduce the size of reactive components needed in the converter; as a result the power density can be increased on top of the power efficiency improvement. Also, ZVS and ZCS involved in the resonant converter operation can greatly reduce dv/dt and di/dt compared to hard-switching converters, thereby reducing the switching stresses and EMI associated.

4.2 Fundamentals of Resonant Converters

Because there are no natural zero crossings existing in a DC power supply, no matter what techniques are used, zero-voltage or zero-current crossings have to be generated to achieve soft switching of semiconductor switches. Circuit resonance based on LC is widely used for this purpose; therefore the terms *resonant converters* and *soft-switching converters* can often replace each other without changing the meaning.

In a broader sense, soft-switching conditions can be satisfied at the power source side, at the semiconductor itself, or at the load side. In fact, a mains power supply has natural zero-voltage crossings, which have been used to create a natural ZVS condition for

TABLE 4.1 Possible Ways that ZVS and ZCS ON and OFF Can Be Achieved

Turning ON	Turning OFF	Method
ZVS! (Needs care) ZCS? (not sure)	*ZVS (Yes, V ~ 0)* ZCS? (not sure)	A parallel capacitor
ZVS? (not sure) *ZCS (Yes, I ~ 0)*	ZVS? (not sure) ZCS! (needs care)	A series inductor
~ZVS (V ~ 0) *ZCS* (not sure)	*~ZVS (V ~ 0)* *ZCS* (not sure)	A parallel clamping diode in conduction
ZVS? (not sure) *ZCS (I = 0)*	ZVS? (not sure) *ZCS (I = 0)*	A series-blocking diode in discontinuous current mode

thyristors in traditional controllable rectifiers, so strictly speaking soft switching may be achieved using an AC link without circuit resonance. However, in a DC-DC converter application there is no such luxury unless a resonant DC-link is specially created; therefore circuit resonance has to be used.

Resonant converters for DC power supplies can be roughly categorized into resonant-switch converters and load resonant converters according to where zero-voltage or zero-current soft-switching condition is created. Depending on whether it is a full resonance or partial resonance, they can also be categorized into full-resonant converters and quasi-resonant converters.

4.2.1 Soft-Switching Mechanism

Depending on the condition of zero-voltage switching or zero-current switching, resonant converters are often called ZVS converter or ZCS converters. Actually switching involves switching on and switching off, which can make the situation more complicated. Table 4.1 summarizes how ZVS or ZCS can be achieved regardless of what specific technique is being used.

Figure 4.1 indicates the conditions in Table 4.1 as applicable to a switch in a resonant converter circuit, and the four possible soft-switching mechanisms are explained below:

1. *Using a capacitor in parallel with the active switch to achieve ZVS OFF.* This is because the capacitor voltage was zero when the switch is ON, and it cannot

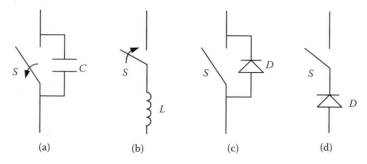

(a)   (b)   (c)   (d)

FIGURE 4.1 Soft-switching configurations of an active power switch: (a) parallel capacitor; (b) series inductor; (c) parallel diode; and (d) series diode.

change simultaneously when the switch is turned OFF. So the voltage remains close to zero during the switching-off period, which is particularly true for modern semiconductor switches as they have a very fast switching time, and therefore quasi ZVS OFF is achieved. However, special care has to be taken to ensure the voltage across the switch is zero (normally by resonance, which will be discussed later) to turn the switch ON, otherwise shorting would occur, which may cause damage to devices. When a parallel capacitor is used, the current flowing through the switch is determined by the circuit and the actual switch used, so it is not sure whether ZCS ON and OFF are achieved.

2. *Using an inductor in series with the active switch to achieve ZCS ON.* This is because the inductor current was zero when the switch is OFF, and it cannot change simultaneously when the switch is turned ON. So the current remains close to zero during switching-off period, and therefore quasi ZCS ON is achieved. However, special care has to be taken to ensure the current flowing through the inductor is zero to turn the switch OFF, otherwise overvoltage would occur, which may cause damage to devices. When a series inductor is used, the voltage across the switch is determined by the circuit and the actual switch used, so it is not sure whether ZVS ON and OFF are achieved.

3. *Using a diode in parallel with the active switch to achieve approximate ZVS ON/ OFF during the diode conduction period.* This is because when the diode is on, its voltage is clamped to its forward on voltage drop, which is small, so quasi ZVS ON and OFF of the active switch can be achieved. The actual current flowing through the switch may be zero because some switches, such as BJTs, cannot conduct negative current through it, or it can be a value sharing with the diode if the switch has bidirectional conductibility like some MOSFTETS.

4. *Using a diode in series with the active switch to achieve ZCS ON/OFF in discontinuous mode.* This is because the diode blocks the current flow in a discontinuous mode. There would be no current flow, so ZCS ON and OFF can be achieved.

It should be noted that the above analysis is based on ideal active switches. In a practical circuit design, the switch parasitic parameters and conductivity may need to be considered. For example, ideally the power losses at ZCS ON is zero, but practically MOSTETs can have large parasitic output capacitances, which can cause large turning on losses at high frequencies. The actual current and voltage across a switch may be zero or a value sharing with the blocking diode depending on the voltage-blocking capability of the switch.

4.2.2 Comparison between Hard-Switching and Soft-Switching Converters

It is well understood that if hard-switching techniques such as PWM are used, the losses in power switches and EMI can be very high, particularly at high frequencies. Soft-switched resonant converters demonstrate clear advantages in these areas, which lead to new converters with high-power efficiency and high-power density. However,

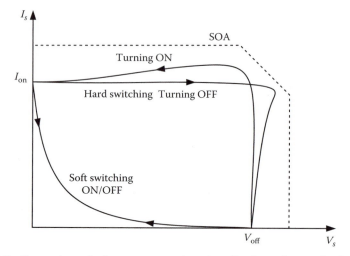

FIGURE 4.2   Comparison of voltage-current trajectories of hard switching and soft switching.

because resonant converters are involved in zero-voltage or -current creation and accurate control of switching at these points, both the main circuit and controller become complex, and often more components are needed; therefore a higher level of understanding and expertise is required from the designer. Also, due to the resonance, normally the peak voltage or current values are higher than their PWM counterparts, which increases the V/I ratings of the devices.

Figure 4.2 illustrates typical voltage and current trajectories of hard switching and soft switching of a power switch. When the switch is on, its voltage is almost zero, and the current is I_{on}; when it is off, the current is zero and voltage is V_{off}. It can be seen that during the hard-switching period, both the voltage and current are high, so the switching power losses can be very large. Overvoltage at switching off and overcurrent at switching on often occur, and it is important to ensure both the on and off trajectories stay within the safe-operating area (SOA) of the switch, otherwise the switch may fail. If the switch is ideally soft switched with either ZVS or ZCS, the switching power losses would be zero. In practice, it cannot be exactly zero but is much lower than hard switching.

A general comparison between hard-switching converters and soft-switching resonant converters is listed in Table 4.2.

TABLE 4.2   Comparison of Soft-Switching Resonant Converters to Hard-Switching Converters

Advantages	Disadvantages
Low switching losses, heat link may not be needed	More component counts to create zero-voltage or current conditions
Low switching stress and EMI	High voltage and/or current ratings for components
High power density due to reduced heat generation and high operating frequency	Complicated main circuit and controller design; parasitic parameters often need to be considered

4.3 Resonant DC-DC Converters

4.3.1 Resonant-Switch DC-DC Converters

The basic structure and operating principal of resonant-switch power converters is no different from traditional hard-switching DC-DC converters. The major change is that LC components are used to resonate at certain periods to offer ZVS or ZCS conditions for the semiconductor switches, so they become "resonant switches."

Depending on the number and configuration of the LC pairs in relation to the active switch and diode, many different types of ZVS and ZVS converters can be configured. Figure 4.3 shows eight possible topologies using only one inductor and one capacitor as a pair of resonant components by Luo and Ye [1]. They also showed 38 possible resonant topologies if three resonant components are used, and 98 for four components. In principle all these configurations may be used in resonant converter design, which can lead to many topological variations of resonant-switch buck, boost, and buck-boost resonant converters. In practice, only simple configurations are used.

Figures 4.4 and 4.5 show two typical ZCS DC-DC buck converter and ZVS DC-DC buck converter circuits respectively. Fundamentally quasi ZCS ON is achieved utilizing the series inductor L_r, and ZVS OFF using the parallel resonant capacitor C_r. The resonance between L_r and C_r generates zero-current and zero-voltage conditions so that ZVS

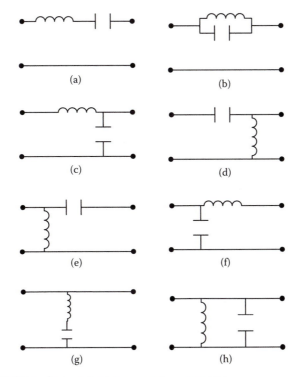

FIGURE 4.3 Eight topologies using two resonant components.

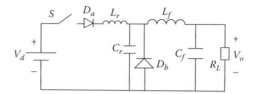

FIGURE 4.4 A ZCS resonant-switch buck converter.

ON and ZCS OFF can be achieved. In practical operation the circuit resonance and zero crossings can be affected by load and parameter variations, so special attention has to be given in designing both the main circuit and controller to ensure that soft-switching conditions are satisfied; otherwise overvoltage or shorting can occur, which may cause damage. Because the resonant period is more or less fixed with given L_r and C_r, the output voltage is regulated by changing the switching frequency [2–3].

Even very simple resonant-switch ZCS or ZVS converters can have many different topological variations. For example, if the series diode D_a in Figure 4.4 is changed to be in parallel with the switch, the circuit would still be able to operate with ZCS, but the current would be in a full-wave mode rather than half-wave; and although the advantage of the converter shown in Figure 4.5 is that the parasitic capacitance of the active switch S can be used advantageously as part of the resonant capacitor C_r, other ZVS configurations exist that can bring in other advantages.

Table 4.3 shows another two ZCS and ZVS quasi-resonant buck converter examples and compares their basic operation. In the case of ZCS, assume in period 1 the load current (not shown in the diagram) is freewheeling through the diode D_b; when the switch is turned ON in period 2, the current flowing through it increases but remains to be very low, so quasi ZVS ON is achieved. After the inductor current reaches the load current in period 3, the diode stops conducting, so C_t and L_t begin to resonate. After the inductor current resonates to zero, the switch can be turned off with ZCS. The diode D_a actually starts to conduct after the current changes the direction, which offers an ideal switching condition. When the resonant voltage across C_t becomes zero, the diode D_b starts to conduct, and the circuit repeats another switching cycle. In the case of ZVS, assume the power switch is on initially and the DC input voltage is applied to C_t and D_b. Because the voltage across C_t cannot change instantaneously, in Period 2 the switch can be turned OFF with quasi ZVS. In period 3, diode D_b starts to conduct, and C_t and L_t start to resonate. In period 4, when the voltage across C_t swings back to V_{in}, the voltage across the

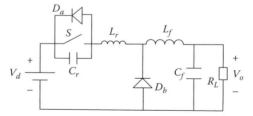

FIGURE 4.5 A ZVS resonant-switch buck converter.

TABLE 4.3 Comparison of ZCS and ZVS Quasi-Resonant Buck Converters

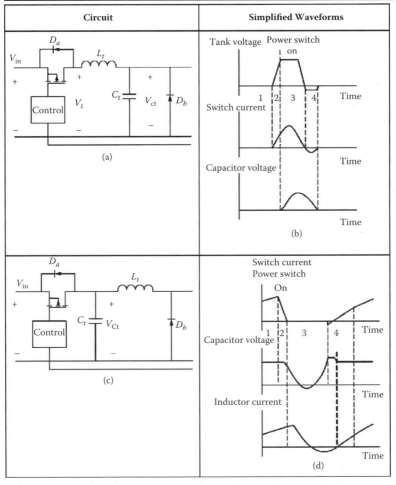

Circuit	Simplified Waveforms
(a)	(b)
(c)	(d)

Source: Reprint from Kularatna, Nihal, *Electronics Circuit Design—from Concept to Implementation*, CRC Press, Boca Raton, FL, 2008. With permission.

power switch becomes zero, so ZVS ON can be achieved. Similar to the situation in ZCS, the body diode D_a would also be conducting, offering an ideal switching condition. After the power switch turns on, the inductor current would increase, D_b would turn off, and the circuit gets ready for the next switching cycle.

In both the cases the resonant frequency is determined by L_t and C_t, although the parasitic parameters of the switch and diode can make the situation more complicated.

4.3.2 Half-Bridge ZVS Resonant Converter Design Example

The TEA1610 produced by Philips is a monolithic integrated circuit controlling for a zero-voltage transition resonant converter. There are two packages available: DIP16

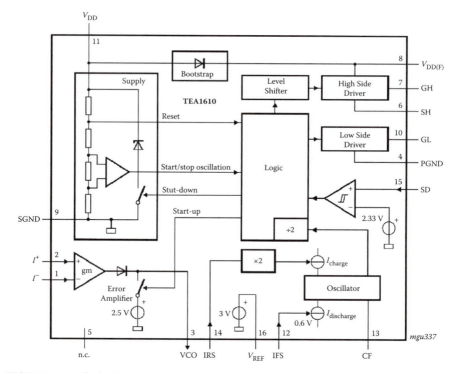

FIGURE 4.6 Block diagram of TEA1610 ZVS controller. (Courtesy of NXP Philips Semiconductors, TEA1610P Product Data Sheet, rev. 03, March 2007.)

(plastic dual in-line package, 16 leads, 300 mil, long body), and SO16 (plastic small outline package, 16 leads, body width 3.9 mm, low stand-off height). The IC can drive two power MOSFETs in a half-bridge configuration. It also includes a level-shift circuit, an oscillator with accurately programmable frequency range, a latched shut-down function, and a transconductance error amplifier [5]. The circuit enables a broad range of applications for different input voltages up to 600V. It is suited to TV and monitor power supplies well. The block diagram of TEA1610 is shown in Figure 4.6.

TEA 1610 can be used to control a half-bridge ZVS resonant power supply using a basic configuration as shown in Figure 4.7. A more detailed circuit diagram is shown in Figure 4.8. At the main circuit side, the half bridge is connected to a series resonant tank consisting of L_r, C_r, and the output, which is series loaded via a transformer with a central tap at the secondary side. A conventional center-tap full-wave rectifier is used to provide a DC output in this example, although other rectifier configurations are acceptable. The leakage inductance of the transformer can be used as the resonant inductance, so the transformer is often purposely designed to have a large leakage. If the leakage inductance is not large enough, then an external inductor has to be used as shown in the diagram. The parallel capacitor Cp in parallel with the lower-side power switch is to achieve quasi ZVS OFF, and it resonates with L_r and C_r when the switch is off, so that zero-voltage switching ON condition can be created. Once the voltage across the switch

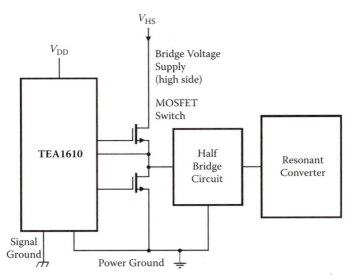

FIGURE 4.7 Basic TEA1610 configuration. (Courtesy of NXP Philips Semiconductors, TEA1610P, Product Data Sheet, rev. 03, March 2007.)

becomes negatively biased, its body diode would turn on, offering a favorable switching condition. Although there is no capacitor directly connected to the high side of the switch, effectively there is a capacitance across it by the input DC power supply pass. There is no direct zero-voltage detection to ensure ZVS, so the resonant circuit and the control range have to be carefully designed to ensure the system can operate safely in the defined load range. The output voltage is monitored, and the controller automatically varies the switching frequency in a closed loop to maintain it to be constant.

TEA1610 has high- and low-side gate drive capability using a bootstrap capacitor. At start-up, after the applied voltage reaches the nominal value of V_{DD}, the low-side power switch is turned-on, and the high-side power switch remains in the nonconducting state. This start-up output state guarantees the initial charging of the bootstrap capacitor (C_{boot} in Figure 4.8) used for the floating supply of the high-side gate drive. During start-up, the voltage on the frequency capacitor (C_f) is zero, which defines the start-up state. The output voltage of the error amplifier is kept constant (typically 2.5 V) and switching starts at about 80% of the maximum frequency when pin V_{DD} reaches the start level. The start-up state is maintained until V_{DD} reaches 13.5 V, the oscillator is activated, and the converter starts operating.

The internal oscillator is a current-controlled oscillator that generates a sawtooth output. The frequency of the sawtooth is determined by the external capacitor C_f and the currents flowing into the IFS and IRS pins. The minimum frequency and the dead time are set by the capacitor C_f and resistors $R_{f(min)}$ and R_{dt}. The maximum frequency is set by resistor $R_{\Delta f}$. The oscillator frequency is exactly twice the bridge frequency to achieve an accurate 50% duty ratio, which is fixed during the operation. The error amplifier is a transconductance amplifier. Thus the output current at pin VCO is determined by the amplifier transconductance and the differential voltage between

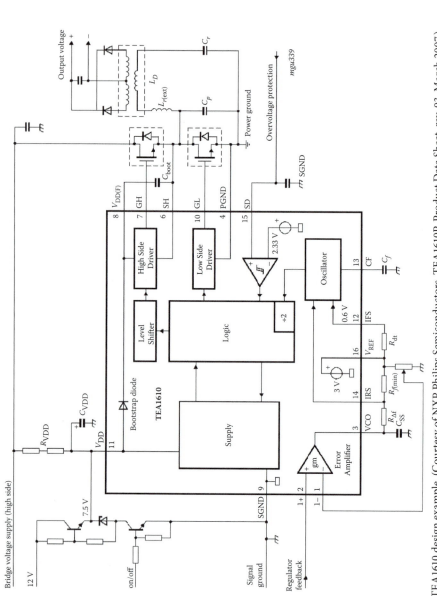

FIGURE 4.8 TEA1610 design example. (Courtesy of NXP Philips Semiconductors, TEA1610P, Product Data Sheet, rev. 03, March 2007.)

input pins I+ and I–. The output current IVCO is fed to the IRS input of the current-controlled oscillator.

The source capability of the error amplifier increases the current in the IRS pin when the differential input voltage is positive. Therefore the minimum current is determined by resistor $R_{f(min)}$, and the minimum frequency setting is independent of the characteristics of the error amplifier. The error amplifier has a maximum output current of 0.5 mA for an output voltage up to 2.5 V. If the source current decreases, the oscillator frequency also decreases, resulting in a higher regulated output voltage. During start-up, the output voltage of the amplifier is held at a constant value of 2.5 V. This voltage level defines, together with resistor $R_{\Delta f}$, the initial switching frequency of the TEA1610 after start-up.

The shutdown (SD) input has an accurate threshold level of 2.33 V. When the voltage on input SD reaches 2.33 V, both power switches switch off immediately, and the TEA1610 enters shutdown mode.

During shutdown, pin V_{DD} is clamped by an internal zener diode at 12.0 V with 1 mA input current. This clamp prevents V_{DD} rising above the rating of 14 V due to low supply current to the TEA1610 in shutdown mode. When the TEA1610 is in the shutdown mode, it can be activated again only by lowering V_{DD} below the V_{DD} (reset) level (typically 5.3 V). The shutdown latch is then reset, and a new start-up cycle can commence.

4.4 Load Resonant Converters for Contactless Power Supplies

4.4.1 Introduction to Inductive Power Transfer

The main objective of an inductive power transfer (IPT) system is to transfer power from a static track loop to some galvanically isolated "pickup" circuits over a large air gap so that wireless/contactless power transfer can be achieved. The air gap gives much freedom to the mechanical movement of secondary power "pickups" but makes IPT a very loosely coupled system. In a practical situation, the track "wire" may be very long while the magnetic "pickup" coils may be quite short so that the actual coupling coefficient can be as low as typically 1% or less, compared with about 95%–98% for transformers and 92% for induction motors [6]. To transfer power from the primary track loop to the secondary pickup coil across an air gap, a high-frequency alternating magnetic field linking the primary and the secondary as illustrated in the dotted line in Figure 4.9 is required. Therefore a power converter is necessary to generate a high-frequency AC current along the track loop. Because the induced voltage in a pickup coil is normally unsuitable to be used directly, a power conditioner is usually needed to maximize the power transfer capacity and regulate the pickup voltage, normally to a constant DC, to drive the load.

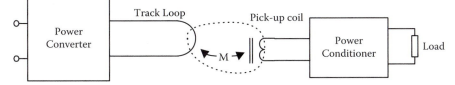

FIGURE 4.9 A basic IPT power supply.

The main purpose of the primary converter is to invert a DC voltage (if it is an AC source, normally it is rectified into DC first) into a high-frequency AC current to generate a high-frequency magnetic field. Because the track loop or coil is normally large and exposed to the outside for easy coupling to the secondary power pickups, the EMI has to be low, which means the current waveform needs to be very good. This, together with the high power efficiency requirement, makes resonant converters desirable for wireless power supply applications.

Because an IPT system has an inherent track inductor and requires high-quality track current waveforms, load resonant converters that utilize the resonance of the track resonant circuit are normally the simplest and most economical choice. When a power pickup circuit is coupled with the primary track inductor, its impedance can be reflected back and modeled as a whole resonant tank. And at the secondary side, the pickup circuit is tuned and rectified to provide a contactless/wireless power supply.

4.4.2 Voltage-Fed and Current-Fed Resonant Converters Topologies

A voltage-fed DC-AC converter/inverter has two basic topologies: full bridge and half bridge, as shown in Figure 4.10. The switching network of the full-bridge topology has four switches, whereas two of these switches are replaced with two suitably large capacitors in a half-bridge topology. As the voltage changes across these capacitors are

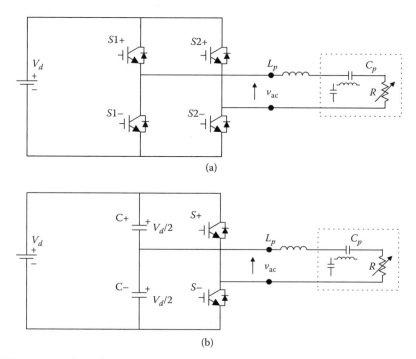

(a)

(b)

FIGURE 4.10   Voltage-fed resonant inverters: (a) full bridge; (b) half bridge.

negligible under steady-state conditions, they serve as voltage sources with half the magnitude of the DC power supply. Therefore the maximum voltage output of the half-bridge inverter is $\pm V_d/2$ compared to $\pm V_d$ in the full-bridge topology

The frequency, magnitude, or phase of the output voltage of a voltage-fed switching converter may be controlled with power switches at their gates. A dead time (also called blanking time) between the turn-on and turn-off of each pair of switches in the same leg is necessary in order to avoid the shorting of the voltage source producing a "short through" failure in the switching devices. The minimum duration of the dead time is determined by the on and off delays of the switching devices, plus a safety factor.

Theoretically, a current-fed resonant converter is the dual of the voltage-fed resonant converter. However, practically there is a big difference. This results from the fact that a current source cannot stand alone naturally like a voltage source without using superconductivity or a closed-loop control. For economic reasons, a large inductor is normally put in series with a voltage source to form a quasi-current source as shown in Figure 4.11. As the current flowing through the inductor is nearly constant at high frequencies under steady-state conditions, it appears like a current source. However, this inductor and voltage source configuration increases the system order and can cause dynamic problems requiring special circuit control and protection methods.

Similar to voltage-fed resonant converters, a full-bridge topology and a push-pull current-fed topology with a phase-splitting transformer dividing the DC current are shown in Figure 4.11. There are not many differences in the performance of these two

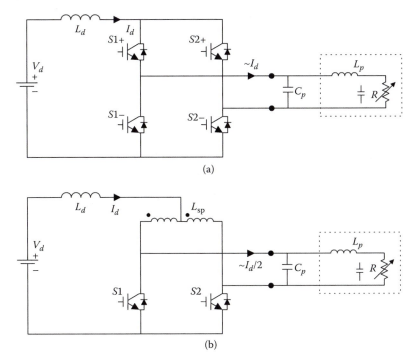

(a)

(b)

FIGURE 4.11 Current-fed resonant inverters: (a) full bridge; (b) half bridge.

topologies; for example, the latter does not require isolated high-side gate drives and doubles the resonant voltage.

Analogous to the protection required to avoid shorting of the voltage source in voltage-fed converters, here the current source must not be broken to avoid the occurrence of high overvoltages. Therefore at least one leg has to be on over the whole period of operation. However, because the switching devices normally turn on faster than they turn off, practically there may be no need to design an overlap time in gate drives of normal switching devices.

A very important aspect needing special attention in the design of voltage- or current-fed inverters is the connection between the switching network and the resonant tank. Because two voltage sources cannot be connected in parallel arbitrarily due to the possibility of shorting the sources, the output of a voltage-fed inverter should not be connected to a voltage-source type of load such as a pure capacitor branch. In consequence, a voltage-fed switching network normally matches series-tuned (or series-parallel-tuned with a series branch in the beginning) types of resonant tanks with at least one inductor being series connected at the input port as illustrated in Figure 4.10. Similarly, for a current-fed inverter, as two current sources cannot be placed in series arbitrarily due to the possibility of creating overvoltage problems, the output of a current-fed switching network should not be connected to a current source type of load such as a load comprising inductive branches only. Thereby, the current-fed inverting network normally matches parallel-tuned (or parallel-series-tuned with a parallel branch first) resonant tanks with at least one capacitive branch connected at the input port as illustrated in Figure 4.11. In the above two situations, a series- or parallel-connected resistor at the input port will also do, but they are seldom used in practice since this introduces high power losses.

4.4.3 A Full-Bridge Voltage-Fed Resonant Converter

As discussed earlier, voltage-fed converters have two basic configurations: a full-bridge topology using four switches and a half-bridge topology where two switches from the full bridge are replaced by two large capacitors. Due to the price drop of the semiconductor switches, the full-bridge topology becomes more popular. Figure 4.12 shows a typical full-bridge voltage-fed series resonant converter for use in IPT applications. It comprises a DC voltage source V_d, an inverting network comprising four switching devices in a full

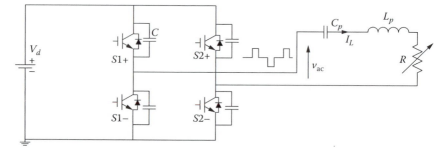

FIGURE 4.12 A voltage-fed series resonant power supply.

bridge, and a series-tuned resonant tank. The track coil is modeled as a lumped inductor L_p, and the load is represented by an equivalent resistor R. A series capacitor C_p is used to tune the track depending on the track length. In a practical application, C_p may be distributed at certain positions along the track so as to limit the maximum voltage on the track. The four capacitors in parallel with the switches are used for soft-switching purposes as will be discussed later. The parasitic output capacitances of the switches are combined into these four capacitors so they are utilized advantageously.

Unlike a current-fed resonant converter whose output AC voltage is mainly determined by the average DC input voltage under ZVS conditions, the output voltage of a voltage-fed inverting network is completely under the control of the switching devices of the inverting network. In consequence, the track current can be regulated by duty cycle control. One control strategy is to shift the phase of the gate signals. In Figure 4.12, the switches S1+ and S1–, S2+ and S2– are complementarily controlled. That is, when S1+ is "on," S1– is "off"; similarly, when S2+ is "on," S2– is "off," and vice versa. If both the upper switches S1+ and S2+ (or both the lower switches S1– and S2–) are "on," the AC output voltage from the inverting network is zero. Otherwise, the output voltage will be either positive V_d or negative V_d, depending on the state of the switches. Because of this, phase-shift duty cycle control can be utilized to regulate the output voltage and consequently the track current. Figure 4.13 shows a situation when the gate signals of S2+ and S2– are lagging S1+ and S1– by 120°. In this case, the output voltage is a PWM square wave with a duty cycle of 2/3. It is obvious that if S1+ (S1–) and S2+ (S2–) are completely in phase, i.e., the phase shift is zero, the output voltage will be zero. On the other hand, if the switching is completely out of the phase, i.e., the phase shift is 180°, then the output voltage reaches its maximum value with a fundamental magnitude (RMS) given by:

$$V = \frac{4}{\pi\sqrt{2}} V_d \qquad (4.1)$$

where V_d is the DC input voltage.

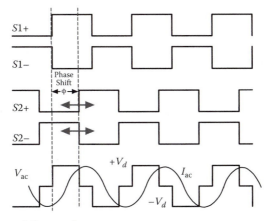

FIGURE 4.13 Phase shift control.

In a more general case where the phase shift angle φ is between 0° and 180°, this voltage can be expressed as:

$$V = \frac{4}{\pi\sqrt{2}} V_d \cdot \sin\left(\frac{\phi}{2}\right) \tag{4.2}$$

or

$$V = \frac{4}{\pi\sqrt{2}} V_d \cdot \sin\left(\frac{\pi D}{2}\right) \tag{4.3}$$

where $D = \varphi/\pi$ is the duty cycle of the output voltage. A duty cycle of 2/3, corresponding to a phase shift of 120°, is a preferable operating point where the harmonic distortion is the lowest for a square waveform output.

In a practical circuit, because the switching devices turn off more slowly than they turn on, a dead time is necessary to prevent momentarily short-circuiting the DC voltage source. Therefore a blanking-out period exists so that the actual PWM phase shift control range is slightly larger than 0° and smaller than 180°. As a result, the maximum output voltage is slightly smaller than the ideal value given in Equation (4.1).

To achieve a constant track current output, a closed-loop control (shown in Figure 4.14) is normally used. A current sensor such as a toroidal current transformer or an LEM device can be used to measure the instantaneous track current. As the measured current changes with time at a relatively high frequency, a process of obtaining the average RMS value or the peak envelope curve of the track current is required. There are many technical options to fulfill this signal-processing task, for example, using phase shift amplifiers or sample and hold techniques. A novel method is to monitor the energy stored in the resonant circuit to estimate the track current. Using this method, simple integral circuits can be used to avoid expensive analog multipliers. After the dynamic track current magnitude has been obtained, it is compared with a current reference and the error is sent to a PI regulator. The output of the regulator is then used for PWM phase-shift control of the inverting network, so that finally the track current can be regulated to the right value as set by the reference.

In a voltage-fed series resonant converter, the track is usually tuned in such a way that its inductance is not completely compensated by the series capacitor but leaves a net inductive residue, so that the track current lags behind the voltage at steady state as shown in

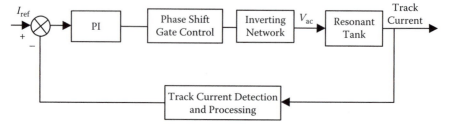

FIGURE 4.14   Closed-loop output track current regulation.

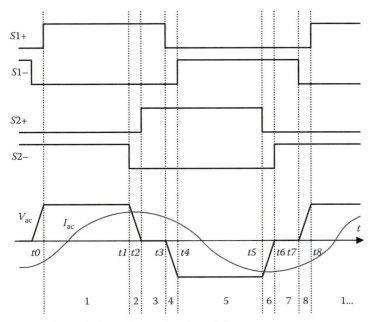

FIGURE 4.15 Soft-switched waveforms of a voltage-fed series resonant converter.

Figure 4.13. In this case, the track current goes through the corresponding body diodes of the switches before the switches are tuned on. Therefore ZVS turn-on is achieved naturally. This is the preferred situation when operating above resonance where ideal turn-on conditions arise for both the switches and diodes. However, when the switches are turned "off," a high current exists at high voltage. If there are no soft-switching capacitors across the switches, the turn-off losses can be high.

Employing parallel soft-switching capacitors helps to eliminate or contain the turn-off losses. When the dead-time considerations are added, the necessary gate control signals and resultant voltage and current waveforms are illustrated in Figure 4.15. The switching process of the active switches under steady-state conditions can be divided into eight states, and their corresponding equivalent circuits are shown in Figure 4.16, where the inductor L_r represents the net residue inductance of the tuned track. These eight states are described in detail below:

1. Assume the first state starts at t_0 when switch S1+ turns on. As both S1+ and S2− are on (S1− and S2+ are off), the output voltage is positive and equals to V_d. Owing to the above-resonance tuning (lagging power factor), the track current lags the fundamental of the driving voltage and changes its direction after a certain time. In the beginning, the current flows through the antiparallel body diodes of S1+ and S2−, and then switches S1+ and S2− themselves begin to conduct after the current changes direction.

2. State 2: t1-t2. S2− turns off at t1 before S2+ turns on at t2, leaving the required dead time. Because of the existence of the capacitor C2−, the voltage across S2− cannot increase suddenly. During the short switch-off time of S2−, the voltage remains

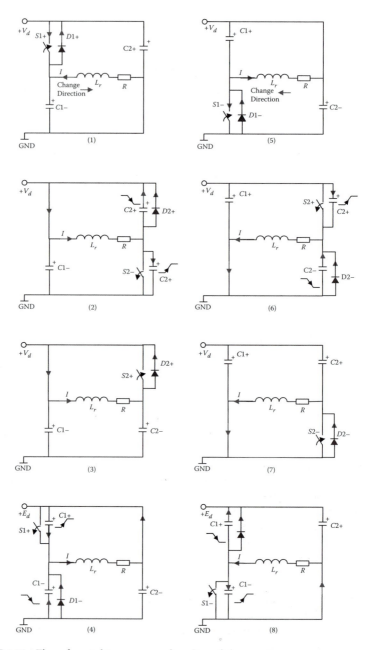

FIGURE 4.16 The soft-switching process of a voltage-fed converter.

almost zero therefore zero-voltage turn-off is approximately achieved. Later, C2+ is discharged and C2– is charged gradually resulting in a ramped voltage change across the track. If the voltage across C2+ is completely discharged corresponding to fully charging C2– from zero to V_d, then D2+ begins to conduct to keep the continuity of the track current. The track voltage drops to near zero (about –1 V).

3. State 3: t2-t3. S2+ is switched on at t2. If D2+ conducts, it clamps the voltage across S2+ to almost zero so that zero-voltage switch-on of S2+ is achieved. In fact, S2+ is also switched "on" at zero current because it does not conduct before the current changes direction. Therefore an ideal "soft" turn-on condition is obtained.

4. State 4: t3-t4. S1+ switches off at t3. Similar to state 2, zero-voltage switch-off of S1+ is achieved because the voltage across C1+ increases slowly. When C1– is fully discharged, D1+ begins to conduct.

5. State 5: t4-t5. S1– is switched on at t4. If D1– conducts, then zero-voltage switch-on of S1– is achieved. As both S1– and S2+ are on, the output track voltage is $-V_d$. In this period, similar to state 1, the current changes direction. In the beginning, the body diodes conducts, then the switches S1– and S2+ begin to conduct after the current changes direction

6. State 6: t5-t6. S2+ is switched off at t5. The process is very similar to states 2 and 4; ZVS during the "off" state is achieved for S2+. When C2– is fully discharged, D2– begins to conduct, offering a zero-voltage turn-on condition for S2–.

7. State 7: t6-t7. S2– is switched on at t6. D2– conducts and clamps the voltage across S2– almost to zero so that zero-voltage switch-on of S2+ is achieved. Similar to other turn-on situations, after S2+ is switched on, it does not begin to conduct immediately since the current has not changed direction.

8. State 8: t7-t8. S1– is switched off at t7. Zero voltage switch-off of S1+ is achieved due to the slow charge up of C1–. When C1+ is fully discharged, D1+ begins to conduct. After this, S1+ can be switched on at zero voltage and the switching cycle repeats from state 1.

From the above analyses it can be seen that due to the charging/discharging of the soft-switching capacitors, the PWM output voltage waveform is trapezoidal rather than square wave. Switching occurs at zero voltage or zero current instants so that the switching losses are essentially eliminated. Moreover, the stresses on the semiconductor devices are alleviated as all the transitions are "soft" at the switching instants. There are essentially no reverse-recovery problems for the diodes. Compared to direct hard-switching methods, an additional advantage of this soft-switching technique is that electromagnetic interference is reduced due to reduced dv/dt, di/dt, and improved resultant track current waveforms.

However, the above analysis also shows that this soft switching is conditional. Specifically, the turn-on of the switches is critical. If the soft-switching capacitors are not completely charged/discharged during the dead time, the residue voltage will be shorted by the switches during turn-on, which may cause the switching devices to fail. The essential issue here is that the track must be tuned above resonance, which ensures that the track impedance is inductive and the current has the right polarity to charge/discharge the capacitors. Also, the inductive energy stored in the track in the residual

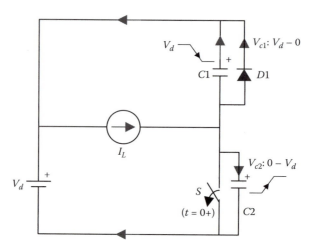

FIGURE 4.17   Equivalent circuit of capacitor charging and discharging during dead time.

inductance needs to be large enough to charge/discharge the capacitors fully to avoid shorting these capacitors during turn-on.

Assuming the track current is constant at I_L during the dead time, the charging/discharging process can be modeled with the simple circuit shown in Figure 4.17. Considering the initial conditions $V_{c1}(0) = V_d$ and $V_{c2}(0) = 0$ when switch S is turned off at time $t = 0$, the voltage across the capacitor C1 during its discharging period (before the diode D starts to conduct) can be expressed as:

$$v_{c1} = V_d - \frac{I_L}{2C}t \qquad (4.4)$$

where $C1 = C2 = C$, and V_d is the DC power supply voltage. If the dead time is given as Δt, from this equation it can be shown that the following condition should be met to guarantee safe soft-switching operation:

$$C < I_L \frac{\Delta t}{2V_d} \qquad (4.5)$$

when designing the soft-switching capacitors. If the capacitance is too large, the track residual inductor will not be able to fully charge/discharge it. As a result, their parallel switches at turn-on will short circuit the capacitors. However, if the capacitance is too small, the capacitor voltages will increase too fast during the turn-on time and the soft switching will be practically impossible. Therefore there is a trade-off between a desirable zero-voltage switching condition and avoiding device failure.

On the other hand, from the track network side proper track tuning is required to achieve soft switching. A basic requirement is that the track network should be able to supply enough inductive energy to charge up the soft-switching capacitors. In practical operation, the parameters of the track inductor and its tuning capacitors may cause some variation in the residue inductance L_r. In addition, load changes also affect the track current value I_L during the dead-time period. Therefore it is a nontrivial task to maintain the soft-switching performance without compromising system security. Dynamic parameter tuning may be a solution, but a more economical approach is to allow the frequency to vary during operation.

4.4.4 A Self-Sustained Current-Fed Resonant Converter

4.4.4.1 Converter Structure

Various current-fed resonant converters with ZVS controllers have been developed and successfully put into practical IPT applications [7,8]. Although external controllers are used in these converters, a careful look into these systems reveals that there is in fact no direct control over the output voltage, current, or frequency. The required track current is actually indirectly obtained via ZVS control. The track current output may have some slight variations in magnitude and frequency, but these variations are so small under normal working conditions that the converters are acceptable for IPT applications. In fact, the main function of the track power supply of an IPT system is to provide high-frequency AC power in a form of magnetic field; thus a constant track current while desirable is not an absolute requirement. The secondary pickups normally have their own control units, which can regulate power flow as required by the load.

Based on the above concept and the particular application, a novel converter that can start up and keep sustained ZVS operation without any additional controllers is proposed. This converter is essentially an autonomous system that gives the same performance as the G1 power supply. As shown in Figure 4.18, it comprises a DC inductor L_d, a phase-splitting transformer L_{sp}, two switching devices S_A and S_B, and a parallel resonant tank. The DC inductor forms an approximate current source from the DC voltage supply V_d provided the inductance is large. The phase-splitting transformer replaces the two top switches of a single-phase full inverting network and essentially divides the DC current into two legs of the inverter network. Compared to the full-bridge topology, a push-pull configuration simplifies the gate drive design (as no isolation is required) and doubles the output resultant resonant voltage. The parallel resonant tank consists of an

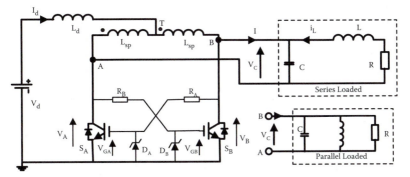

FIGURE 4.18 Proposed DC-AC converter without external controllers.

inductor L and a capacitor C, with an equivalent load resistor R that is series connected. In other applications such as DC-DC converters [9,10], the load can also be connected in parallel as illustrated in the lower dotted block of Figure 4.18.

For this current-fed converter (Figure 4.18), ZVS not only minimizes switching losses and EMI but also is crucial for safe operation of the circuit. If ZVS fails, then the resonant capacitor C will be shorted by the active switches and their body diodes, which may cause the switching devices to fail. In some topologies, additional diodes are placed in series with the active switches S_A and S_B to prevent the shorting and allow for non-ZVS operation, but these diodes cause voltage drops and power losses so they are not commonly used.

Unlike conventional converters with two function blocks comprising the main circuit and the controller, Figure 4.18 shows a novel approach of driving the switching devices without any additional controllers. Both the power and signals needed for the gate drive are obtained directly from the voltages across the main switching devices. The circuit is so simple that only a resistor and a zener diode are used for each switch. The voltage rating of the zener diodes D_A and D_B are chosen according to the switching requirements of the devices, for example, 4.7–5.6 V for most low-voltage threshold MOSFETs and 12–15 V for IGBTs. The current limiting resistors R_A and R_B are designed according to the resonant voltage level and the current rating of the zener diodes.

4.4.4.2 Self-Sustained Operation Analysis

In principle, autonomous operation of the converter shown in Figure 4.18 is based on the oscillatory property of the proposed topology. The critical conditions of the series-loaded parallel-resonant tank have been investigated, and it has been found that the minimum bounds on Q to ensure start-up and steady-state ZVS operation of such series-load converters are 2.54 and 1.86 respectively [6, 8]. Theoretically there is no limit on Q for a standard parallel-loaded tank (see Figure 4.18) as no DC offset voltage exists across the load resistor.

An advantageous feature of the proposed resonant converter is that it can start up automatically. At turn-on, switches S_A and S_B are initially "off." Once the DC source is switched on, a DC voltage will be exerted across the switches. In consequence both the switches are turned on and the current in the DC inductor increases, resulting in some energy storage in the DC inductor, which has been proven beneficial for boosting the circuit oscillation [6]. Due to the existence of parameter differences and external disturbances, the voltages across the active switches cannot be exactly the same in a practical circuit. The lower voltage, say V_A, will provide a lower gate drive voltage V_{GB} in the other leg. Consequently, S_B will turn off, resulting in a higher voltage drop V_B, which will further increase the gate drive voltage V_{GA} and decrease the voltage V_A. This positive feedback will quickly (typically within several ms) lead to complete resonant operation with ZVS.

Figure 4.19 shows typical steady-state voltage and current waveforms of the converter obtained from a PSpice simulation. It can be seen that the gate drive signal of one switch is in fact the capped voltage across the other switch. Switch transitions occur approximately at the zero voltage points with a maximum error of a zener diode voltage drop (4.7 V). The resultant resonant voltage waveform is very good as shown.

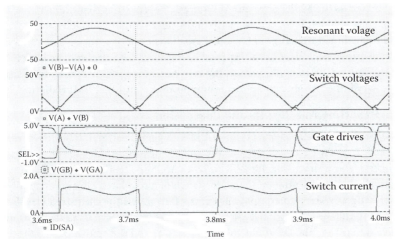

FIGURE 4.19 Typical simulated waveforms of the proposed converter.

If the resonant voltage is below the zener diode voltage level, the gate drive voltage becomes very small and both switches tend to switch off. This should be strictly prohibited for current-fed resonant converters, because the DC current has to continue to avoid the occurrence of dangerously high voltages. Fortunately, the proposed converter can protect the overvoltage automatically. From Figure 4.19 it can be seen that if the voltage V_B is lower than the zener diode voltage 4.7 V, the gate voltage V_{GA} starts to drop. When V_{GA} reaches the threshold voltage of about 3 V, switch S_A starts to turn off, causing a rapid voltage rise that turns the other switch S_B on very quickly. As a result, the DC current follow continues and the occurrence of a high overvoltage is avoided. Since in practice, switches turn on faster than they turn off, and no external control loop delay exists in this converter, it is impossible for the dynamic voltage to become too high, although small glitches occur in the voltage waveforms of V_A and V_B during the switch transitions as shown in Figure 4.19.

The gate drive waveforms shown in Figure 4.19 have a reasonably good rising edge because they are directly driven by the large resonant voltages across the main switches, which quickly rise above 5 V. However, the falling edge of the gate drive lasts longer due to the large resistance in the discharge path of the gate input capacitor and the Miller effect. Theoretically a higher resonant voltage and higher frequency will make both the rising and falling edge of the gate drive signal sharper if only a simple resistor and zener diode circuit is considered, but practically the need for larger limiting resistances R_A and R_B and stronger Miller effect at high voltages may prevent the devices from turning off. At low voltages, all these effects are small so that circuit oscillation can be sustained and the circuit functions well. An advantageous factor of this circuit is that the turn-on process is faster so that the current is quickly diverted away, which helps the switch to turn off. Figure 4.19 shows that the tail current of the switch (I_D) lasts a very short time. Also, the Miller effect is small during turn-on, but it helps to reduce the gate drive voltage during the fall period of the resonant voltage in the second half of the waveform. The gate voltage can become negative (as shown in Figure 4.19) because the zener diode becomes forward biased as the Miller capacitor discharges. This ensures that the switch is turned off completely before it is turned on again in the next switching cycle.

It should be noted that although some distortions exist over the switch voltages during circuit transitions, they hardly appear in the final resonant voltage waveforms as shown in Figure 4.19. In consequence, the resultant resonant current in the track loop contains very low harmonic components and produces minimal EMI radiation.

4.4.4.3 Experimental Results and Discussion

A prototype converter of the form of Figure 4.18 with parameters of $L = 200\ \mu H$, $C = 0.47\ \mu F$, $L_d = 0.2$ mH, $L_{sp} = 2$ mH, and $R_A = R_B = 1\ k\Omega$, and a zener diode of 5.1 V/1 W, was built and tested in the laboratory using 100 V/8 A/0.4 Ω MOSFETS. The converter can start up automatically by just turning on the main switch of a regulated DC power supply without employing any start-up equipment. Figure 4.20 shows the measured waveforms of the switching voltage over switch S_A (a half wave), the track current (a full wave), and the gate drive signal V_{GA}. The DC input voltage V_d is 24 V, and the equivalent load resistance R on the track is 1 Ω. It can be seen from this result that switching occurs approximately at the zero-voltage crossing points as expected, and the sinusoidal waveform of the resonant current is perfectly acceptable.

The above test was based on a series-connected load (see Figure 4.18). A parallel-connected load gave very similar results. In both the cases, the converter could start up automatically and keep self-sustained ZVS. The resonant voltage and current vary proportionally to the DC input voltage, and it has been observed that the gate drive waveforms improve with increasing the DC input voltage. Nevertheless, the increase is

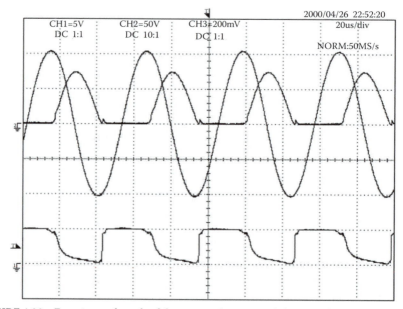

FIGURE 4.20 Experimental result of the proposed converter (Ch1: Gate drive voltage: 5 V/div, Ch2: Switch voltage: 50 V/div, Ch3: track current: 2 A/div).

limited by the power ratings of the resistors and zener diodes of the gate drive circuit so that operating at high voltages is impractical, as will be discussed later. Moreover, the operating frequency of the converter can be adjusted by simply varying the capacitor or inductor of the resonant circuit, and this can even be done during operation. Under such variations, the converter automatically adapts to the new operating conditions with dynamic ZVS operation. There is no danger of damaging the switching devices owing to the inherent overvoltage protection property of the converter discussed earlier. Because of the elimination of any external controllers, the system delay and the total component count is greatly reduced. This helps to increase the maximum possible operating frequency and power density, as well as improve the power efficiency and reliability at reduced cost. However, as with the G1 and variable frequency power supplies based on free oscillation and energy injection control, the operational frequency of this converter is dependent on the load so that frequency stability problems can occur. As such the design of a complete IPT system using such converters requires care to achieve the required power transfer capacity.

Apart from the above example converter based on the current-fed parallel-resonant G1 supply, there are other options to achieving a self-sustained switch-mode operation that are worth further exploration in the future. One potential application of this type of converter is in the development of higher-frequency converters (say 1–10 MHz) aiming to significantly increase the power density. It is known that the design of gate drives becomes difficult at such high frequencies for normal low-voltage drive circuits. Surprisingly, it can be easier for the self-sustained converters to achieve desirable gate drive waveforms because the required signals and power are integrated and internally supplied from the main circuit where high-frequency voltages or currents are available. To overcome the shortcomings of this type of converter regarding the lack of control flexibility, other control mechanisms such as varying circuit parameters (rather than via gate drive control) may be used to regulate the final output as required.

4.4.4.4 Gate Drive Improvement

The self-sustained converter is shown to be both simple and effective. However, the autonomous simple gate drive circuit does not give ideal gate drive waveforms. Also, at high voltages the current limiting "dropper" resistors and zener diodes of the gate drive circuit become lossy, which will reduce the overall power efficiency of the power converter.

A possible solution is to have a voltage-controlled nonlinear resistor as shown in Figure 4.21(a). If the resistance of this variable resistor decreases as the voltage across it decreases, then the problems discussed above can be solved automatically. In high-voltage periods, the resistance is high so that the power losses are low. Conversely, during the low-voltage transition periods, the resistance is low so that both the rising edge and falling edge of the gate drive signal can be improved. Unfortunately, no suitable resistor with such a nonlinear property has been found in the commercial market. Normal thermally variable resistors are far too slow for this application.

Alternatively, adding additional components, such as a speed-up capacitor in parallel with the current-limiting resistor as shown in Figure 4.21(b), can provide a phase advance and reduce the transition delay. However, the dilemma is that a small capacitance has little effect, but a large capacitance makes the gate drive voltage drop too early

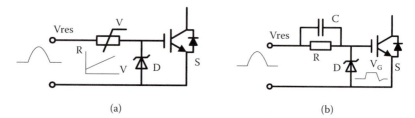

FIGURE 4.21   Passive gate drive circuits for the self-sustained resonant converter: (a) using a voltage-controlled variable resistor; (b) using a speed-up capacitor.

before the actual zero-voltage crossing due to the fact that the resonant voltage starts to drop in the second half of the "off" period. As noted, turning the devices off too early interrupts the DC current flow and causes high overvoltages, which may damage the switching devices. As a result, more sophisticated charging and discharging circuits are needed to achieve desirable gate drive waveforms for high-voltage applications.

To further improve the gate drive waveforms and power losses, PLL (phase-locked loop) and direct ZVD (zero-voltage detection) techniques were investigated to build more efficient IPT power supplies. In these two cases, although gate controllers are used essentially they only help to follow zero-voltage crossings of the resonant voltage, so the current is still autonomous.

Figure 4.22 is a block diagram showing how the PLL technique can be used to control a resonant circuit. In this diagram the VCO (voltage-controlled oscillator) unit is set at a predetermined frequency by a resistor R_T and a capacitor C_T, say 40 kHz, when its input voltage is set at half the DC supply voltage V_{cc} by two equal resistors R. The output from the VCO is divided by 4 to obtain the system nominal frequency (e.g., 10 kHz here) and a phase shift of 90° using a normal D flip-flop circuit. Then the output signal is divided into two complementary signals and sent to an IGBT gate drive circuit (such as ICL 7667) to drive the two main switches of the resonant circuit. From the ZVS frequency analysis of the current-fed resonant converter undertaken, it has been shown that the switching frequency has to be varied to achieve ZVS operation. For this reason, the resonant voltage (v_{res}) of the main circuit is detected and fed back to change the actual operating frequency, which can be slightly higher or lower than the nominal frequency. The resonant voltage may be measured directly across the two main switching devices; however, the voltage measured will be very high so a voltage divider with a very large turn-down ratio has to be used, and this is not desirable to cater to a wide range of operating voltages. Alternatively, a one-turn coil added to the phase-splitting transformer of the converter (see Figure 4.18) can be employed as a very simple and reliable voltage sensor. The

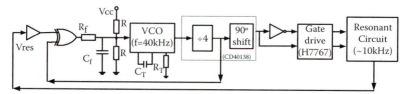

FIGURE 4.22   PLL gate drive technique.

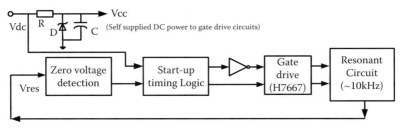

FIGURE 4.23 Direct ZVD gate drive technique.

measured resonant voltage is converted to a square waveform using a high-gain amplifier and is then compared with the output of the VCO (after the frequency divider) using an XOR logic circuit. These two signals should be equal in frequency but have a 90° phase shift as required for the XOR detection circuit. After an RC low-pass filter, the output pulses from the XOR will control the input voltage of the VCO, thus varying the actual gate driving frequency to the point where a stable ZVS frequency is obtained.

Unlike the PLL gate drive, which is based on frequency shift control, Figure 4.23 shows the block diagram of a direct ZVD technique. The advantage of this technique over PLL is that after the resonant voltage is measured, the zero crossings of the resonant voltage are detected instantly using a zero-voltage detection circuit, and the output is used to control the gate drive directly. However, the current-fed resonant converter has a start-up problem because the resonant voltage does not go to zero naturally in the beginning. A complete new technique is used to overcome this problem. On starting, both switches are turned on at the same time so that the current in the DC inductor increases rapidly. When enough energy is stored in the inductor compared to the energy in the resonant circuit in normal operation, one switch is turned off and the circuit then completes a first half-cycle followed by normal operation with direct ZVS control. To achieve this desirable condition, start-up timing logic is added as shown in Figure 4.23. Also, to avoid an additional DC power supply for the gate control circuitry and form a completely self-sustained converter, a simple zener diode configuration is used although the initial power buildup process has to be considered to make the circuit work properly. If the ramp-up delay of a practical input DC supply is taken into account, an overshoot-free soft start-up can be achieved. Consequently, the system can start up simply by turning on the main switch of the input power supply without employing any additional equipment.

Figure 4.24 shows the measured steady-state resonant voltage (across one of the main switches) and the gate signal waveforms of a practical 80 A/10 kHz current-fed parallel resonant IPT power supply using the direct ZVD technique. The DC input voltage is 240 V and the DC current can be up to 60 A, corresponding to a maximum load of about 15 kW. The main circuit of the converter is the same as the self-sustained converter (essentially the G1 power supply) shown in Figure 4.18. The track inductance is 125 μH, and its parallel tuning capacitance is about 2μF. It can be seen that ZVS operation is achieved and both the rising and falling edges of the gate drive waveform are significantly improved owing to the adoption of the specially designed MOS gate drive circuit (ICL 7667).

A PLL circuit was also built to control a power converter with the same specifications as that of the direct ZVD. Because the same gate drive chip was used, very similar gate

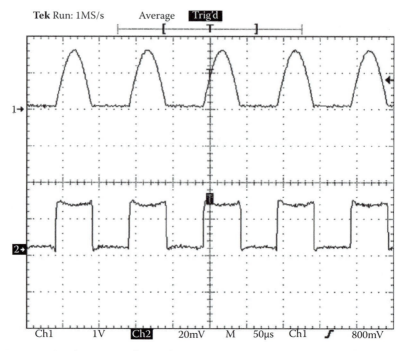

FIGURE 4.24 Steady-state waveforms of an IPT power supply using direct ZVD (Ch1: Resonant voltage across the switch: 500 V/div, Ch2: Gate drive signal: 10 V/div).

signal waveforms were obtained. To avoid repetition, Figure 4.24 shows the resonant voltage and the resultant track current of this IPT power supply. In fact, it has been observed that the track current waveforms of the two supplies are basically the same because they achieve very similar steady-state ZVS performance. However, a close look at the resonant voltage waveforms shows that the switching of the ZVD scheme is slightly slower due to the circuit delay (which is mainly caused by the practical comparator used). In comparison, the average ZVS error of the PLL circuit can be smaller under ideal steady-state conditions because it has an integral control loop and can add in some phase advance to the gate drive signals. However, this is true only when the feedback loop of the PLL is tuned accurately. Practical circuit parameter and voltage variations can introduce errors in the PLL circuit, and as a result the actual difference between the steady-state ZVS errors of the two IPT power supplies is barely perceptible.

Apart from being simple and cost-effective, the most outstanding advantage of the direct ZVD circuit over the PLL is its dynamic performance. The PLL circuit has a predetermined frequency, and its internal "integral" process causes relatively slow frequency response so that dynamic ZVS is impossible. As a result, additional blocking diodes have to be put in series with the main switches for safe operation, and special care has to be taken in designing the PLL circuit to ensure the frequency can be captured by the PLL during start-up and load transients. Conversely, the ZVD circuit simply follows the zero voltage crossings with limited predictable errors during the whole

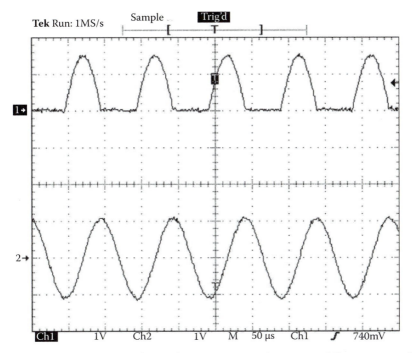

FIGURE 4.25 Steady-state waveforms of an IPT power supply using PLL (Ch1: Resonant voltage across the switch: 500 V/div, Ch2: Track current: 100 A/div).

process so that the series-blocking diodes are optional. Although frequency stability problems may occur under some extreme conditions as discussed previously, during normal working conditions the circuit oscillation continues and the direct ZVD operation is very reliable. The actual operating frequency can be varied simply by changing the resonant inductor or the parallel-tuning capacitor without any modification of the gate drive circuit. The basic concept employed here is essentially the same as the simple self-sustained converter proposed earlier, but the gate-drive performance is significantly improved.

Using the method discussed above, an 80 A/10 kHz IPT power supply based on direct ZVD has been built and overshoot-free dynamic ZVS start-up and steady-state performance has been achieved. Figure 4.25 shows a measured result of the resonant voltage across one of the two main switches and the gate drive signal of the other switch. It can be seen that the voltage across the switch, which is equal to the DC input voltage, ramps up gradually in the beginning when both the switches are off. After a very short DC current buildup period (where both the switches are on, shown with the first short pulse of the gate drive signal in Figure 4.26) before the DC voltage reaches its maximum value, the circuit goes into ZVS operation very quickly and smoothly without any overshoots. As a result, the stresses and losses of the switching devices are minimized. This feature is particularly useful for applications where frequent "on" and "off" control of the converter is required. An IPT pulse

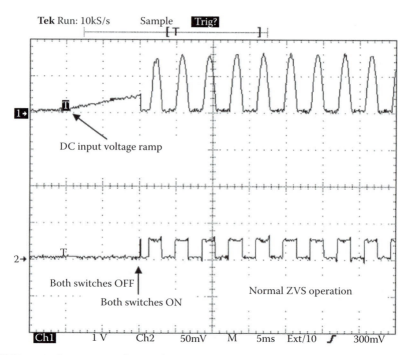

FIGURE 4.26 Start-up waveforms of a practical IPT power supply using direct ZVD (Ch1: Resonant switch voltage: 500 V/div, Ch2: Gate drive signal: 10 V/div).

battery-charging system controlled from the primary side is a good example of such an application.

References

1. Luo, Fang Lin, and Hong Ye. 2004. *Advanced DC/DC converters*. Boca Raton, FL: CRC Press.
2. Mohan, N., T. M. Undeland, and W. P. Robbins. 1995. *Power electronics—converters, applications, and design*. New York: Wiley.
3. Ang, Simon, and Alejandro Livia. 2005. *Power-switching converters*, 2nd ed. Boca Raton, FL: CRC Press.
4. Kularatna, Nihal. 2008. *Electronics circuit design—from concept to implementation*. Boca Raton, FL: CRC Press.
5. NXP Philips Semiconductors. 2007. TEA1610P, TEA1610T zero-voltage-switching resonant converter controller. Product Data Sheet, rev. 03, March.
6. Hu, Aiguo Patrick. 2009. *Wireless/contactless power supply—inductively coupled resonant converter solutions*. Saarbrücken, Germany: VDM Verlag.
7. Green, A. W., and J. T. Boys. 1994. 10 kHz inductively coupled power transfer—concept and control. *IEE Power Electronics and Variable Speed Drives Conference*, PEVD, Pub. 399, 694–99.

8. Boys, J. T., A. Hu, and G. Covic. 2000. Critical Q analysis of a current-fed resonant converter for ICPT applications. *IEE Electronics Letters* 36 (17, ISSN 0013-5194, August): 1440–42.

9. Luo, Fang Lin, and Hong Ye. 2006. *Synchronous and resonant DC/DC conversion technology, energy factor, and mathematical modeling.* Boca Raton, FL: CRC Press.

10. Kazimierczuk, Marian K., and Dariusz Czarkowski. 1995. *Resonant power converters.* New York: Wiley.

5

Control Loop Design of DC-to-DC Converters

Kosala Kankanamge-Gunawardane

5.1 Introduction

In our previous chapters on linear and switching regulators, our discussions did not consider the detailed aspects of the feedback loop used in the topology. In a well-designed DC-DC converter, irrespective of the load behavior, under all load currents the power supply should be free of any stray oscillations, and the DC rails should not fluctuate beyond specified margins. In a real-world, practical power supply design, when we consider the nonideal components such as op-amps, smoothing capacitors (with finite ESR), and inductors and transformers with parasitics, the design of the control loop becomes a challenge. This is further complicated by the finite frequency response properties of BJTs and MOSFETS, which have junction capacitances and other complex secondary elements. The subject of stability of a power supply that pertains to the closed-loop frequency response of the DC-DC converter has received much attention during the past quarter century. This chapter provides a summary of essential theory and different practical approaches to get the best control loop design without a rigorous mathematical approach.

5.2 Feedback Control and Frequency Response

For most designers, feedback control loop stability is shrouded by a cloud of mystery. Although most designers understand the problem of unwanted oscillations of a power supply, many use trial-and-error procedures or fancy mathematical models that require computing resources. In this section we discuss feedback loop stability, with some practical insights into the theory and suggesting some useful practical procedures for refining the design process.

A system that maintains a prescribed relationship between the output and the reference input by comparing them and using the difference as a means of control is called a feedback control system. Feedback control systems are often referred to as closed-loop control that uses the feedback control action in order to reduce the system error.

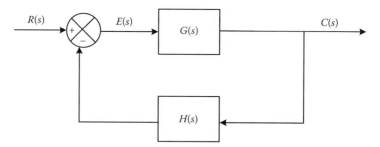

FIGURE 5.1   Closed-loop system.

A typical block diagram of a closed-loop system is shown in Figure 5.1. Closed-loop transfer function of this is given by

$$\frac{C(s)}{R(s)} = \frac{G(s)}{1 + G(s)H(s)} \tag{5.1}$$

where R(s) is the input, output C(s) is obtained by multiplying the transfer function G(s), and H(s) is the feedback element form the output. More details can be found in Ogata [1].

Frequency response is the steady-state response of a system to a sinusoidal input. In frequency response analysis, frequency of the input signal is varied over a certain range and we study the resulting performance. In order to analyze control systems, frequency response analysis is a reliable approach that utilizes the data obtained from the measurements on the physical system without deriving the mathematical models. Frequency response methods are most powerful in the conventional control theory. An advantage of frequency response analysis is that this approach is simple in general and can be made accurately by use of readily available sinusoidal signal generators and common measurement equipment.

5.3 Poles, Zeros, and S-Domain

The frequency response can be analyzed by representing the gain as a function of the complex frequency s. The relationship between the input and output of a system, which is called the transfer function T(s), is derived as

$$T(s) = \frac{V_o(s)}{V_i(s)} = \frac{N(s)}{D(s)} \tag{5.2}$$

Once this function is derived, by replacing s with $j\omega$ we can evaluate its frequency behavior. In general, the transfer function T(s) in its own form (without substituting $s = j\omega$) can reveal many useful details about the stability of the circuit. The important thing here is to recognize that this function has a gain and phase associated with it. In such an equation the roots of the numerator $N(s) = 0$ are called zeros of the system, while the roots of the denominal $D(s) = 0$ are called the poles of the system. In general, function $T(s)$ can be expressed in many different forms, including:

$$T(s) = a_m \frac{(s - Z_1)(s - Z_2)\dots(s - Z_n)}{(s - P_1)(s - P_2)\dots(s - P_n)} \tag{5.3}$$

and

$$T(s) = \frac{a_m s^m + a_{m-1} s^{m-1} + \dots + a_0}{s^n + b_{n-1} s^{n-1} + \dots + b_0} \tag{5.4}$$

In Equation (5.3), Z_1, Z_2, ... , Z_m are called transfer function zeros or transmission zeros, and P_1, P_2,..., P_n are called transfer function poles or the natural modes of the network. The n of the transfer function is called the order of the network.

For example a first-order transfer function, which has a single pole, can be written as

$$T(s) = \frac{a_0}{s - P_0} \tag{5.5}$$

where P_0 is called the pole frequency. (Sometimes first-order systems may involve one zero as well.)

A second-order transfer function, which has two poles, can be written as

$$T(s) = \frac{a_0}{(s - P_0)(s - P_1)} \tag{5.6}$$

where P_0 and P_1 are the pole frequencies. (Sometimes second-order systems may involve one or two zeros.) For example, a low-dropout regulator has three poles and one zero, which has a third-order transfer function, which will be discussed later in this chapter.

5.4 Stability Using Bode Plots

Bode plots are convenient tools because they contain all the information necessary to determine if a closed-loop system is stable. Bode diagrams consists of two graphs, a logarithmic of the magnitude of sinusoidal transfer function and the phase plot, which is plotted against the frequency on logarithmic scale as shown in Figure 5.2(b). In order to construct the Bode plot, first determine the transfer function and arrange the equation in the form of Equation (5.3).

A good place to start is the simple RC circuit, as in Figure 5.2(a), and its gain and phase plot. For this simple single-pole circuit, the transfer function is given by

$$G(s) = \frac{V_{out}(s)}{V_{in}(s)} = \frac{1}{1 + sRC} \tag{5.7}$$

and the pole frequency is given by

$$f = -\frac{1}{2\pi RC} \tag{5.8}$$

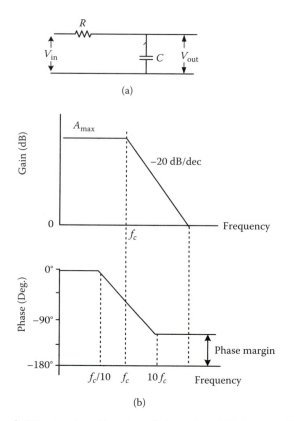

(a)

(b)

FIGURE 5.2 Simple RC network and its gain and phase plots: (a) RC circuit; (b) gain and phase plot.

Equation (5.7) shows an important result, that a pole will cause the transition of the gain plot from 0 to −1 at a frequency of $f_c = 1/2\pi RC$. At this corner frequency f_c, the asymptote breaks and the slope is −6 dB/octave or −20 dB/decade. In general, a pole in a transfer function will cause a transition from +1 to a 0 slope, or 0 to −1, or −1 to −2, or −2 to −3, etc., with a gain change of −6 dB/octave (or −20 dB/decade). This is associated with a phase shift of −90°/octave or −45°/decade. Zeros are the points where the Bode plot breaks upward, causing an opposite change of slopes with leading phase shifts. More details in constructing Bode plots can be found in Ogata [1] , Chryssis [2], and Sedra [3].

5.5 Linear Regulators' Feedback and Loop Stability

As discussed in Chapter 1, all closed-loop-type linear regulator configurations use a feedback loop to hold the output voltage constant. In analyzing linear regulators, loop gain will be accounted as the magnitude of the voltage gain that the feedback signal experiences as it travels through the loop. In order to maintain corrected regulated output voltage and a stable loop response, it is important to note that negative feedback must be used for a linear regulator. Figure 5.3(a) illustrates the block diagram of typical

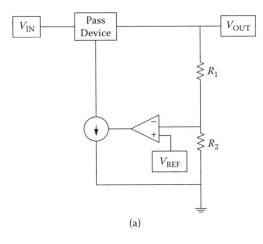

(a)

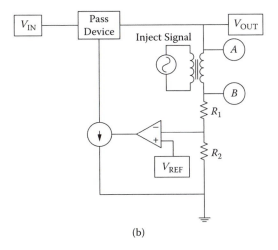

(b)

FIGURE 5.3 (a) Simplified case of Figure 1.5(b) linear regulator; (b) modeling technique to determine the loop gain of a linear regulator.

linear regulator that applies this concept. The resistor divider, error amplifier, and pass element form a *closed loop*. The output voltage, V_{OUT}, provides a feedback voltage through the resistor divider, to the inverting input of the error amplifier. The bandgap reference output (V_{REF}) is a high-precision, fixed voltage that is tied to the noninverting input of the error amplifier. The error amplifier, essentially an op-amp, then makes fraction of output equal to V_{REF} by sourcing a ground current to the base of the pass element. The pass element, in turn, supplies sufficient output current to keep V_{OUT} at a certain value.

The regulated output voltage, therefore, is defined as

$$V_{OUT} = V_{REF} \cdot \left(1 + \frac{R_1}{R_2}\right)$$

(5.9)

This equation illustrates that the feedback is negative, and the adjustment of the output voltage resulting from the feedback is in opposite polarity to the "original" change to the output voltage. This means when there is a certain fluctuations in the output voltage by means of a rising or a falling of the output voltage (due to the changes in input voltage), the negative-feedback loop will respond to force it back to the nominal value.

In the analysis of feedback response in order to model the actual fluctuations on the output voltage, through a transformer a small sinusoidal signal is coupled to the feedback path. This small-signal sine wave is injected into the feedback path between points A and B and is used to modulate the feedback signal as shown in Figure 5.3(b). The AC voltages at A and B are measured and used to calculate loop gain. Initially, the small signal has a value of ΔV_A at point A, and a value of ΔV_B at point B with respect to ground. The signal at point B then travels through the loop and eventually arrives at point A, with a value of $\Delta V_{B'}$. Although ΔV_A and $\Delta V_{B'}$ have the same magnitude (unity gain), there is a difference in phase (in degrees) between the two.

If the error amplifier creates an ideal negative feedback to the loop, ΔV_A would lag $\Delta V_{B'}$ by $-180°$ as shown in Figure 5.4(a). However, because of the pass element's built-in

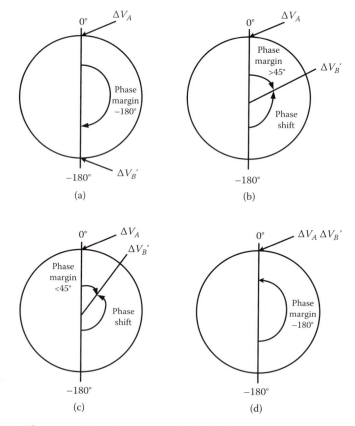

FIGURE 5.4 Phase map: (a) perfect negative feedback—loop stable; (b) phase margin >45°—loop stable; (c) phase margin <45°—loop unstable; (d) perfect positive feedback—loop unstable.

capacitance, it introduces a phase shift that reduces the perfect −180° phase difference by a value between 0° and −180° (counterclockwise, with −180° as the starting point) as shown in Figures 5.4(b), 5.4(c), and 5.4(d). A stable loop typically needs at least 45° of phase margin. Phase margin (Φ) is a positive number (clockwise) between 0° and 180°, and is expressed as

$$\Phi = 180° + \text{phase shift} \tag{5.10}$$

where phase shift is a negative number that is counterclockwise between 0° and −180°.

To keep the loop stable, the phase margin must be no less than 45°. If a 45° phase margin cannot be achieved by the intrinsic architecture of the linear regulator, some form of compensation, either internal or external, is required.

Phase shift and phase margin can be calculated easily using Bode plot analysis by means of poles and zeros. A pole as shown in Figure 5.5 is defined as a point where the slope of the gain curve changes by −20 dB/decade with respect to the slope of the curve prior to the pole. Each additional pole will contribute to increase the negative slope by the factor N (−20 dB/decade), where N is the number of additional poles. A pole also introduces a −90° phase shift (counterclockwise) from the frequency one decade below

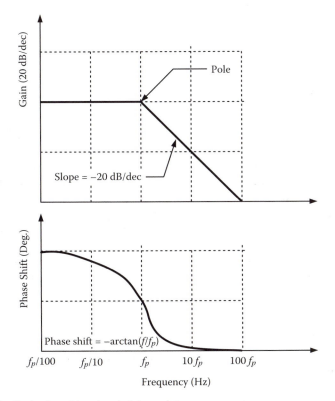

FIGURE 5.5 Bode plot with pole gain/phase plot.

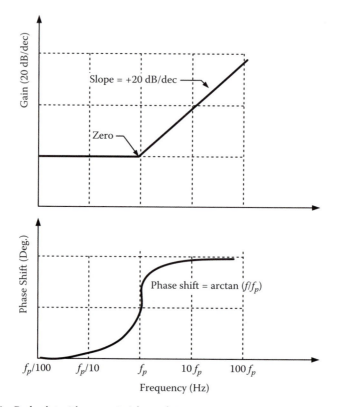

FIGURE 5.6   Bode plot with zero gain/phase plot.

pole frequency to the frequency one decade above pole frequency, with a −45° phase shift at pole frequency.

A zero is defined as a point where the gain changes by +20 dB/decade with respect to the slope prior to the zero as shown in Figure 5.6. Similar to the case of poles, the change in slope is additive with additional zeros. The phase shift introduced by a zero varies from 0 to +90°, with a +45° shift occurring at the frequency of the zero. The most important thing to observe about a zero is that it is an antipole, which is to say its effects on gain and phase are exactly the opposite of a pole.

This is why zeros are intentionally added to the feedback loops of LDO regulators so that they can cancel out of the effect of one of the poles that would cause instability if left uncompensated.

5.5.1 Bode Plot Analysis

In order to analyze the effect of stability using poles and zeros in a Bode plot, let's consider an example of three poles and one zero as shown in Figure 5.7.

It is assumed that the DC gain is 80 dB, with the first pole occurring at 100 Hz. At that frequency, the slope of the gain curve changes to −20 dB/decade.

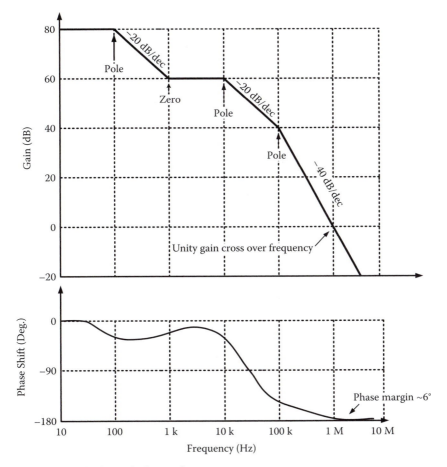

FIGURE 5.7 Bode plot with phase information.

The zero at 1 kHz changes the slope of the gain to +20 dB/decade and the result of 0 dB/decade until the second pole at 10 kHz, where the gain curve slope returns to −20 dB/decade. The third and final pole at 100 kHz changes the gain slope to the final value of −40 dB/decade. It can also be seen that the unity gain (0 dB) crossover frequency is 1 MHz. The unity gain crossover frequency is sometimes referred to as the loop bandwidth where the gain of the loop reaches zero.

In order to get the phase plot, the phase shift at each frequency point was calculated based upon summing the contributions of every pole and zero at that frequency.

The first pole, next zero, and the second pole contribute their full phase shifts of −90°, +90, and −90° respectively, resulting in a net phase shift of −90°. The final pole is exactly one decade below the 0 dB frequency. Using Equation (5.10) for pole phase shift, this pole will contribute −84° of phase shift at 1 MHz. Added to the −90° from the two previous poles and the zero, the total phase shift is −174°, which results in a phase margin of 6°. Therefore the loop is not stable and it would either oscillate or ring severely.

5.5.2 Stability Analysis

A linear regulator with an NPN pass transistor (Figure 1.5[b]) is in the common collector configuration. For this configuration, considering the intrinsic factors of the pass transistor (at this stage we neglect the intrinsic factors of the error amplifier) the small signal equivalent circuit can be drawn as shown in Figure 5.8.

The load current of the regulator can be written as

$$I_L = i_B + g_m V_\pi - \frac{V_0}{r_0} \tag{5.11}$$

Assuming r_0 is an infinitely larger value, Equation (5.11) becomes

$$I_L = i_B + g_m V_\pi \tag{5.12}$$

Based on Figure 1.5(b), Thevenin equivalent circuit parameters can be derived as

$$V_T = \frac{R_Y}{R_X + R_Y} V_0 = k V_0$$

and $R_T = R_X \mathbin{/\!/} R_Y$. Therefore

$$V_{R_{in}} = \frac{R_{in}}{R_{in} + R_T} k V_0 \tag{5.13}$$

and

$$V_x = A_{OL} \frac{R_{in}}{R_{in} + R_T} k V_0 \tag{5.14}$$

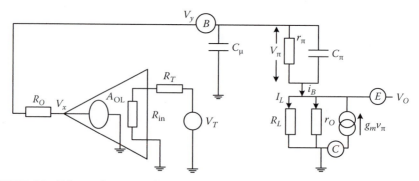

FIGURE 5.8 Effects of intrinsic factors of an NPN pass transistor in the linear regulator stability.

Assume R_0 is a negligible resistance, so that $V_x = V_y$. Therefore i_b can be written as

$$i_B = \frac{V_y - V_o}{r_\pi} + \frac{V_y - V_o}{1/sC_\pi r_\pi} \tag{5.15}$$

Substituting i_b and $V_\pi = V_y - V_0$ into Equation (5.11) gives

$$\frac{V_o}{I_L} = \frac{1}{\left[\left(A_{OL}k \middle/ \left(1 + \frac{R_T}{R_{in}}\right) \right) - 1 \right]} \frac{r_\pi}{[1 + g_m r_\pi + sC_\pi r_\pi]} \tag{5.16}$$

According to Equation (5.16) there is a pole that inherits from the intrinsic parameters of the pass transistor. Similarly, considering the error amplifier's intrinsic capacitances it can be shown that there is another pole due to the NPN linear regulator.

The following discussion shows the stability analysis for some practically available linear regulators configurations. The derivations are based on a similar approach to the above analysis.

5.6 Stability

5.6.1 Stability of Standard NPN Linear Regulator

The standard NPN regulator is based on the NPN Darlington pass transistor with PNP driver as shown in Figure 5.9.

As the pass transistor of the NPN regulator is connected in a circuit configuration known as the common collector, it inherits the low output impedance, which is an important characteristic of all common collector circuits. Therefore a power pole (P_{PWR}),

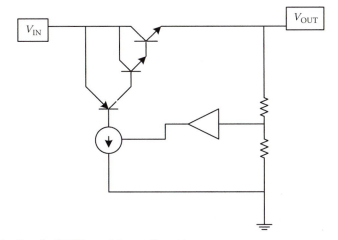

FIGURE 5.9 Standard NPN regulator configuration.

created by the pass element, occurs at a higher frequency because of the pass element's relatively low output impedance.

$$f(P_{PWR}) = 1/2\pi \cdot R_{PWR} \cdot C_{PWR} \tag{5.17}$$

where R_{PWR} and C_{PWR} are the pass element's resistance and capacitance.

The dominant pole (P_{INT}), created by the intrinsic elements of the regulator, occurs at a very low frequency due to very high output impedance of the regulator especially due to the relatively high capacitance.

$$f(P_{INT}) = 1/2\pi \cdot R_{INT} \cdot C_{INT} \tag{5.18}$$

where R_{INT} and C_{INT} are the regulator's intrinsic resistance and capacitance.

According to the Bode plot analysis as shown in Figure 5.10, there are two poles in the standard NPN regulator. Still, P_{PWR} can occur at a frequency below the 0 dB crossover point, which may create a less than 45° phase margin. In this case, the loop becomes unstable, which creates a need for compensation (Graphs shown in dotted lines show the compensated case, with a phase margin greater than 45°. Figure 5.10).

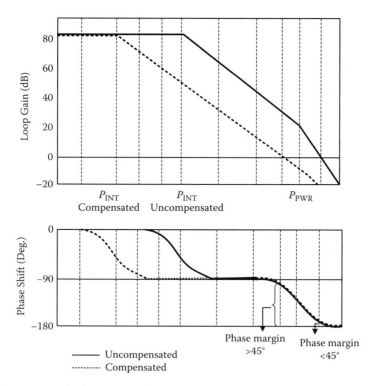

FIGURE 5.10 Standard NPN regulator Bode plot (uncompensated and dominant pole compensation).

In case of instability, the standard NPN regulator employs an internal compensation known as dominant pole compensation, which is achieved by adding intrinsic capacitance to the regulator so that P_{INT} moves to an even lower frequency. Therefore the standard NPN regulator doesn't need an external compensation due to the superior frequency response of the NPN pass device.

5.6.2 Stability of Quasi-LDO Regulator

The quasi-LDO, as shown in Figure 5.11, uses an NPN pass device, and it is in the common-collector configuration, which means its output impedance is relatively low.

However, because the base of the NPN is being driven from a high-impedance PNP current source, the regulator output impedance of a quasi-LDO is not as low as the standard NPN regulator with an NPN Darlington pass device. Therefore P_{PWR} occurs at a frequency lower than that of the standard NPN regulator, and usually below the 0 dB crossover point. The result is a less than 45° phase margin, where compensation is needed for loop stability (Figure 5.12).

Because of the relatively low frequency of P_{PWR}, the NPN pass transistor regulator cannot be stabilized by using only dominant pole compensation. Instead, a zero must be placed between P_{INT} and P_{PWR} to increase the phase margin to at least 45° (Figure 5.12). This is accomplished by using an external compensation method, such as the addition of an output capacitor next to V_{OUT}. The frequency of the added zero (Z_{COMP}) is defined by

$$f(Z_{COMP}) = 1/(2\pi . ESR . C_{OUT}) \tag{5.19}$$

where ESR is the equivalent series resistance of the output capacitor.

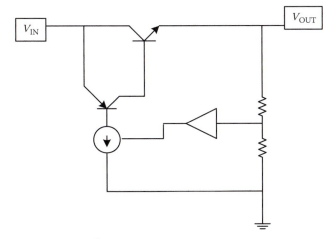

FIGURE 5.11 Quasi-LDO configuration.

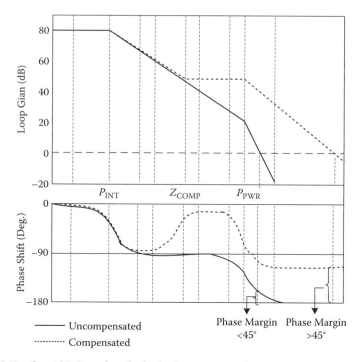

FIGURE 5.12   Quasi-LDO regulator bode plot (uncompensated and dominant pole compensation)

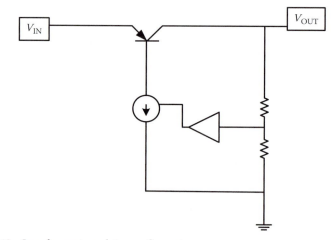

FIGURE 5.13   Low dropout regulator configuration.

Because P_{PWR} of the NPN pass transistor regulator still occurs at a relatively higher frequency, output capacitor selection is fairly easy. As long as $f(Z_{COMP})$ is lower than $f(P_{PWR})$, C_{OUT} can be small, and ESR is not critical.

5.6.3 Stability of LDO Regulator with a PNP Pass Transistor

Compared to other linear regulator configurations, PNP pass transistor regulator requires more careful selection of an output capacitor.

Its pass element, the PNP transistor, is in a common emitter configuration, which exhibits high output impedance, so P_{PWR} occurs at a frequency lower than in the other linear regulator configurations. Furthermore the impedance of the load, created by load resistance and output capacitance, becomes a significant contributor to loop stability by adding another low-frequency load pole (P_L) to the Bode plot. The frequency of P_L is expressed as

$$f(P_L) = 1/(2\pi \cdot R_{LOAD}C_{OUT}) \tag{5.20}$$

Typically, R_{LOAD} and C_{OUT} are higher than the regulator's intrinsic resistance (R_{INT}) and capacitance (C_{INT}), which makes P_L occur at a frequency lower than $f(P_{INT})$. With three poles the PNP pass transistor regulator is not stable, even with dominant pole compensation. A zero must be added to compensate for this instability, which can be accomplished by the addition of an output capacitor next to V_{OUT}. Particularly, the location of the zero in the frequency space is critical, because it translates to a careful matching of the output capacitor's capacitance and ESR.

The zero must occur somewhere between P_{INT} and P_{PWR} (Figure 5.14). Because P_{PWR} occurs at a relatively low frequency, the "space" between P_{INT} and P_{PWR} is narrow; hence, the choice of $f(Z_{COMP})$ is narrow and is demonstrated as

$$f(P_{INT}) < f(Z_{COMP}) < f(P_{PWR})/10 \tag{5.21}$$

Equation (5.21) shows that Z_{COMP} must occur above $f(P_{INT})$, and at least one decade below $f(P_{PWR})$. This is because Z_{COMP} needs the full one decade above $f(Z_{COMP})$ to fully realize its +90° phase shift. Based on Equation (5.21) in order to maintain loop stability at any given capacitance value, the ESR must satisfy the following:

$$R_{INT}C_{INT}/C_{OUT} > ESR > 10R_{PWR}C_{PWR}/C_{OUT} \tag{5.22}$$

If the ESR is either too high or too low, it is out of range. The ESR is too high when the following condition exists:

$$ESR > R_{INT}C_{INT}/C_{OUT} \tag{5.23}$$

In this condition, Z_{COMP} occurs at a frequency below $f(P_{INT})$. Additionally, because of the narrow space between P_L and P_{INT}, Z_{COMP} can be within one decade of $f(P_{INT})$. This

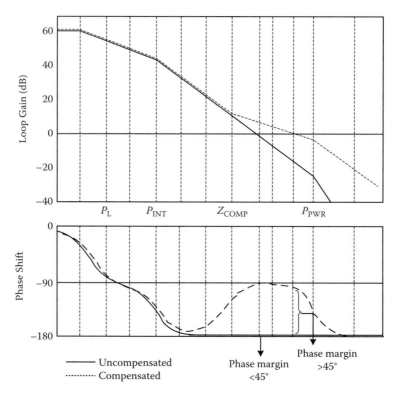

FIGURE 5.14 LDO regulator Bode plot (uncompensated and dominant pole compensation).

prevents Z_{COMP} from realizing its full +90° phase shift. As a result, the phase margin may not rise above 45° and, hence, the loop is unstable.

On the other hand, ESR is too low when the following is true:

$$ESR < 10R_{PWR}C_{PWR}/C_{OUT} \tag{5.24}$$

Here, Z_{COMP} occurs within one decade below f(P_{PWR}), which prevents it from realizing its full +90° phase shift. Therefore the phase margin will not rise above 45°, and the loop, again, is unstable.

The LDO manufacturer provides a set of curves that define the boundaries of the stable region, plotted as a function of load current (Figure 5.15). Furthermore the capacitor used at the output should have some stability within the operational temperature ranges. Figure 5.16 shows typical aluminum electrolytic capacitor characteristics over frequency and temperature.

More details on the stability of linear regulators can be found in King [5], O'Malley [6], Simpson [4,7], Goodenough [8], Kwok and Mok [9], Chava and Silva-Martinez [10], Schiff [11], Rincon-Mora [12–14], Deng [15], and Lee [16].

There are novel modifications to the basic LDO architecture from Analog Devices' ADP330X series LDO regulator family in which careful selection of output capacitor

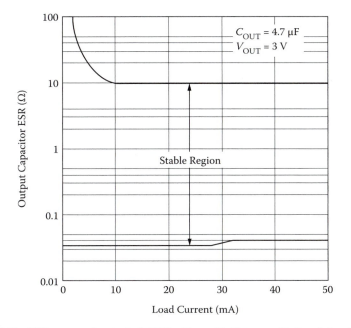

FIGURE 5.15 ESR range of a typical LDO. (From C. Simpson, National Semiconductor, Application Note 1148, May 2000.)

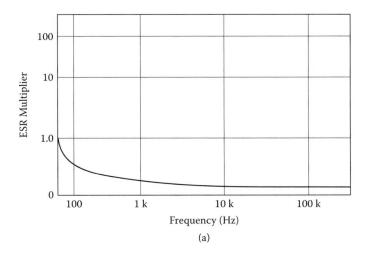

FIGURE 5.16 Behavior of typical aluminum electrolytic capacitors at different frequencies and temperatures: (a) frequency behavior; (b) ESR change with temperature; (c) capacitance change with temperature. (From ON Semiconductor, Application Note SR003AN/D. With permission.)

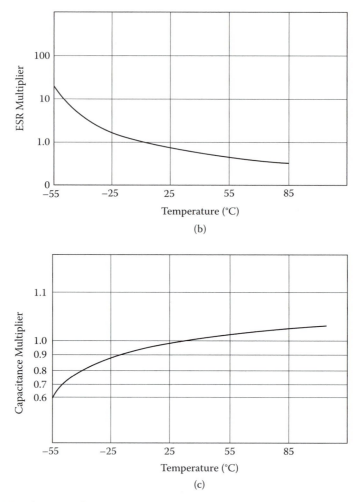

FIGURE 5.16   (Continued).

is not necessary required for these models. These regulators are also known as the any-CAP® family, so named for their relative insensitivity to the output capacitor in terms of both size and ESR [17].

Leung et al. [18] have analyzed and shown that the conventional frequency compensation cannot effectively stabilize a low-voltage LDO with a two-stage error amplifier. Further they emphasized that the conventional technique is parameter dependent and very sensitive to external effect so is limited to certain operation conditions and not the optimum. In order to stabilize low-voltage LDOs, they have developed a novel approach based on nested Miller compensation, called pole control frequency compensation (PCFC), which is a systematic approach and improves the stability of the low-voltage LDO under load current and temperature.

Conventional ESR frequency compensation is not an optimal method to maintain stability for a wide range of loading current as the pole at the output node changes significantly with different loading conditions. Kwok and Mok [9] have introduced a tracking zero to cancel the output pole so that the frequency response becomes independent of the load current.

5.7 Feedback Loop and Stability of Switch-Mode Power Supplies

Any switching regulator also may be treated as a closed-loop feedback control system. In analyzing the feedback loop and stability, the same Bode plot-based analyzing procedure of linear regulators is applicable to switching regulators as well.

Automatic regulation of the output voltage against fluctuations can be achieved with negative feedback. The difference between output voltage and a reference is sampled and used to control the duty cycle. Negative feedback reduces the duty cycle to lower the output voltage and vice versa.

For a closed loop of a switching supply, as shown in Figure 5.17, the loop consists of two typical blocks: the modulator and the error or feedback amplifier. (The case shown is a simple buck converter only, but a similar simplified block diagram can be developed for any other configuration.)

The speed and accuracy of voltage regulation depends on the total voltage "gain" around the complete switch, filter, feedback amplifier, and the duty cycle of the control loop as a function of frequency. The loop gain can be measured by injecting a small AC

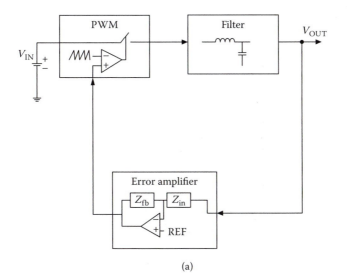

(a)

FIGURE 5.17 Control loop as applicable to a switching power supply.

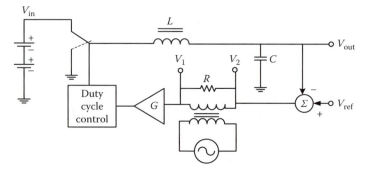

FIGURE 5.18   Estimation of the control loop gain of SMPS.

signal into the feedback loop using an isolation transformer as shown in Figure 5.18 and measuring the ratio of the amplified AC error voltage (V_2) to the injected signal voltage (V_1) versus frequency.

To achieve overall system stability and adequate phase margin, the amplifier gain combined with the modulator gain should produce an overall gain plot that crosses the unity gain (0 dB) line at the desired crossover frequency as shown in Figure 5.19 (similar to the case of linear regulators) [19,20].

All off-the-line PWM switching power supplies consist of a modulator, an error amplifier, an isolation transformer, and an output LC filter. In single-port direct duty cycle control PWM power supply topologies, a voltage V_c is applied to the control port of the PWM comparator, and it is compared to a saw tooth voltage of constant amplitude

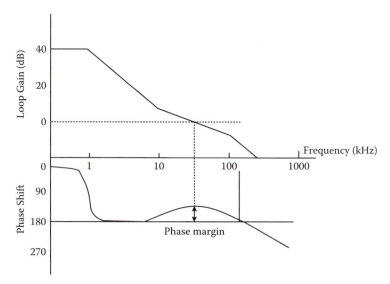

FIGURE 5.19   Bode plot of a SMPS.

V_s to change the comparator's output duty cycle from 0 to 1. The resulting duty cycle of the drive waveform then varies as

$$\delta = \frac{V_C}{V_S} \tag{5.25}$$

The gain of the buck family (forward, push-pull, and bridge) converters is given by

$$\frac{V_{out}}{V_{in}} = \frac{N_s}{N_p}\delta = \frac{N_s}{N_p}\frac{V_c}{V_S} \tag{5.26}$$

where $\frac{N_s}{N_p}$ is the transformer secondary-to-primary turns ratio and V_{in} is the transformer primary voltage.

The control-to-output voltage DC gain of a PWM power supply is

$$\frac{\partial V_{out}}{\partial V_C} = (dcgain) = \frac{N_s}{N_p}\frac{V_{in}}{V_S} \tag{5.27}$$

Equation (5.27) shows the gain H(s) of the system. For more details, refer to Chryssis [2].

An error amp based on the op-amp can be designed to have any pole-zero combination to change the Bode plot to attain unconditionally stable characteristics of the circuit. The op-amp circuit can have different configurations, and in general there are three common types used in practical switching power supply environments. Table 5.1 shows the simplified circuit configurations and the associated transfer functions. In error amplifier configurations, such as in Table 5.1, break or corner frequencies are predetermined by the design objectives. Type 1 is a simple RC low-pass filter with a single pole; type 2 is with a pole-zero pair; and type 3 is with two pole-zero pairs. In type 3, loop crossover should occur between f_2 and f_3 for better stability. Another mathematical technique useful in this process is called the K-factor method, which is a mathematical tool for defining the shape and characteristics of the transfer function; more details can be found in Chryssis [2].

In practice, the overall loop can be complicated, and in the final stage of a design project, the design team can make use of some tests to measure the loop transfer function. One test is to inject a signal into the loop and then measure the loop transfer function. However, due to monolithic ICs that do not allow injection of the signal in the best location, one may have to compromise with an "achievable method." Figure 5.20 shows the case from Venable [21], where the desired versus achievable methods are indicated for a simplified case of a computer power supply. In Figure 5.20, for the case of a common PWM control IC type UC 3844, two cases of feedback paths exist (a fast loop via resistor R5, and a slow loop via R7). Figures 5.20(a) to 5.20(f) show these measurements in this typical example [21]. One practical difficulty with such measurements is to select an injection transformer with wide frequency response. Although some companies sell such transformers, they can be expensive. One solution proposed is to modify an off-the-shelf current transformer and use it with a typical network analyzer- based measurement setup. Figure 5.21 shows a typical injection transformer setup with a network analyzer [22]. Figures 5.21(b) and 5.21(c)

TABLE 5.1   Different Error Amp Types Usable in Switching Power Supplies

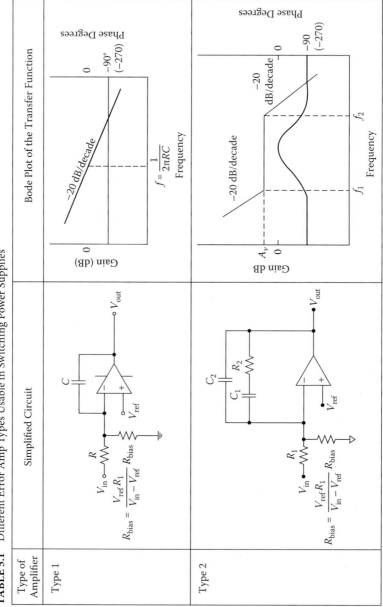

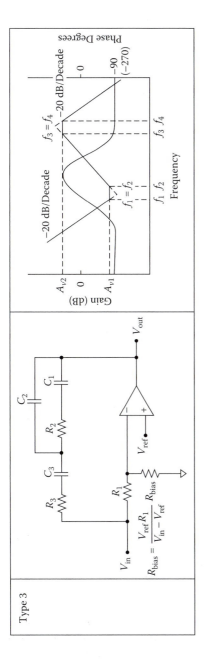

Type 3

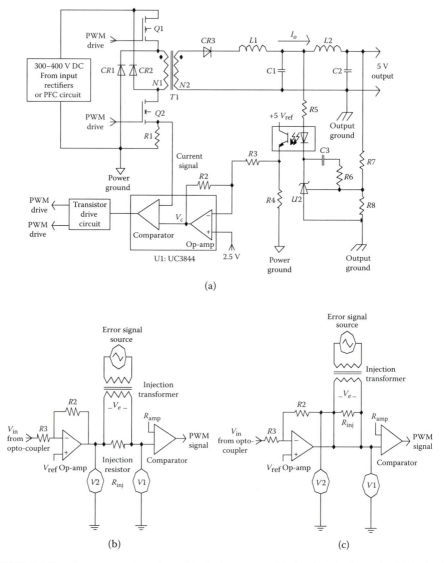

FIGURE 5.20 Loop measurements and typical response: (a) simplified schematic; (b) desired injection method; (c) achievable injection method; (d) transfer function of slower loop (measured in series with R7); (e) transfer function of fast loop (measured in series with R5); (f) transfer function of entire loop (measured in series with R2). (Courtesy of *PCIM*. Source: [21])

show a typical frequency response curve for a commercial current transformer, such as the PE-51687 from Pulse Engineering [22]. Williams [23] discusses practical guidelines for an iterative procedure for easy frequency compensation using a test setup. Hesse [24] provides some analytical aspects of a battery-powered buck converter example. Gain equalization aspects of flyback converters are discussed in Sandler [25].

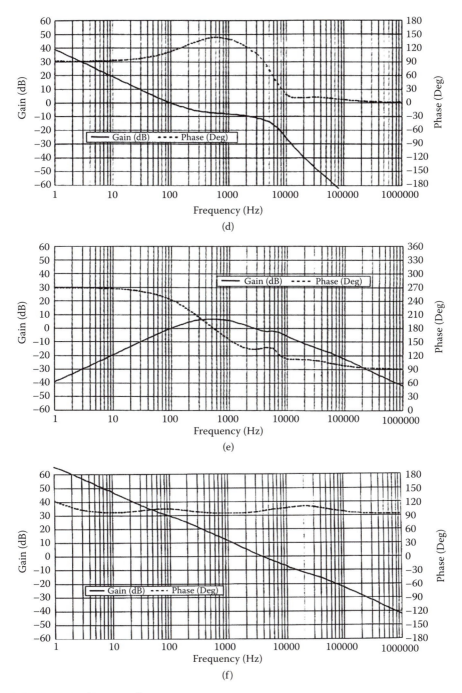

FIGURE 5.20 (Continued).

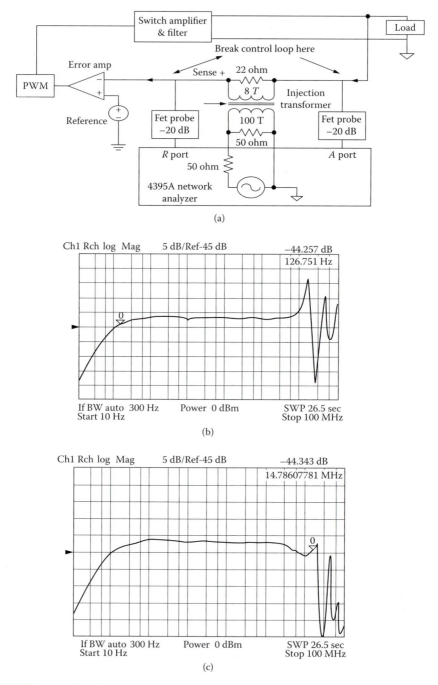

FIGURE 5.21 Typical injection transformer setup for loop-gain measurements: (a) setup; (b) frequency response of the PE-51687 transformer with eight-turn secondary; (c) frequency response for the same with 22 Ω on the secondary as in (a). ([22]Courtesy of Power Electronics Technology.)

5.8 Digital Control

Over the past eight years, the concept of digital control in power supplies has become a frequent topic of discussion, while microcontroller and DSP suppliers were introducing low-cost programmable controllers. In switching power supplies, the analog control concepts are used to control the "on" and "off" states of a power switch, using PWM or soft-switching techniques. If the function of the analog control circuits can be duplicated inside the software algorithms of a digital controller block, so that the power switch is driven by the intended PWM waveform, we can conceptually achieve fully digital control. Figure 5.22 indicates the concept in a simplified form for the case of a compound buck-and-boost configuration; the topology is shown in Figure 5.22. In a typical example based on the above digital control concept [26], three basic steps are performed:

1. Analog output is sampled and converted to digital format using an ADC.
2. Processed digital information (analog input based) is subjected to the digital equivalent of the transfer function (which resides within the software algorithms).
3. Output of the transfer function drives the power switches to control the output voltage.

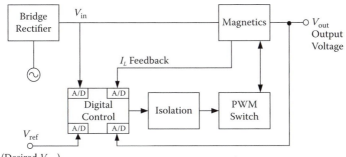

(a)

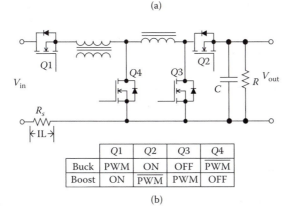

(b)

FIGURE 5.22 Digital control concept used in SMPS with buck-and-boost mode possibilities: (a) concept of digital control; (b) topology control; (c) compound buck-boost flowchart; (d) basic flowchart. (Source [26], Courtesy of *PCIM*)

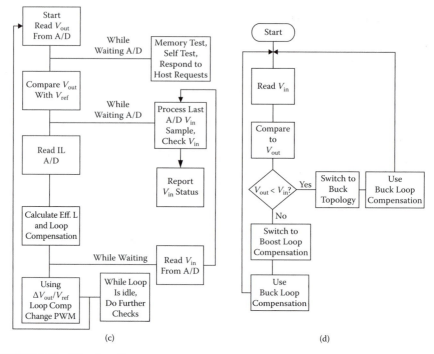

FIGURE 5.22   (Continued).

In Figure 5.22(a), the signal I_L (the current in the inductors) is also fed into the digital controller, whereas input voltage is fed at another ADC input, which makes the system decide on the buck or boost requirement.

Figure 5.22(c) shows the flowchart applicable to the system for the selection of the buck or boost mode and the overall power conversion-related flowchart (Figure 3.50[d]). In such a system, many other aspects of overall control, such as efficiency management, diagnostics, and interactive communications with the power supply, can be easily achieved [26]. Given the conceptual approach in the above example, the digital control concept in power supplies may have four different levels [27]:

- Level I—adds simple functions that are difficult to achieve with analog components (e.g., a ramping PWM waveform for the soft-start function).
- Level II—secondary management function around a traditional analog circuit. The digital controller monitors the output parameters and uses existing external controls to enhance the functionality of the power supply, though the control loop is still analog.
- Level III—a higher level of integration where the switcher is integrated with the microcontroller, but the implementation of the feedback loop is still analog.
- Level IV—complete digital control where all parameters are digitized and analyzed by the controller to provide appropriate outputs.

This usually requires a DSP with high-speed ADCs and PWM outputs. Figure 5.23 shows the concept and implementation aspects of a digitally controlled buck converter

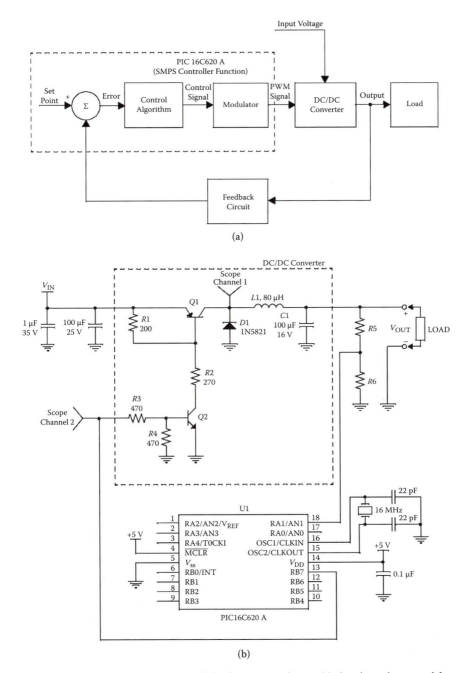

FIGURE 5.23 Digital control in a simple buck converter design: (a) absorbing the control function into the microcontroller; (b) circuit; (c) flowcharts. (Courtesy of Microchip Technology, Inc.)

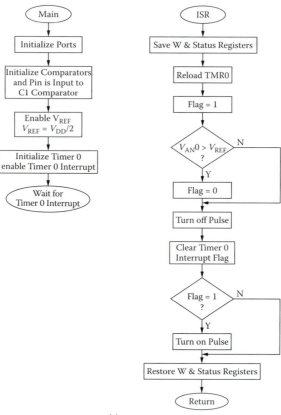

(c)

FIGURE 5.23   (Continued).

based on a low-cost microcontroller such as the PIC 16CXXX from Microchip Technology. Figure 5.23(a) shows the concept of absorbing the control function into the digital controller [28]. Figures 5.23(b) and 5.23(c) show the final circuit and applicable flowcharts for this pulse-skipping modulator-based design. The main and interrupt subroutines for timing generators are shown in the flowcharts.

Using a simple low-pin-count microcontroller and a suitable driver IC, a more compact digitally controlled switcher can be developed. Figure 5.24(a) shows a digital power converter schematic based on a PIC microcontroller. Figure 5.24(b) shows the development environment for the PIC controller, including the in-circuit debugger (ICD) [29].

For more advanced control requirements, microcontroller manufacturers such as Microchip Technology have introduced 16-bit digital signal controllers (DSCs) with DSP functionality, low pin count, and low cost. A block diagram of a typical DSC from Microchip Corp. is shown in Figure 5.25(a). Figure 5.25 shows the concept of implementing a synchronous buck converter, with the approximate timing involved. In such

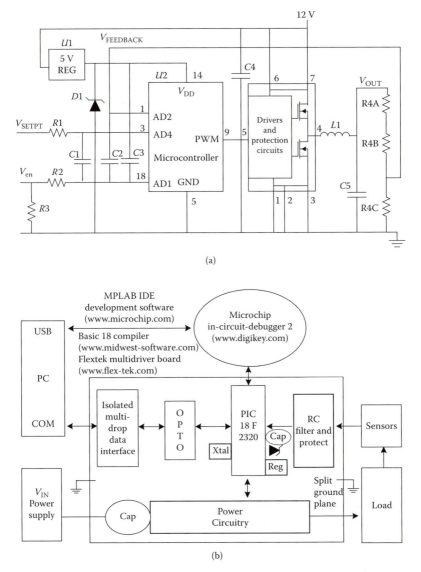

(a)

(b)

FIGURE 5.24 A PIC microcontroller-based power converter using a half-bridge driver IC: (a) circuit; (b) development environment. (Source: (28) Courtesy of *Power Electronics Technology*.)

a system, sampling triggers and ADC capability are important so as not to miss critical parameters, as in Figure 5.25, where asynchronous ADC sampling was used [30].

Analog comparators available within the DSC should be used for current limiting so that these comparators provide benefits that are not practicable or desirable to perform directly in the digital control loop (Figure 5.25[d]). With the cost of DSPs dropping to $2 or less, use of a DSP in the control loop is becoming a viable option [31].

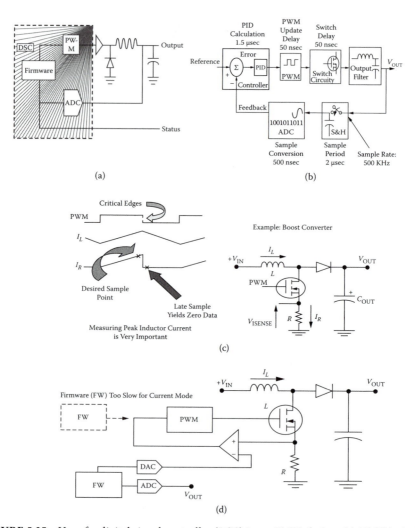

FIGURE 5.25 Use of a digital signal controller (DSC) in an SMPS design: (a) DSC block diagram; (b) example of a synchronous buck implementation; (c) need for asynchronous ADC sampling; (d) analog comparator used for current limiting outside the loop. (Source: [30] Courtesy of *Power Electronics Technology*.)

Caldwell [32], Kris [33], Etter and Fosler [34], Hagen and Freeman [35], and Pandola [36] provide some useful information for designing digital controllers for switching supplies, including the frequency response of the power stage [35]. Figure 5.26 shows a DSC-based approach for designing a full-bridge version with PFC where two DSC chips are used. Bramble and Holden [37] discuss details of a digital control system for a buck converter based on an 8051-compatible microcontroller.

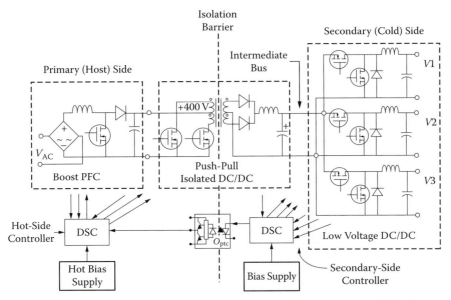

FIGURE 5.26 DSC-based approach for PFC-based full-bridge configuration with dual processors. (Courtesy of Microchip Corp.)

5.9 Control Modes of Switch-Mode Converters

In developing the control circuits for a DC-DC converter, there are two important practical steps: selection of the control IC and design of the feedback loop. In this section, a brief introduction to control ICs is provided. The primary function of the control IC in a switch-mode power supply is to sense any change in the DC output voltage and adjust the duty cycle of the power switches to maintain the average DC output voltage constant. In general, an oscillator within the IC allows the designer to set the basic frequency of operation. A stable, temperature-compensated reference is also provided within the IC. There are two basic modes of control used in PWM converters: voltage mode and current mode.

5.9.1 Voltage-Mode Control

This is the traditional mode of control in PWM converters. It is also called single-loop control, as only the output voltage is sensed and used in the control circuit. A simplified diagram of a voltage-mode control circuit is shown in Figure 5.27. The main components of this circuit are an oscillator, an error amplifier, and a comparator. The output voltage is sensed and compared to a reference. The error voltage is amplified in a high-gain amplifier. This is followed by a comparator, which compares the amplified error signal with a sawtooth waveform generated across a timing capacitor.

The comparator output is a pulse-width modulated signal that serves to correct any drift in the output voltage. As the error signal increases in the positive direction, the

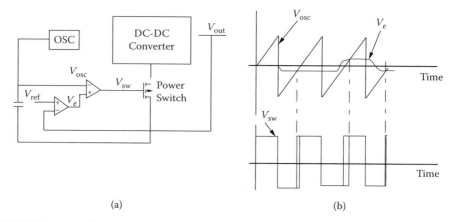

(a)   (b)

FIGURE 5.27   Voltage-mode control: (a) block diagram; (b) associated waveforms.

duty cycle is decreased, and as the error signal increases in the negative direction, the duty cycle is increased. The voltage mode control technique works well when the loads are constant. If the load or the input changes quickly, the delayed response of the output is one drawback of the control circuit, as it senses only the output voltage. Also, the control circuit cannot protect against instantaneous overcurrent conditions on the power switch. These drawbacks are overcome in current-mode control.

5.9.2 Current-Mode Control

The current control mode is a multiloop control technique that has a current feedback loop in addition to the voltage feedback loop. This second loop directly controls the peak inductor current with the error signal rather than controlling the duty cycle of the switching waveform. Figure 5.28 shows a block diagram of a basic current-mode control circuit. The error amplifier compares the output to a fixed reference. The resulting error signal is then compared with a feedback signal representing the switch current in the current-sensing comparator. This comparator output resets a flip-flop that is set by the oscillator. Therefore switch conduction is initiated by the oscillator and terminated when the peak inductor current reaches the threshold level established by the error amplifier output. Thus the error signal controls the peak inductor current on a cycle-by-cycle basis. The level of the error voltage dictates the maximum level of peak switch current. If the load increases, the voltage error amplifier allows higher peak currents. The inductor current is sensed through a ground-referenced sense resistor in series with the switch.

The disadvantages of this mode of control are loop instability above 50% duty cycle, a less than ideal loop response due to peak instead of average current sensing, and a tendency toward subharmonic oscillation and noise sensitivity, particularly at very small ripple currents. However, with careful design using slope compensation techniques [38,39], these disadvantages can be overcome. Therefore current-mode control becomes an attractive option for high-frequency switching power supplies. Some special problems, such as pulse-skipping oscillations, and solutions are discussed in Dobrenko [40].

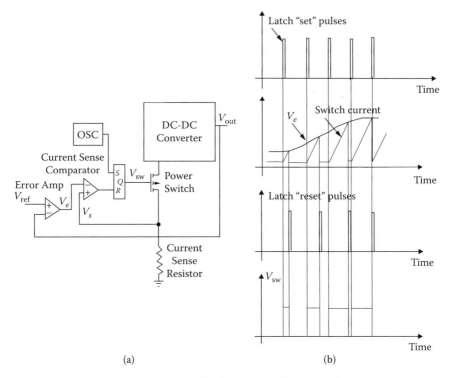

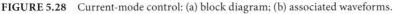

(a)   (b)

FIGURE 5.28   Current-mode control: (a) block diagram; (b) associated waveforms.

5.9.3 Hysteretic Control

For processors such as Pentiums™, current requirements are on the order of 20–30 A in desktops and similar systems. Other requirements are extremely low-output ripple voltages (on the order of only 50–100 mV, at most) with step load current transients on the order of 30–50 A/μs. In such situations the speed of the controller becomes very critical. Table 5.2 indicates the core voltage requirements of an old 300 MHz Pentium processor. A relatively newer technique using a hysteretic controller or ripple

TABLE 5.2   Design Constraints for a 300 MHz Pentium™ II Processor

Parameter	Typical Value
Core voltage (V_{core})	2.8 V
Static voltage tolerance	+100 to −60 mV
Dynamic voltage tolerance	±140 mV
Maximum process core current ($I_{CC(max)}$)	14.2 A
Typical processor core current ($I_{CC(typ)}$)	8.7 A
Step current slew rate	30 A/μs
Input current di/dt	<0.1 A/μs

Note: From E. Vosicher, *PCIM*, January 2000. With permission.

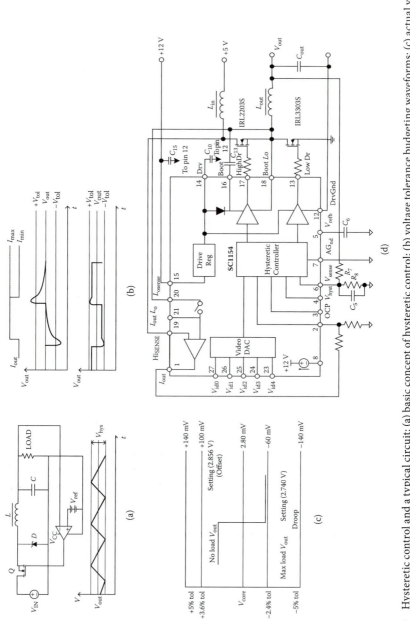

FIGURE 5.29 Hysteretic control and a typical circuit: (a) basic concept of hysteretic control; (b) voltage tolerance budgeting waveforms; (c) actual voltage tolerancing for the case of the Intel VRM 8.3 specification; (d) a typical circuit. (Source [41]: Courtesy of Power Electronics Technology.)

regulator has become prominent. Ripple regulation combines the advantages of voltage-mode regulation and current-mode solutions to power supply regulation. Voltage-mode regulation is noted for reliable operation within a specified window, but it is slow in responding to transient current demands, and loop compensation is difficult to implement. Current-mode regulation offers better transient response than voltage-mode regulation but at the expense of additional losses due to current-monitoring resistors in the circuit.

A ripple regulator responds quickly to step current demands, and it is power efficient. In addition, a well-designed hysteretic controller keeps the ripple-regulated VO within the specified window, maintaining general conditions required by a power-hungry CPU. Figure 5.29(a) shows the basic concept of a hysteretic-controlled buck converter that compares the actual output voltage to a reference signal corresponding to a desired output. Within the V_{hys} margins set by the Schmitt trigger/comparator, the output voltage will ramp up and down. One important aspect of this approach is that the frequency of operation is not constant and depends on the input-output voltage differential, inductor value, and ESR of the output capacitor. The typical range of frequencies is from about 150 to 700 kHz. The graph in Figure 5.29(a) shows the shape of the output ripple current flowing through the output capacitor. The hysteretic control concept is easy to implement, but EMI control may be difficult due to the unpredictable noise spectrum from the variable operational frequency. Figure 5.29(b) shows voltage tolerance budgeting waveforms at the application of a typical load (upper waveform with no droop compensation; lower with droop compensation). Figure 5.29(c) shows the actual voltage tolerances of a hysteretic control-based power supply where some offset voltages are applied above and below the nominal core voltage. In this design approach using adjustable values of offset and droop values for no-load and maximum-load conditions, respectively, variation of the output voltage is maintained within the specified limits of the processor. For example, SC1154 (from Semtech Corporation) and TPS 5211(from Texas Instruments) provide the design capabilities for the Intel VRM Spec 8.3. These controllers allow the output voltage to be adjusted from 1.3 V to 3.5 V, depending on a five-bit DAC output, which is a common requirement of VRM specifications. A guide to design calculations is available in Vosicher [41] and Nowakowski and Hodson [42]. Figure 5.29(d) shows a typical power supply based on a hysteretic controller such as SC1154.

References

1. Ogata, K. 2006. *Modern control engineering*. Englewood Cliffs, NJ: Prentice Hall.
2. Chryssis, G. C. 1989. *High frequency swiching power supplies*. New York: McGraw-Hill.
3. Sedra, A. S., and K. C. Smith. 2004. *Microelectronic circuits*, 5th ed. New York: Oxford University Press.
4. Simpson, C. 2000. Linear regulators: Theory of operation and compensation. Application Note 1148. National Semiconductor, May.
5. King, B. M. 2000. Understanding load transient response of LDO. *Analog Application Journal*, November. http://focus.ti.com/lit/an/slyt151/slyt151.pdf

6. O'Malley, K. 2001. Compensation for linear regulators—on semiconductor application note.

7. Simpson, C. A user's guide to compensating low dropout regulators. http://www.national.com/assets/en/appnotes/f10.pdf

8. Goodenough, F. 1996. LDO controller handles 250A/μs load transients. *Electronic Design*, 162.

9. Kwok, K. C., and P. K. T. Mok. 2002. Pole-zero tracking frequency compensation for low dropout regulator. *IEEE International Symposium on Circuits and Systems* 4:735–38.

10. Chava, C. K., and J. Silva-Martinez. 2002. A robust frequency compensation scheme for LDO regulators. *IEEE International Symposium on Circuits and Systems* 5:825–28.

11. Schiff, T. 2000. Stability in high speed linear LDO regulators. Application Note. ON Semiconductor, October.

12. Rincon-Mora, G. A. 1996. Current efficient low voltage low dropout regulator. *Electrical Engineering*, Georgia Institute of Technology.

13. Rincon-Mora, G. A., and P. E. Allen. 1998. A low-voltage, low quiescent current, low drop-out regulator. *IEEE Journal of Solid-State Circuits* 33 (1): 36–44.

14. Rincon-Mora, G. A., and P. E. Allen. 1998. Optimized frequency shaping circuit topologies for LDO's. *IEEE Transactions on Analog and Digital Signal Processing* 45 (6): 703–8.

15. Deng, Q. 2007. An LDO primer—a review on regulation, stability and compensation. *EngineerIT,* February pp 37–42.

16. Lee, B. S. 1999. Understanding the stable range of equivalent series resistance of an LDO regulator. *Analog Application Journal* (November): 14–16.

17. *Practical design techniques for power and thermal management.* 1998. Analog Devices Technical Reference Books. Englewood Cliffs, NJ: Prentice Hall.

18. Leung, K. N., P. K. T. Mok, and W. H. Ki. 1999. A novel frequency compensation technique for low-voltage low-dropout regulator. IEEE *International Symposium on Circuits and Systems* 5:102–5.

19. Bruce Carsten Associates, Inc. 2007. Seminar Notes CD-ROM, February.

20. Bull, C., and C. Smith. 2003. Integrated building blocks for dual-output buck converter. *Power Electronic Technology*. Oct 1, 2003, http://powerelectronics.com/mag/power_integrated_building_block/

21. Venable, D. 1997. Testing and stabilizing power supply feedback loops. *PCIM*, Sept, 8.

22. Mannas, S. 2004. Analysis of closed-loop DC-DC converters. *Power Electronics Technology*, October 1 pp. 42–46.

23. Williams, J. 1987. Regulator IC speeds design of switching power supplies. *EDN*, 193.

24. Hesse, K. 2001. Battery powered applications require high performance buck converter ICs. *PCIM*, January 20.

25. Sandler, S. 2006. Gain equalization improves flyback performance. *Power Electronics Technology*, July, 46.

26. Vinsant, R., J. DiFore, and R. Clarke. 1994. Digitally controlled SMPS extends power system capabilities. *PCIM*, June, 30–37.
27. Duvenhage, F. 2004. The role of digital control in power supplies. *Power Electronics Technology*, 74.
28. Dharmawaskita, H. 2002. DC/DC converter controller using a PICmicro microcontroller. Application Note AN216. Microchip Technology, Chandler, AZ.
29. Caldwell, D. 2004. Microcontroller enables digital control in SMPS. *Power Electronics Technology*, February, 30.
30. Hutchings, B. 2006. Achieving high performance, reliable digital power supplies. *Proceedings of the PET Conference*, October 24–26 (PES01-CD ROM).
31. Choudhury, S., and M. Harrison. 2003. DSPs simplify digital control implementation of SMPS. *Power Electronics Technology*, July, 40.
32. Caldwell, D. 2005. Power goes digital. *EDN*, August 18, 75.
33. Kris, B. 2006. DSCs ease migration to digital control. *Power Electronics Technology*, November (supplement), 3.
34. Etter, B., and R. Fosler. 2006. Digital power control enables system identification. *Power Electronics Technology*, November (supplement), 6.
35. Hagen, M., and D. Freeman. 2006. Digital control measures in system response. *Power Electronics Technology*, November (supplement), 12.
36. Pandola, M. 2006. Explore the lesser known benefits of digital power. *Power Electronics Technology*, November (supplement), 16.
37. Bramble, S., and P. Holden. 2006. Digital feedback controls supply voltage accurately. *Power Electronics Technology*, January, 22.
38. Modeling, analysis and compensation of the current mode converter. 1993. Application Note U97. Unitrode Corporation, Merrimack, NH, 9-51–9-57.
39. A 25W off-line flyback switching regulator. 1993. Application Note U-96A. Unitrode Corporation, Merrimack, NH, 9-47–9-51.
40. Dobrenko, D. 2002. Solving the problem of pulse skipping oscillations. *Power Electronics Technology*, October, 28.
41. Vosicher, E. 2000. Hysteretic controller fits processor needs. *PCIM*, January, 28.
42. Nowakowski, R., and L. Hodson. 2000. Hysteretic controller IC enables PC power supply to meet advanced CPU requirements. *PCIM*, July, 74.

6

Power Management

6.1 Introduction

Practically all electronic systems work with DC power rails, which are usually derived from a commercial AC utility source or a limited-energy battery pack. Due to reasons related to political, environmental, and consumer pressures, saving energy, a green environment, electromagnetic compatibility, and safety aspects drive the power supply designer to be very optimistic about the overall structure of a power supply subsystem. With more feature-packed processor-based systems becoming the norm, particularly in portable systems, multiple DC power rails are required. Another important aspect of the modern power management systems is the transient response of the power rails for load currents with fast slew rates, due to common situations such as the processors transitioning between sleep and wake-up modes.

6.2 Design Approaches and Specifications

A low-voltage power supply subsystem must fulfill four essential requirements: isolation from the mains, change of voltage level, conversion to a stable and precise DC value, and energy storage. In the modern world of power-hungry products with mixed power supply rail values where a battery pack or another limited-capacity alternate energy source is used as the primary or the secondary source, a few additional requirements need to be considered. These are energy-saving aspects, quality of the output with respect to fast load transients, electromagnetic compatibility issues, protection and supervisory aspects, packaging aspects, and communication interfaces.

For the power supply, the design team has two basic choices: buy or build. The final decision must consider overall cost and the time to deliver. However, when a system becomes quite complex and requires multiple power rails and critical power management, design and build becomes the choice.

6.2.1 Centralized Power Architecture versus Distributed Power Architecture

Traditional power distribution techniques in a system have relied on a centralized architecture where all the required power rails are derived from a single high-power

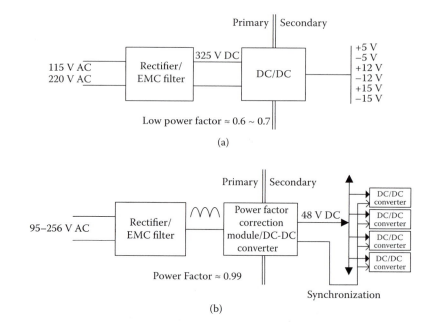

(a)

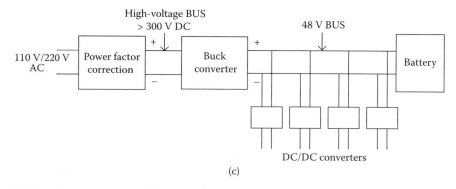

FIGURE 6.1 Comparison of the centralized power architecture and the distributed power architecture (DPA): (a) centralized power architecture; (b) distributed power architecture; (c) a DPA system with 48 V intermediate bus with battery backup.

switch-mode power supply (SMPS), as shown in Figure 6.1(a). These conventional power supplies use an unregulated DC power supply based on a rectifier and smoothing capacitors followed by a DC-DC converter with or without power factor correction (PFC) blocks to provide different DC power rails.

The distributed power architecture (DPA) approach is to distribute power at an intermediate "medium" voltage throughout the system and convert power locally. Distribution becomes more efficient with lower currents, and local power conversion occurs at lower power levels with the many different voltage levels required by individual circuit blocks. DPA systems are typically power factor corrected, have high efficiency (greater than 90%), and consist of a DC bus within the overall system. The selected DC bus voltage is determined by the safety requirements. Safe electrical low voltage (SELV) requires less than 60 V, and a common value for telecom systems is 48 V. Figure 6.1(b) shows an example of a universal AC input (95–265 V AC) power factor-corrected DPA system with multiple DC-DC converter modules to derive individual power rail values. Davis [1] and Curatolo [2] provide an overview of DPA concepts, including system cost aspects; Cassani et al. [3] provide some design details of a system with a 36 V intermediate-bus voltage for DC-DC converter inputs. Figure 6.1(c) shows a DPA system with a 48 V intermediate bus with a battery backup possibility [4]. Smith [5] provides key considerations for DPA systems, and Hemena and Malik [6] provide a case for a personal computer/server situation. Several important technical considerations in the DPA approach are partitioning of the load, determining the intermediate-bus voltage value, end-to-end efficiency considerations (on the basis of AC line to load), and technology selection [2]. When the overall line-to-load efficiency (LTLE) is a serious design consideration, designers should carefully account for the efficiencies of individual power supply blocks [7].

Designing power solutions for line cards in information systems that handle multiple tasks at high speeds is complex. As these boards must process large amounts of data, they incorporate multigigahertz microprocessors, dual-logic application-specific integrated circuits (ASICs), and other high-performance devices. These new-generation devices can require two operating voltages per chip: one for the processor core (about 2.5 V or 1.8 V and rapidly moving downward) and the other for input/output (I/O) devices, which is higher (about 2.5–3.3 V). Until recently, these power supply requirements were addressed using multiple single-output isolated DC-DC converters. However, with rising currents and declining voltages, along with tighter regulation tolerances and faster load current slew rates, multiple isolated converter solutions are not as effective in such applications. These isolated converters are not space efficient and can cause higher ohmic losses along long interconnections. Under such conditions, maintaining high overall efficiency and tight point of load (POL) regulation becomes challenging. To alleviate these issues, now there are integrated building blocks that provide DPA architecture-based solutions for POL requirements. An example from International Rectifier (part number iP1201/iP1202), where an intermediate-bus architecture (IBA) is based on an 8 V rail, is discussed in Bull and Smith [8]. Figure 6.2(a) illustrates the concept in this approach suitable for line cards up to about 150 W power consumption. Figure 6.2(b) illustrates implementation of a POL using iP1202 [8].

6.2.2 Selection of DC-DC Converter Techniques

In the modern world of electronics, the three different basic approaches available for the process of DC-to-DC conversion are the linear approach, the switching approach, and the charge pump approach.

In a practical system, one can mix these three techniques to provide a complex, but elegant, overall solution with energy efficiency, effective silicon or PCB area, and noise and transient performance to suit different parts of an electronic system. Switching-type DC-DC converters—once the clear choice for 5 V systems—suffer

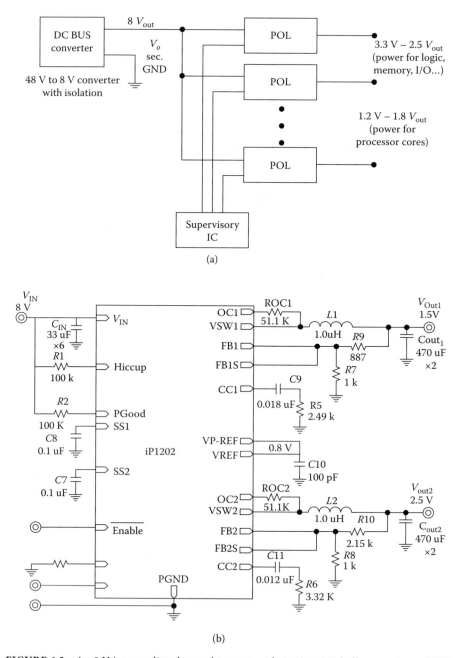

FIGURE 6.2 An 8 V intermediate-bus architecture with 48 V to 8 V bulk converter and POL stages: (a) concept; (b) a complete schematic for a dual-output buck converter-based POL using iP1202. (Source: [8], Courtesy of *Power Electronics Technology*, 1998.)

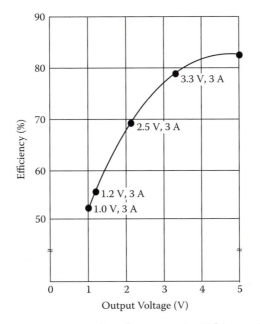

FIGURE 6.3 Efficiency versus output voltage for a compact switching regulator family—a typical example. (Courtesy of *EDN Magazine*.)

lower efficiency at lower voltages. In linear regulators, if the control circuits are designed with a low-power approach, efficiency is approximately given by $(V_{out}/V_{in}) \times 100$. Compared to a high-frequency switching technique-based SMPS solution, LDO regulator ICs (or simply LDOs) are faster in responding to load current changes, produce less noise, and are more compact on a PCB or as an integrated solution for a complete silicon solution. Often, combining a switcher and an LDO makes more sense in electronic systems where the DC rail voltages are less than 5 V or 3.3 V and a combination of many different low voltages are within the same system. This is illustrated in Figure 6.3, which is derived from the characteristic curves for the Power Trends PT6305 integrated switching regulators [9]. The graph indicates that the efficiency drops from approximately 79% for a 3.3 V output model to 56% for a 1.2 V output device. This is mostly due to the rectification losses at very low output voltages. Industry is continuously attempting to move forward with higher efficiencies with the switchers, by employing synchronous rectification and resonant conversion and using nonsilicon power semiconductors.

Charge pumps, switched capacitors, flying capacitors, and inductorless converters are all different names for DC-DC converters that use a set of capacitors rather than an inductor or transformer for energy storage and conversion. For many years, designers have used charge pumps for DC-DC conversion in applications for which the regulation tolerance, conversion efficiency, and noise specifications are not very stringent. As

TABLE 6.1 Comparison of Popular DC-DC Converter Techniques

Feature	LDOs (and) very low drop out regulators (VLDO)	Charge Pump Converters	Switching Regulators
Design complexity	Low	Moderate	Moderate to high
Cost	Low	Moderate	Moderate
Noise	Lowest	Low	Low to moderate
Efficiency	Low to moderate	Moderate to high	High
Thermal management	Poor to moderate	Good	Best
Output current capability	Moderate	Low	High
Requirement of magnetic parts	No	No	Yes
Limitations	Cannot step up	V_{in}/V_{out} ratio	Layout considerations

discussed below, these circuits use capacitors combined with switches to boost or invert the input voltage, and they do not occupy more PCB or silicon area to implement as a single-chip converter [10]. Recent generations of charge pumps have become viable DC-DC conversion methods for cellular phones, portable wireless equipment, notebook computers, and PDAs, where high-density DC-DC conversion is necessary and PCB area is at a premium [10].

In a practical system, the power supply designer has the possibility of combining these techniques. For an effective overall solution taking all specifications and cost into account, combining a large-capacity bulk SMPS in tandem with LDOs and charge pumps becomes a very effective approach. Table 6.1 provides a comparison of the three popular DC-DC converter techniques. In portable application design, the designer should be careful when selecting techniques, and in many situations the method can be a mix of the three techniques discussed.

6.2.3 Switched Capacitor DC-DC Converters (Charge Pump Converters)

Switched capacitor (charge pump) converters use capacitors rather than inductors or transformers to store and transfer energy. The most compelling advantage is the absence of inductors, which have greater component size, more EMI, greater layout sensitivity, and higher cost. Compared with other types of voltage converters, the switched capacitor converter can provide superior performance in applications that process low-level signals or require low-noise operation. These converters offer extremely low operating current—a useful feature in systems where the load current is either uniformly low or low most of the time. Thus for small handheld products, light-load operating currents can be much more important than full-load efficiency in determining battery life. The basic operation of switched capacitor voltage converters is shown in Figure 6.4 [11].

When the switch is in the left position, C_1 charges to V_1 (see Figure 6.4[c]). The total charge on C_1 is given by $q_1 = C_1 V_1$. When the switch moves to the right position, C_1

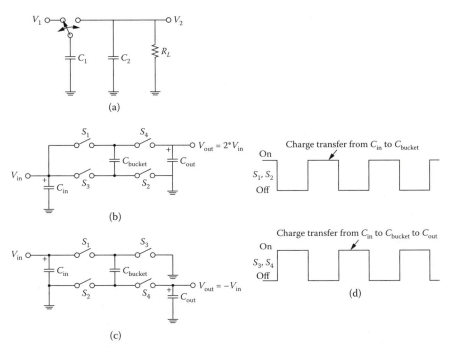

FIGURE 6.4 Switched capacitor converters: (a) basic principle of operation; (b) doubler; (c) inverter; (d) timing.

discharges to V_2. The total charge on C_1 is now given by $q_2 = C_1 V_2$. The total charge transfer is given by $q = q_1 - q_2 = C_1(V_1 - V_2)$.

If the switch is cycled at a frequency f, the charge transfer per second, or the current, is given by

$$I = fC_1(V_1 - V_2) = (V_1 - V_2)/R_{eq} \qquad (6.1)$$

where R_{eq} is given by $1/fC_1$. The reservoir capacitor C_2 holds the output constant. A basic charge pump can work as a doubler or an inverter, as shown in Figures 6.4(b) and 6.4(c), respectively. Figure 6.4(d) shows the switch drive waveforms for the two cases. Some variations of the basic doubler exist for which the output voltage is about 1.33—1.5 times the V_{in} value [10]. In some cases, output voltage can be programmed with an external resistor divider. Charge pumps can be either regulated or unregulated.

After initial charge-up transient conditions and when a steady-state condition is reached, the charge pump capacitor has to send only a small amount of charge to the output capacitor on each switching cycle. The amount of charge transferred depends on the load current and the switching frequency. During the time the bucket (pump) capacitor is charged by the input voltage, the output capacitor, C_{out}, must supply the load current. The load current flowing out of the C_{out} causes a drop in the output voltage that corresponds to a component of the output voltage ripple. Higher switching frequencies allow smaller capacitors for the same amount of droop. There are, however, practical limitations on the switching speeds and switching losses, and switching frequencies are usually limited to few hundred kHz [11].

Recent generations of charge pumps offer improved specifications and have become a viable DC-DC conversion method for many portable appliances where high-density converters are necessary and circuit area is limited. Two common charge pump types are hysteretic and fixed frequency. Figure 6.5a shows the concept of hysteretic control in charge pumps. With this technique, an output voltage that falls below the reference voltage enables the oscillator. During the first clock cycle, the bucket capacitor charges to the input voltage. During the next cycle, the total charge, consisting of C_{bucket} and C_{in}, transfers to the output capacitor. This cycle repeats until the output voltage reaches the upper hysteretic threshold, at which point the comparator disables the oscillator. The internal comparator continues to enable/disable the charge pump switches based on the output level [10]. The example shown is for an SC1517-5.

In a fixed-frequency type with linear regulation, shown in Figure 6.5(b), the internal oscillator runs at a fixed frequency when the device is not shut down. The charge pump provides an unregulated voltage to an internal linear regulator that adjusts this voltage to a fixed output. The device achieves regulation by using an internal comparator that senses the output voltage and compares it with an internal reference while adjusting the gate drive to the internal pass MOSFET for fixed output voltage. The oscillator frequency that controls the charge pump is usually outside the sensitive frequency spectrum of cellular communication bandwidths. Unlike the hysteretic types, this type can restrict any generated switching noise to noncritical bandwidths [10]. For details, see Khorshid [10], Williams and Huffman [11], Arimoto [12,13], and Vitchev [14].

6.2.4 Power Management Concepts

Modern electronic product design requires dealing with different low-voltage DC rails for the longest battery life, thermal management, EMC compliance, and PCB area optimization. Modern multigigahertz-order processors and peripherals generate fast load current transients on their multiple DC rails, and users expect longer run times from batteries. Therefore in designing the DC power supply subsystem, one should consider the following:

- Battery pack or energy source-related aspects
- Load-partitioning aspects to minimize the number of DC rails to be used
- Dealing with power factor, harmonics, and other EMC issues
- Packaging and thermal issues
- Transient protection of the power supply and the product
- Issues related to swapping of modules
- Electrical isolation requirements
- Effects of fast transient loading at the output rails and the stability of the converter blocks

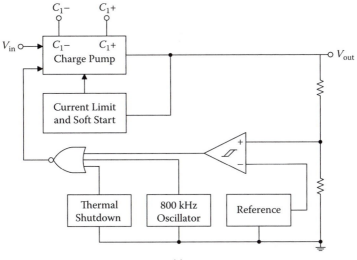

(a)

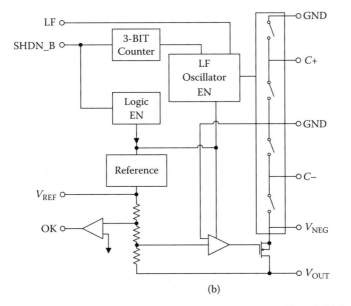

(b)

FIGURE 6.5 Different types of charge pumps: (a) hysteretic control based; (b) fixed-frequency type. (Courtesy of *EDN Magazine*, © 2000, Reed Business Information, a division of Reed Elsevier. All rights reserved. [10])

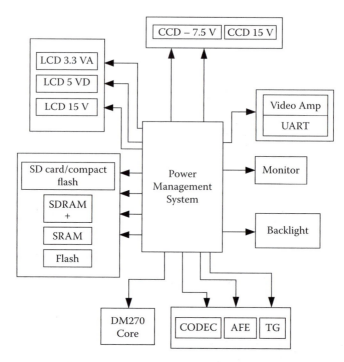

FIGURE 6.6   Digital still camera system components. (Courtesy of *Power Electronics Technology.* [16])

To achieve the above, designers have access to a wide variety of technologies and power management IC families, battery management techniques, architectures, and standards. The important consideration is to deal with all of the above in an integrated manner and in a cost-effective way for a "power management solution" [15]. If the overall system consumes more than 50 W with several DC rail requirements, one has to first carefully analyze the load and have an overall view of its DC rail voltages and the transient behavior of the load. Let's look at a few examples, such as a digital still camera and a cellular phone. Figure 6.6 depicts the system-level block diagram of a digital still camera, based on a TMS320DM270 programmable DSP-based media processor ("DM 270") [16]. The input power source for a case like this will be an Li-ion or Li-polymer rechargeable battery pack. If this particular system operates from an Li-ion battery, the input operating voltage will be from 3.6 to 4.2 V. This operating range is a critical factor in selecting the power supply topologies in the design process. In this kind of product, there are different mixes of blocks, such as the processor, memory, analog front ends (AFEs), video amplifier, motor, liquid crystal display (LCD), CCD module, and backlight for the LCD.

Table 6.2 shows each system component and its power requirements. A few concepts used in such a system include:

- DC rails for low-noise analog circuits need be separated from digital block DC rails.
- System components frequently turned on and off need to be separately powered from the continually running circuit blocks.

TABLE 6.2 Partitioning of the System Loads

System Component	Indication in Figure 3.4	Bus Voltage (V)	Typical Bus Current (mA)	Typical Bus Power (mW)	Efficiency of an LDO-Based DC Supply (%)	Efficiency of a Switcher (%)	Weighted Efficiency of an LDO (%)	Weighted Efficiency of a Switcher (%)
SD card/compact flash	3.3 V I/O	3.3	30					
SDRAM		3.3	125					
SRAM		3.3	140					
Flash		3.3	30					
Total I/O (3.3 V)		3.3	325	1072.5	89.2	93.0	34.03	35.48
DM 270 core	1.5 V_Core	1.5	185	277.5	40.5	91.0	4.00	8.98
DM 270 Codec	3.3 VA	3.3	5					
Timing generator		3.3	80					
Analog front end		3.3	15					
Total analog circuits (3.3 V)		3.3	100	330	89.2	93.0	10.47	10.92
LCD	3.3 VA_LCD	3.3	20	66	89.2	93.0	2.09	2.18
LCD	5 V LCD	5	20	100	n/a	94.0	n/a	3.34
LCD	15 V_LCD	15	6	90	n/a	85.0	n/a	2.72
CCD	15 V_CCD	15	6	90	n/a	85.0	n/a	2.72
CCD	7.5 VN_CCD	-7.5	-6	45	n/a	80.0	n/a	1.28
Video amp	5 VA	5	3					
UART		5	10					
Total analog (5 V rail based)		5	13	65	n/a	94.0	n/a	2.17
Lens motor	<2.7 V	2.7	250	675	73	95.0	17.52	22.81

Note: Adapted from Ref. [16].

- 3.3 V rails (a total of eight) can be grouped into three separate sets of 3.3 V rails.
- Secure digital (SD) card memory, synchronous dynamic random access memory (SDRAM), static random access memory (SRAM), and flash memory on the DM 270 can all be powered by the same bus.
- The codec, analog front end (AFE) and video timing generator require low-noise LDOs.
- The 3.3 V rail for the LCD needs to be separated because it goes on and off during use.

Appropriate supply topology selection depends on the input voltage, output voltage, noise, efficiency, cost, and space. The last three items usually compete with each other. The first two restrict the choice of the topology. If the output voltage is always lower than the input voltage, a buck converter or an LDO will work. Otherwise, a buck-boost or a single-ended primary inductance converter (SEPIC) topology will work. A typical Li-ion battery's voltage profile varies from a high value of 4.2 V (at full charge) to 3.0 V when fully discharged. Between these limits, the battery maintains approximately 3.7 V. All selections need to be based on these values. In terms of efficiency, it is important to consider the overall efficiency for prolonged battery life, and for this reason a special approach is to use the *weighted efficiency* for each power rail, as indicated in Table 6.2. Weighted efficiency is calculated by multiplying each efficiency value by the typical bus power divided by total estimated power ratio. In real-world operation, the example here can have different modes of operation, and for each mode one has to develop a table and analyze and estimate the best options. Then it is necessary to have an estimate of the percentage of the time the camera spends in each mode. For lower-noise considerations, even if an LDO solution is considered (in lieu of a switcher), weighted efficiency indicates otherwise for items such as the DM270_Core. For the 5 V analog bus for the video amp, low-noise operation may be mandatory, but at lower battery voltages such as 3.5 V or lower, an LDO will cease to operate. In this situation, a switcher is to be selected and then followed up with an LDO. After defining the power supply requirements, the designer can start choosing individual ICs required for the system. Figure 6.7 indicates part of the solution using a Texas Instruments TPS 65010 power management IC, which has several switcher outputs and two LDOs. In addition, a TPS 61120 IC will be required. For details, see Day [16] and the data sheets of the relevant ICs [17,18].

Let's briefly discuss the case of powering a modern cellular phone. By 2007, most cellular phones allowed many features in addition to voice communication. A trend was for cellular phones to act as MP3 players or to add a micro hard disk or a very large amount of silicon memory. Figure 6.8 shows a generic block diagram of a 2005-generation cellular phone. In much older generation phones the main power rail was 3.3 V, but the newer-generation chipsets use a 1.5 V main power rail because the majority of large-scale integrated circuits (LSICs) operate with voltage rails of 1.5 V or less. Examples are baseband chipsets running from 1.375 V rails and video-processing DSP chips running from 1.2 V rails. In these situations it is possible to mix very low dropout (VLDO) linear regulators with conversion efficiencies from 80% to 90% [19]. With the common acceptance of 600 mAh Li-based cell phone battery packs, and dealing with packaging problems, thermal

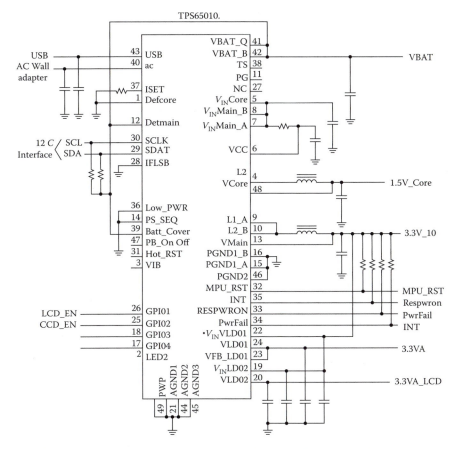

FIGURE 6.7 Part of a power management (PM) solution for the DSC from Texas Instruments [Source: [16]].

management, and noise issues, power management of the product becomes quite critical and the designer has to make a well-informed and critically analyzed approach [19]. A simple example is the case of lower than 50% efficiency of LDO solutions for converting 3.3 V to 1.8 V compared to the situation of modern-generation power rail requirements from 1.5 V to 1.2 V, which can be supported by an LDO solution with efficiencies of about 80%. For more discussion on the practical design considerations for portable wireless products, see Armstrong [19] and Maxim Integrated Products [20].

6.3 Specifying DC Power Supply Requirements

Let's start with the simple fact that the DC supply subsystem is expected to provide a constant set of output voltages at a maximum set of load currents. Given this requirement to be derived from an AC input rail such as 230 V/50 Hz or 120 V/60 Hz or an energy source such as a battery or a fuel cell, the designer has to start the list of specifications with input voltage, output voltages, and respective load currents. Then we can add

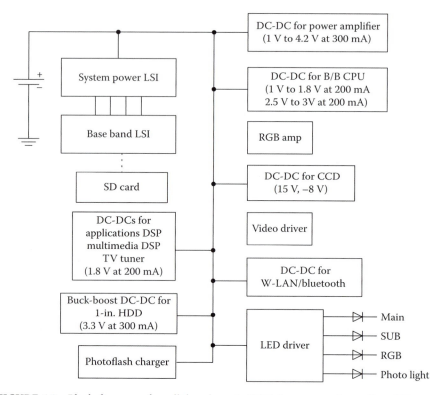

FIGURE 6.8 Block diagram of a cellular phone in 2005 shows many low-voltage DC supply requirements. (Source: [19], Courtesy of *Power Electronics Technology.*)

as much secondary information as possible. The more requirement specifications we list, the easier it is to narrow down the available options.

Design specifications act as the performance goal that the ultimate power supply must meet in order for the product to meet its overall performance specification. When developing the specification, the power supply designer must keep in mind what is a reasonable requirement and what is an idealistic requirement. Most specification-related parameters are measurable using common test setups under different environmental conditions. These specifications can be grouped into several subsets, as shown in Table 6.3.

In developing these specifications, the designer should have a clear idea of the load requirements and steady-state and transient behaviors. In a very simple case where load consumes a few watts to about 50 W in a single- or dual-rail requirement, the load can be supplied by a simple linear or switching supply where only a few of the above items need to be specified. Many complex loads require multiple rails, power management and green design concepts, transient loading conditions, tight space or weight requirements, and cost restrictions, and the designer may have to start with the generalized concept of a power supply, as in Figure 6.9. In the case of an AC-powered situation, concepts in

TABLE 6.3 Power Supply Specifications

Subset	Item	Remarks
Input specifications	Nominal input voltage	Product is expected to work around this voltage most of the time.
	Range of input voltages	Product is expected to withstand this range of fluctuations.
	Frequency (for AC input systems) or total energy available from a battery pack (in mAh or Ah)	In the case of a battery pack, an off-the-line charger may be designed for the input frequency.
	In-rush current	Important for the start-up conditions.
	Voltage transients	Important for the reliability of the power supply and the load for reliable operation.
	Permissible harmonics or power factor	Governed by various standardization bodies.
	Fusing	Speed of fusing is based on the I^2t rating of the device.
Output specifications	Nominal output voltage	The load is expected to operate at this voltage.
	Average and peak currents	RMS values to be used.
	Turn-on delay	• Capacitor/inductor energy storage dependent at the time of initial switch-on. • In multiple-rail output situations, carefully timed sequencing may be necessary.
	Stability over a specified period	Based on the age of the components.
Regulation specifications	Load regulation	• Variation of the output voltage versus current. • Specified as a percentage or graphically shown for different input voltages.
	Line regulation	• Variation of the output voltage in response to changes in the input line. • Specified as a percentage or graphically shown for different load currents.
	Hold-up time	Amount of time the output remains within usable limits when the input source is disconnected temporarily.
	Output voltage temperature coefficient	The stability of the output voltage rails. Dependent primarily on the reference source stability and the temperature tolerances of the output sampling chain resistors.

(*continued*)

TABLE 6.3 (CONTINUED)   Power Supply Specifications

Subset	Item	Remarks
Regulation specifications (Contd.)	Transient specifications • Overshoot, undershoot, and settling time • Step response of the output for sudden changes of load current	• Overshoot, undershoot, etc., are dependent on the control loop behavior. • Step response is very important in dealing with complex high-current processor loads.
	Output impedance	Represents the Thevenin equivalent resistance of the power supply output.
	Ripple and noise limits	These specifications may be very significant for the reliable operation and accuracy of analog- and mixed-signal circuitry.
Protection conditions	Over- and undervoltage limits at the output	These specifications indicate the safe operation limits of the load.
	Output current limit	Maximum load current expected.
	Thermal limits	To avoid excessive temperature conditions within the product or the power supply.
Power conversion specifications	Overall efficiency	In battery-powered products and green designs, this is a critical specification.
	Thermal dissipation	Determines the need for cooling and packaging limitations.
Safety and regulatory agency specifications	Isolation requirements • Dielectric withstand voltage • Insulation resistance	• For the safety of the user, a power supply should have galvanic isolation between the AC power input side and the load side. • Insulation resistance is usually specified for the transformers involved in the design.
	RFI/EMI requirements • Conducted EMI • Radiated RFI	• Conducted EMI specifies the line-filtering requirement. • Radiated RFI affects the physical layout and enclosures.
Power management requirements	Energy conservation and green design aspects	Particularly important in high-power loads.
	Sequencing and resetting of the output rails	Critically important specification in multiple-rail situations.

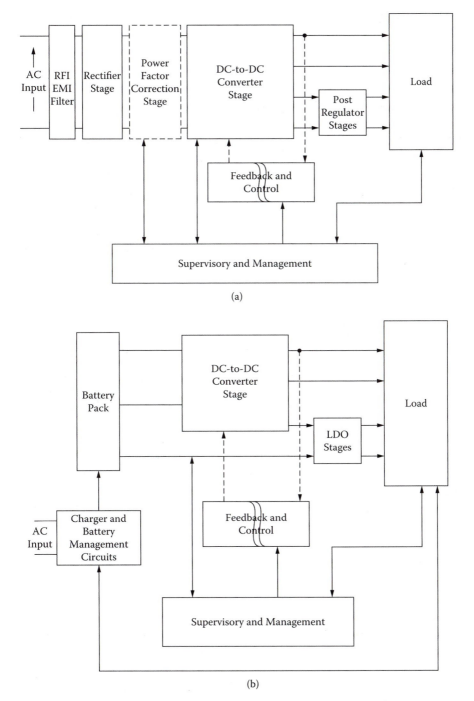

FIGURE 6.9 Overall design approach to a complete power supply subsystem: (a) based on an AC input source; (b) based on a battery pack with a charging subsystem.

Figure 6.9(a) apply, whereas for battery-powered products the concepts in Figure 6.9(b) apply. Brown [21] and Rubadue [22] provide useful details on specifications and design concepts. A discussion of battery management for longest run time and standby time is presented in Chapter 8.

6.4 Loading Considerations

Load connected to a power supply can be as simple as a single-rail requirement, which can be easily met by a simple linear or switching power supply. In extreme cases, the load may consist of several complex processors or other mixed-signal loads that may require multiple power rails, specialized power management aspects, and ultra-low-voltage DC rails that consume 100 A or more. Some processor loads may demand digitally controlled adjustable power rails for effective power management. In communication subsystems, the load may demand extra low-noise and low-voltage power rails. Designers should have an initial estimate of the load requirements and an idea of the nature of the load in general. The key considerations include:

- The number of voltage rails and regulation aspects
- The nature of the load transients and stability considerations
- Efficiency and power management aspects
- Protection requirements
- Noise and EMC considerations and regulatory requirements

For the simple cases of single- or dual-rail power requirements, there is a choice of three-terminal linear regulator chips; these are low cost and easy to implement, with excellent noise and drift characteristics. The most useful property is their speed of response to transient loads. The only major disadvantage of these solutions is their low efficiencies, which in general range between 30% and 50%. There are ways to improve the efficiencies of linear regulators by manipulating the rectifier circuits in the input stages using silicon-controlled rectifiers [23]. Many common loads can tolerate slower responses and greater amounts of high-frequency noise. For such simple requirements, there are switching regulator solutions where the equivalent of a three-terminal linear IC solution is provided by integrated switching regulator (ISR) techniques by companies such as Power Trends. ISRs are able to provide buck, boost, or inverting voltage values from a single DC bus supply such as 5 V [24]. Figure 6.10 indicates a DPA solution based on an intermediate-bus architecture (IBA) of 5 V. Another fully packaged switching solution for high-current-capability DC rails is the "brick converter," where a wide range of voltages (0.9 V to 48 V) is possible at currents up to a few tens of amperes [25, 26]. Figure 6.11 shows the relative sizes of quarter, eighth, and sixteenth brick sizes.

Further to examples given in Section 6.2, powering portable devices such as palm computers, pose different issues. Some of these are powered by a few AA cells from which different voltages need to be generated. The typical power source for a palm computer is a disposable alkaline cell, and using such a cell creates other design challenges, such as generating higher voltages (such as a 5 V rail from two alkaline cells of 3 V) efficiently, generating LCD bias generators (typically –24 V), and generating miscellaneous

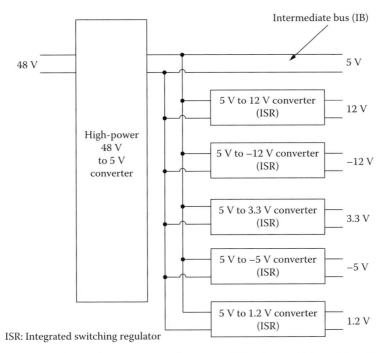

FIGURE 6.10 ISR devices from Power Trends, Inc. in a DPA solution. (Adapted from Narveson, B., *Electronic Design*, January 6, 1997, 137.)

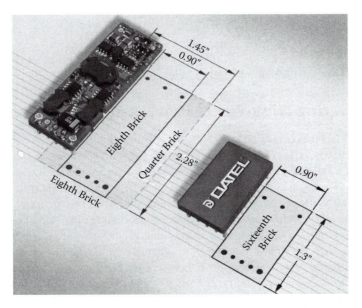

FIGURE 6.11 Brick converter examples. (Courtesy of *Power Electronics Technology*.)

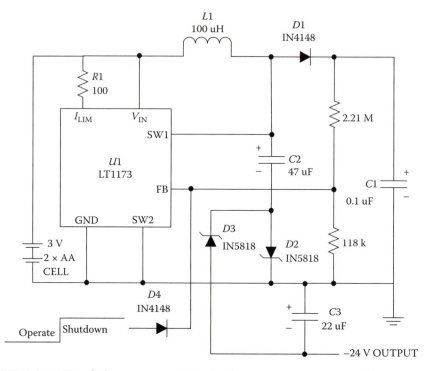

FIGURE 6.12 Use of a boost converter IC and a charge pump to convert 3 V from two alkaline cells to generate a +24 V and −24 V LCD bias supply. (Reprinted with permission of *EDN Magazine*, © 1993, Reed Business Information, a division of Reed Elsevier. All rights reserved.)

lower voltages such as 3.6 V, 2.4 V, and other values [27]. In such cases, boost converters, such as the LT 1173 from Linear Technology, with charge pump configurations (to invert the +24 V to −24 V) can be used [27]. See Figure 6.12.

6.5 Powering High-Power Processors and ASICs

Advanced microprocessors and ASICs are power hungry and can consume as much as 100 A from power rails of 1.0 V or less. When the Pentium™ range processors entered the market in the mid-1990s (with only a few hundred megahertz clock speeds), their power consumption was a few amps to more than 10 A from voltage rails of 1.8 V to 3.3 V. In CMOS digital circuits, the power consumption is proportional to V^2, and this fact encourages chip designers to develop processors that operate with lower rail voltages [28]. Two other important facts about high-end digital components are that they require multiple-rail voltages for efficiency and speed, and the load currents can have slew rates easily up to 100 A/μs. To achieve these rates, most modern high-power processors have digital command signals based on four- to five-bit code to command a voltage regulator module (VRM) to output different voltages to power the processor. This concept is shown in Figure 6.13, where the processor outputs a

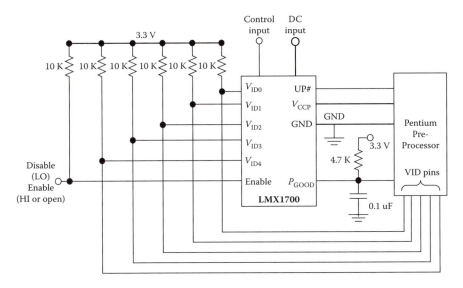

FIGURE 6.13 A voltage regulator module under the VID command signals (four or five bits) from the processor.

four- or five-bit code via a special set of pins (called voltage identification [VID] pins) that command a VRM to change the voltage output from about 0.8 V to about 3.5 V in steps of 100 mV or 50 mV. For more details, see Mannion [29]. The VRM is capable of adjusting its own output voltage under the command code bits from the processor within the range of values specified by the processor. An early example of this is the LXM1700 from LinFinity [29]. The example in Figure 6.13 is for a five-bit VID pin code from the processor. With the load demands of extremely high current slew rates, the VRM should be capable of responding quickly with low-ESR capacitors at the output. More of these design aspects are discussed in later sections. The PCB track inductances can jeopardize the required slew rate; thus designers must pay special attention to PCB layout.

To power Intel, AMD, and other high-end processors, the concept of VRMs has created a special set of power modules coming under the VRM specifications series. For details, see Brown [30], Gentchev [31], and Wong et al. [32]. Most of the VRMs used on processor boards are powered by the 5 V rail of the PC power supply ("silver box").

In most processor-based systems, power rails have typical value combinations such as 1.8 V, 2.5 V, or 3.3 V, and these rails need to be properly sequenced for the reliable operation of a system. There are different possibilities of sequencing depending on the application, as shown in Figure 6.14. Common themes are sequential power-up, ratiometric method, and output tracking (or simultaneous or coincidental power-up).

In sequential power-up (Figure 6.14a), the system turns on the core voltage, and when it reaches the voltage set point it turns on the second I/O rail. This technique can be used to delay the start-up of the second rail at some predetermined time after the first rail is turned on. In this technique, interlocks can be introduced, as shown in

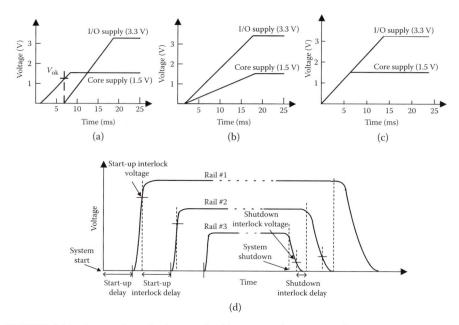

FIGURE 6.14 Sequencing of voltage rails: (a) sequential power-up; (b) ratiometric method; (c) output tracking (simultaneous or coincidental); (d) sequential tracking with interlocks.

Figure 6.14(d). Switching off can be based on the same principle in the reverse order [33]. In Figure 6.14(b), the ratiometric technique is shown. The two rails are turned on simultaneously, reaching regulation at their respective set points at the same time. In this method, the two rails are controlled with different slew rates. Figure 6.14(c) shows the output tracking or simultaneous power-up scheme where the two rails reach their output voltages at the same rate, and once the core voltage is reached, it remains constant, while the I/O rails reach the 3.3 V value [34].

6.5.1 Advance Configuration and Power Interface (ACPI) Specification

In the early days of personal computing, there was no power management. Over the history of 40 years in single-chip microprocessors, power consumption has grown exponentially as evidenced by details in Table 6.4, based on Intel processors. Around 1989, Intel shipped processors with technology to allow the central processing unit (CPU) to slow down, suspend, or shut down part or all of the system platform, or even the CPU itself, to preserve and extend battery life. In 1992, an early version of PC power management techniques based on the advanced power management (APM) specification was introduced. With the need for energy saving simultaneously with the instantly available PC (IAPC), an advance version of the APM was introduced, called the advanced configuration and power interface (ACPI). The history and technical details related to this process are available in Kolinski et al. [35].

TABLE 6.4   Development of Microprocessors by Intel

Processor	Year	Power Consumption (Watts)	Clock Frequency (MHz)	Number of Transistors	Feature Size of Transistors (Microns (μm))
4004	1971	0.5	0.75	2.3k	10
8008	1972	0.5	0.5-0.8	3.5k	10
8080	1974	0.5	2	6k	6
8086	1978	2	5–10	29k	3
80286	1982	3	6–12	134k	1.5
Intel 386	1985	1–1.5	16–25	275k	1.5-1.0
Intel 486	1989	0.3 –2.5	25–100	1.2M	1-0.6
Pentium	1993	8–17	60–300	3.2–4.5M	0.8-0.35
Pentium Pro	1995	29–47	166–200	5.5M	0.6-0.35
Pentium II	1997	17–43	233–450	7.5M	0.35-0.25
Pentium III	1999	14–44	450–1000	9.5–28M	0.25-0.18
Pentium IV	2000	21–115	1400–3800	42–178M	0.18-0.065
Pentium M	2003	5–27	1300–2130	77–140M	0.13-0.090
Core	2006	6–31	1000–1860	152M	.065
Core 2 Duo	2006	10–65	1060–3160	167–410M	.065-.045
Core i7	2008	45–130	2660–3330	731M	.045
Atom	2008	1.4–13	800–1860	47M	.045

The ACPI specification provides a platform-independent, industry-standard approach to operating system-based power management. The ACPI is the key constituent in operating system power management (OSPM). OSPM and ACPI apply to all classes of computers, including handheld, notebook, desktop, and server machines. In ACPI-enabled systems, the basic input/output system (BIOS), hardware, and power architecture must use a standard approach that enables the operating system to manage the entire system in all operational situations. From a computer power system designer's viewpoint, ACPI power management means generating and managing a multitude of voltages on the motherboard and riser cards with no user intervention to enable the processing of audio, video, and data streams. ACPI-compliant computers require the generation of these multiple voltages at various current ratings as the system transitions between sleep states. The ACPI defines six possible discrete system operating states, which are referred to as S0 to S5, in order of highest to lowest power consumption. For more details on the ACPI and ACPI power controllers, see Kolinski et al. [35] and Lakkas and Duduman [36].

6.5.2 Multiple-Output Power Supplies versus Point of Load (POL) Approach

Traditionally, the approach for a multiple-rail output requirement was a multirail switching supply based on a multiple-winding high-frequency transformer. A classic example of this approach is in the desktop computer power supply (known as the "silver box"), where a transformer-isolated forward-mode topology is used. In modern devices such as

TABLE 6.5   Comparison of Multiple Output Flyback Designs versus a POL Architecture

Feature/ Performance	Multiple-Output Flyback Design	POL Architecture
Complexity/Component count	Low	Medium to high
Flexibility and adaptability	Rigid	Easy to adapt to changing requirements
Performance (general)	Low	Medium to high
Accuracy	4–7%	1–5%
Transient response	Slow	Fast
Efficiency	70–77%	80–90%
Sequencing	Difficult	Easy
Cost	Low	Low to medium
Energy Star compliance	Difficult	Easy
Regulatory approval	Long	Short
Time to market	Slow	Fast

set-top boxes, many higher-voltage rails are also required, in addition to the usual 3.3 V, 2.5 V, 1.8 V, 1.5 V, and other lower-DC rails [37]. Traditionally such requirements can be fulfilled by utilizing an off-the-line flyback power supply. However, in powering set-top boxes and similar systems, flexibility, Energy Star™ compliance, power sequencing, and other attributes can be easily approached by a point of load (POL) regulation, compared to a traditional multirail output approach based on flyback design. Table 6.5 compares these two approaches from a designer's viewpoint.

A POL solution for a system such as a set-top box is easily developed starting with an off-line converter, an intermediate-bus voltage (IBV) such as 12 V, in addition to a core supply of 3.3 V or similar. Then using the IBV the designer can combine efficient synchronous converters, ordinary switchers, LDOs, inverters (positive to negative rail converters), or charge pump type converters. A typical example is shown in Figure 6.15. More details can be found in [37].

Using the concepts discussed above, many new single-chip power management solutions have started coming into the marketplace. Some examples are discussed in [38, 39].

6.5.3 External Power Supplies and New Energy Standards

For certain types of consumer products, by using an external or a wall plug-in supply, designers can gain board space, eliminate a major source of heat, and reduce EMI noise. In addition, this may even provide the option of freeing the designer from the burden of getting safety approvals for the product. Until the late 1980s, wall plug-in supplies were limited to less than 25 W, but as of 2005 this capability has grown to about 250 W. Most low-power, older versions were linear regulator based, but recent energy-saving and related standards worldwide have pushed these products to adopt switch-mode designs. It is important for designers to note the recent standards released by the California Energy Commission (CEC), Environmental Protection Agency (EPA), and European Union (EU) regarding efficiency requirements and the nameplate output power ratings, as summarized in Figure 6.16. For more details on external power

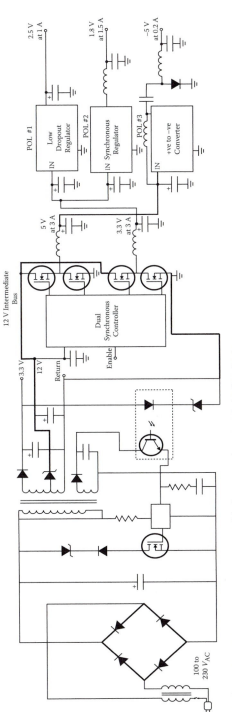

FIGURE 6.15 POL design approach suitable for set-top boxes and similar devices.

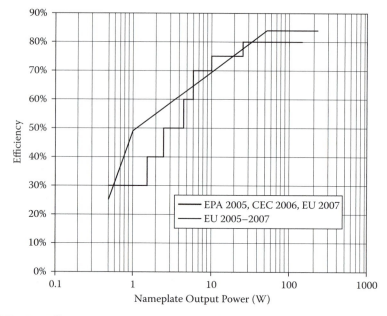

FIGURE 6.16   Efficiency requirements for external power supply curves based on CEC, EPA, and EU recommendations. Courtesy of *Power Electronics Technology*.

supplies, international power converter requirements, and recent energy-saving standards, see Everett [40], Forrester [41], and Jovalusky [42].

In the latter part of 2009, regulators in the United States and EU have been busy proposing new minimum efficiency standards for off-the-line power supplies. This mostly covers external power supplies such as mobile phone chargers. Around 70% of minimum acceptable efficiency during device operation, as well as 250–300 mW of no-load dissipation, expected by these regulatory bodies is currently met by the industry. By 2011 the European Commission is expected to lower the no-load dissipation down to 150 mW. One difficult target may be the 30 mW no-load consumption limit agreed upon by LG Electronics, Motorola, Nokia, Samsung, and Sony-Ericsson. Table 6.6 summarizes the future scenario of efficiency targets based on these "star-ratings." Five stars mean a 30 mW no-load consumption, the hard target to achieve in the near future.

TABLE 6.6   New Star Ratings and the No-Load Consumption of Chargers

No Load Consumption Limits	No of Stars
Less than 30 mW	*****
30 mW to 150 mW	****
150 mW to 250 mW	***
250 mW to 350 mW	**
350 mW to 0.5 W	*
≥.5W	No stars

Power management chip design companies are currently spending heavy on R&D efforts to achieve these regulatory requirements by introducing new ICs and measurement techniques. One example is in [43]. Another example of optimizing magnetic for the AC-DC adaptors with Energy Star is detailed in [44].

6.5.4 Load Switches and Switching Off Parts of the Power Supply Rails

With manufacturers and consumers both trying to be conscious of energy conservation, particularly for standby power (sometimes called the vampire power), switching unused parts of the circuits has become very common in portable products. In order to achieve this under processor control, various load switches have been introduced by power component manufacturers. Some discussion on this is in [45].

6.5.5 PMBus and Digital Control Protocol for Power Converters

Over the last 10 years many board designers have moved to intermediate-bus architecture (IBA) discussed in the early parts of the chapter. This is due to processor-based products, particularly the portables, as well as large systems such as data center power units that have multiple on-board DC-DC converters to generate the diversity of power rails needed by different silicon devices. With many power supply manufacturers and IC houses launching digitally programmable POL converters, a common control bus and an open-standard architecture for power converter control was required. As a result power management bus (PMBus) architecture was proposed in the early part of the new millennium, with a fully defined command language and a transport and a physical interface. The protocol was founded by a coalition of power supply and semiconductor manufacturers, and in 2005 revision 1.0 of the protocol was placed in the public domain. Development of the protocol and the promotion of the standard was transferred to a special interest group known as the System Management Interface Forum.

PMBus is not a standard for AC-DC power supplies or DC-DC converters. It only facilitates communication with a power converter or another device. The PMBus transport layer is based on the Version 1.1 of the system management bus (SMBus) (discussed briefly in Chapter 8), which is based on the industry favorite I^2C bus. A typical bus implementation is shown in Figure 6.17(a). It uses the wired-AND connection of all devices on the bus to provide arbitration in the event of bus contention, and it is electrically similar to the I^2c bus. Its master device is not based on proprietary silicon.

PMBus communications are based on a simple command set as depicted in the summary of Figure 6.17(b). Every packet contains an address byte; followed by a command byte; zero, one, or more data bytes; and an optional packet error code (PEC). More details are in [46]. Figure 6.17(c) depicts the flexibility of the PMBus specification, in power supply control parameters. It summarizes the conceptual view of how these commands are used. PMBus usage in power converter units in data centers allows better efficiency and power management for optimal control of energy usage [47].

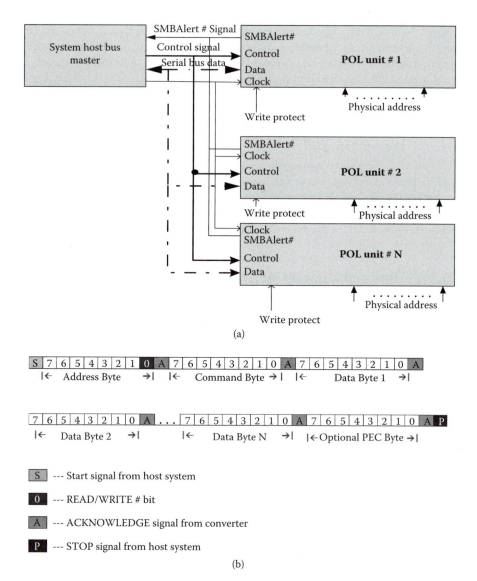

(a)

(b)

S --- Start signal from host system

0 --- READ/WRITE # bit

A --- ACKNOWLEDGE signal from converter

P --- STOP signal from host system

FIGURE 6.17 PMBus protocol and connections: (a) PMBus bus-master connections to POL converters; (b) typical host to salve information transfer; (c) command settings allowed. (Reproduced from White, B., *Power Electronics Technology*, September 2005, 14–19. With permission.)

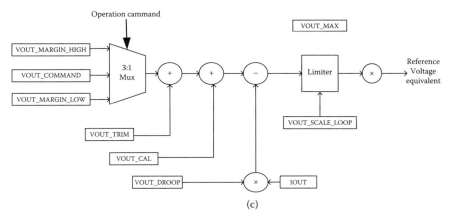

(c)

FIGURE 6.17   (Continued)

References

1. Davis, S. 1989. High frequency power conversion in the next decade. *PCIM*, April, 18.
2. Curatolo, T. 2000. High density power components add flexibility to distributed power design, *Electronic Design*, June 12, 129.
3. Cassani, J. C., J. J. Hurd, D. R. Thoms, R. G. Hodgins, and H. A. Wittlinger. 1993. 80 W, 1 MHz supply employs current controlled PWM power IC. *PCIM*, January, 7.
4. Small, C. H. 1994. Distributed power takes center stage, *EDN*, April 28, 54.
5. Smith, C. 2000. The top 10 keys to successful distributed power architecture design. *PCIM*, April, 120.
6. Hemena, W., and R. Malik. 2000. A distributed power architecture for PC industry. *Proceedings of the PCIM Conference*, October 1.
7. Alderman, A. 2003. Distribution vs. efficiency: Power architect's dilemma. *Power Electronics Technology*, October, 42.
8. Bull, C., and C. Smith. 2003. Integrated building blocks for dual-output buck converter. *Power Electronics Technology*, October, 68.
9. Travis, B. 1998. Linear vs. switching supplies: Weighing all the options. *EDN*, January 1, 40.
10. Khorshid, O. 2000. Selecting charge-pump DC/DC converters. *EDN*, August 17, 115.
11. Williams, J. & Huffman, B. 1988, Switched capacitor networks simplify DC-DC converter design, *EDN*, November 24, 171.
12. Arimoto, K. 2000. Efficient power ICs help mobile products harness battery power. *AEU* (Japan), July, 73.
13. Arimoto, K. 2001. Charge pumps shine in portable designs. Application Note 669. Maxim Integrated Products, Sunnyvale, CA.
14. Vitchev, V. 2006. Calculating essential charge pump parameters. *Power Electronics Technology*, July, 30.
15. Lam, C. 2006. Power-management techniques for multimedia mobile phones. *EDN*, April 13, 55.

16. Day, M. 2003. Integration saves time and board space. *Power Electronics Technology*, October, 64.

17. TPS65010 data sheet. Texas Instruments, Dallas, TX. Available at http://www. datasheetcatalog.com/datasheets_pdf/T/P/S/6/TPS65010.shtml.

18. TPS61120 data sheet. Texas Instruments, Dallas, TX. Available at http://www. datasheetcatalog.com/datasheets_pdf/T/P/S/6/TPS61120.shtml.

19. Armstrong, T. 2005. Wireless products signal new uses for VLDOs. *Power Electronics Technology*, July, 38.

20. Maxim Integrated Products. 2001. Linear regulators in portable applications. Application Note 751, May 24. Available at http://www.maxim-ic.com/appnotes. cfm/appnote_number/751.

21. Brown, M. 2001. *Power supply cookbook*. London: Newnes.

22. Rubadue, J. 2003. Powering your core voltage. *EDN*, September 27, 69.

23. Williams, J. 1989. Astute designs improve efficiencies of linear regulators. *EDN*, August 17, 151.

24. Narveson, B. 1997. How many isolated DC-DC converters do you really need? *Electronic Design*, January 6, 137.

25. Strassberg, D. 2002. Tiny titans: Choose 'em and use 'em with care. *EDN*, May 2, 41, 2002.

26. Bindra, A. 2003. Eighth brick gains momentum, sixteenth gets proposed. *Power Electronics Technology*, March, 56.

27. Pietkiewicz, S. 1993. Specialized circuits condition power for Palm top computers. *EDN Asia*, August, 55.

28. Goodenough, F. 1997. Advanced microprocessors demand amperes of current at <2V. *Electronic Design*, January 20, 44.

29. Mannion, P. 1997. VRMs: Technological first aid, but for how long? *Electronic Design*, January 20, 31.

30. Brown, S. 1997. Microprocessor controls its dedicated voltage regulator module. *PCIM*, April, 66.

31. Gentchev, A. 2000. Designing high-current VRM compliant CPU power supplies. *EDN*, October 26, 155.

32. Wong, P.-L., F. C. Lee, P. Xu, and K. Yao. 2002. Critical inductance in voltage regulator modules. *IEEE Transactions on Power Electronics* 17:485.

33. Cooper, D. 2004. Power management for high performance. *Power Electronics Technology*, July, 33.

34. Thornton, C. 2003. Auto-Track™ feature simplifies simultaneous power supply voltage sequencing. Analog Application Brief No. 6. Texas Instruments, Dallas, TX.

35. Kolinski, J., R. Chary, A. Henroid, and B. Press. 2001. *Building the power-efficient PC*. Santa Clara, CA: Intel Press.

36. Lakkas, G., and B. Duduman. 2001. The ACPI advantage for powering future generations of computers. *EDN*, September, 91.

37. Perzow, J. 2002. Point-of-load- regulation adds flexibility to set-top-box design. *EDN*, June 27, 73–78.

38. Morrison, D. 2004. Power architecture tackles power management complexity. *Power Electronics Technology*, February, 62.

39. Davis, S. 2010. Mobile electronic products employ integrated power management. *Power Electronics Technology*, September, 31–36.
40. Everett, C. 1987. External power supplies eliminate more than just heat from your design. *EDN*, April 15, 107.
41. Forrester, S. 1995. International power converter requirements. *PCIM*, December, 8.
42. Jovalusky, J. 2005. New energy standards banish linear supplies. *Power Electronics Technology*, March, 42.
43. Mugee, M. 2009. Achieving five-star energy efficiency in cell phone chargers. *Power Electronics Technology*, September, 10–13.
44. Sun, J. 2010. Optimizing power supply adaptor design. *Power Electronics Technology*, July, 27–35.
45. Davis, S. 2010. New generation load switch ICs cut standby power. *Power Electronics Technology*, April, 45–46.
46. White, B. 2005. PMBus offers open-standard digital power management. *Power Electronics Technology*, September, 14–19.
47. Griffith, B. 2008. PMBus takes command of data center power issues. *Power Electronics Technology*, April, 14–18.

7

Off-the-Line Switching
Power Supplies

7.1 Introduction

The basic theory of the switching power supply has been known since the 1930s. Practical switch-mode power supplies (SMPSs) have existed since the 1960s, starting with components designed for other uses and thus poorly characterized for SMPS use. Early SMPS systems were running at very low frequencies, usually from a few KHz to about 20 KHz. Early SMPS designs justified the SMPS techniques for higher output power ratings, usually above 500 W.

Toward the mid-1980s most SMPS units were available for much lower power ratings with multiple output rails, while oscillator frequencies were reaching the upper limit of 100 KHz. Today most switchers operate well above 500 KHz, and newer techniques allow the systems to run at several MHz, making use of new magnetics, surface mount components, and the resonant conversion techniques. The rapid advancement of microelectronics in the last two decades (1975–1995) has created a necessity for the development of sophisticated, efficient, lightweight power supplies that have a high power-to-volume (W/in^3) ratio with no compromise in performance. The high-frequency switching power supply meets these demands. Recently it has become the prime powering source in the majority of modern electronic systems. The trends associated with the switch-mode power supplies for the electronic products and systems are (a) reaching direct off-the-line design approach, (b) utilizing higher frequencies, (c) increasing output rating/volume, (d) minimizing the components, and (e) increasing reliability.

7.2 Building Blocks of a Typical Off-the-Line Switching Power Supply

Figure 6.9(a) depicts the basic block diagram of a typical off-the line SMPS. This block diagram depicts the AC line RFI filter block, the supervisory and power management blocks, and the requirement of I/O isolation in the overall system. (Note the isolation indicated in the feedback and control block, which is essential for this purpose.) The EMI/RFI filter can be either part of the power supply or external to it, and it is generally designed to comply with national or international specifications, such as the FCC class A or class B and VDE-0871. Within the past two decades, because of the emphasis placed on power quality issues, power factor correction (PFC) as applicable to the nonlinear behavior of the input current waveform of a rectifier has become an important issue. For power supplies with output capacities greater than 75 W, PFC is mandatory in some jurisdictions. Input/output isolation is essential to off-the-line switchers. The isolation used within different stages may be optical or magnetic, and it should be designed to comply with Underwriters Laboratories (UL)/Canadian Standards Association (CSA) or Verband Deutscher Electronotechniker (VDE)/International Electrotechnical Commission (IEC) safety standards. The UL and CSA require 1000 V AC isolation voltage, whereas VDE and IEC require 3750 V AC. Consequently, any step-down power transformer or high-frequency switching transformer within the DC-DC converter stage has to be designed to the same safety isolation requirements.

7.3 Rectifier Section

In older linear power supplies, a step-down transformer was used with low-voltage rectifier circuits. In off-the-line switching power supplies, a line voltage capability rectifier set is used directly off the AC line without any low-frequency line isolation transformer between the AC mains and the rectifiers. Because in most of today's electronic equipment the manufacturers are generally addressing an international market, power supply designers must use an input circuit capable of accepting many different line voltages, normally 90 to 130 V AC or 180 to 260 V AC. Figure 7.1 shows such a circuit using a

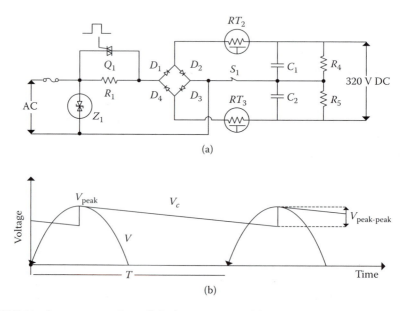

(a)

(b)

FIGURE 7.1   Input section of an off-the-line power supply: (a) overall arrangement; (b) ripple voltage calculations.

voltage doubler technique. When the switch S_1 is closed, the circuit may be operated at a nominal line of 110 V AC. During the positive half cycle of the AC, capacitor C_1 is charged to the corresponding peak voltage, approximately 160 V DC, through diode D_1. During the negative half cycle, capacitor C_2 is charged to 160 V DC through diode D_4. Thus the resulting DC output is the sum of the voltages across $C_1 + C_2$, or 320 V DC. When the switch is open, D_1 to D_4 form a full-bridge rectifier capable of rectifying a nominal 230 V AC line and producing the same 320 V DC output voltage. In many universal input systems, there is an automatic changeover capability built into the design.

Two important aspects of the representative circuit in Figure 7.1(a) should be noted. One is the energy storage in capacitors C_1 and C_2. The other is the resistor R and special components such as negative temperature coefficient (NTC) thermistors for the purpose of inrush current limiting. Figure 7.1(b) illustrates the simplified case of a half-wave rectifier where the charging and discharging of the capacitor is simplified by a sawtooth assumption for the discharge during the negative half of the AC cycle. During this process, approximate average DC output voltages for the half-wave or full-wave cases are given by

$$V_{DC} = \sqrt{2}V_{rms} - \frac{I_{Load}}{2fC} \qquad \text{(7.1a) [Half wave]}$$

$$V_{DC} = \sqrt{2}V_{rms} - \frac{I_{Load}}{4fC} \qquad \text{(7.1b) [Full wave]}$$

where V_{RMS} is the RMS AC input voltage, I_{load} is the average load current, f is the line frequency, and C is the value of the effective smoothing capacitor. The RMS ripple voltage is given by

$$V_{ripple(rms)} = \frac{I_{load}}{2\sqrt{3}fc} \qquad \text{(7.2a) [Half wave]}$$

$$V_{ripple(rms)} = \frac{I_{load}}{4\sqrt{3}fc} \qquad \text{(7.2b) [Full wave]}$$

For more details, see Smith [1, 2]. It is important to note that the above equations are based on a simplified approach only, and a more accurate and detailed treatment is available elsewhere [3].

For diodes, maximum forward rectification current capability, peak inverse voltage (PIV) capability, and the surge current capability (to withstand the peak current associated with turn-on) are the most important specifications. Proper calculation and selection of the input rectifier filter capacitors are very important, because this will influence performance parameters such as low-frequency AC ripple at the output power supply and the holdover time. Normally, high-grade electrolytic capacitors with high ripple current capacity and low ESR need to be used with a minimum working voltage of 200 V DC. Resistors R_4 and R_5 (Figure 7.1[a]) provide a discharge path when the AC supply is switched off.

7.3.1 Fuses

Even the selection of the fuse for the input section needs to be done based on proper specifications. Fuses are categorized by three major parameters: current rating, voltage rating, and, most important, "let-through" current, or I^2t rating. The current rating of a fuse is the RMS value or the maximum DC value that it must exceed before blowing. The voltage rating of a fuse is not necessarily linked to the supply voltage. Rather, the fuse voltage rating is an indication of the fuse's ability to extinguish the arc that is generated as the fuse element melts under fault conditions. The voltage across the fuse element under these conditions depends on the supply voltage and the type of circuit. For example, a fuse in series with an inductive circuit may see voltages several times greater than the supply voltage during the clearance transient.

The I^2t rating of a fuse is defined by the amount of energy that must be dissipated in the fuse element to cause it to melt. This is sometimes referred to as the prearcing let-through current. To melt the fuse element, heat energy must be dissipated in the element more rapidly than it can be conducted away. This requires a defined current and time product. The heat energy dissipated in the fuse element is estimated in watt-seconds (or joules), or I^2Rt for a particular fuse with an internal resistance of R. As the fuse resistance is a constant, this is proportional to I^2t, normally referred to as the I^2t rating for a particular fuse or the prearcing energy. The I^2t rating categorizes fuses into the more familiar slow-blow, normal, and fast-blow types. It should be noted that the I²t energy can be as much as 20 times greater in a slow-blow fuse of the same DC current rating. For example, a 10 A fuse can have an I^2t rating ranging from 5 A²s for a fast fuse to 3000 A^2s for a slow fuse. The selection of fuse ratings for off-line SMPSs is discussed by Billings [4]. For high-power semiconductors such as power diodes and transistors, manufacturers indicate a value of I^2t, from 10 ms (for 50 Hz) or 8.3 ms (for 60 Hz), that should not be exceeded. Comparing this value with the fuse I²t permits us to verify the protection [5, 6].

7.3.2 Inrush Current Limiting

An off-the-line switching power supply may develop extremely high peak inrush currents during turn-on unless it incorporates some form of current limiting in the input section. These currents are caused by the charging of the filter capacitors, which at turn-on present a low impedance to the AC lines, generally limited by the ESR plus the total input resistance within the charging path. If no protection is employed, these surge currents may approach very large values to blow the input fuses.

Several methods are widely employed in introducing a high impedance to the AC line at turn-on. Figure 7.2 illustrates a few common methods. In Figure 7.2(a), an NTC

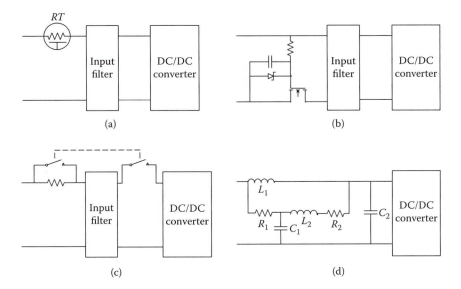

FIGURE 7.2 Power supply in-rush current limiter techniques: (a) thermistor technique; (b) MOSFET-based approach; (c) resistor relay; (d) inductor based. (Source: [7] Courtesy of *Power Electronics Technology*.)

thermistor limits the inrush. Initially, the thermistor resistance is high, which limits the inrush. As the inrush current flows through the thermistor, it heats up and the resistance decreases for normal operation. Figure 7.2(b) indicates a power semiconductor (MOSFET)-based technique. The FET limits the inrush by turning on slowly as its gate capacitor charges. Figure 7.2(c) shows an approach using a resistor and a relay. The input filter charges through the resistor until the relay is commanded to connect the filter to the converter and short the resistor. Each of these has its advantages and drawbacks [7]. Figure 7.1(a) illustrates the combination of thermistor and series power semiconductor (such as a triac)-based techniques, which is very popular in computer power supplies.

One of the self-limiting techniques is to use an input inductance to limit inrush current, as in Figure 7.2(d). This filter implements the required damping using inductor L_2 and resistors R_1/R_2 in parallel to the main DC-carrying inductor L_1. The value of L_1 is typically 5 to 10 times larger than L_2. More details of this technique are discussed by Bell [7], giving attention to the design of the magnetics.

7.3.3 Power Factor Correction Blocks

Power factor correction aligns the current waveform with the input voltage waveform. If the waveforms are not aligned, the power factor (PF) is less than 1; if they are aligned, the PF is unity. Figure 7.3 illustrates this. Figure 7.3(a) illustrates a typical non-PFC case where the nonsinusoidal current waveform does not align with the voltage waveform.

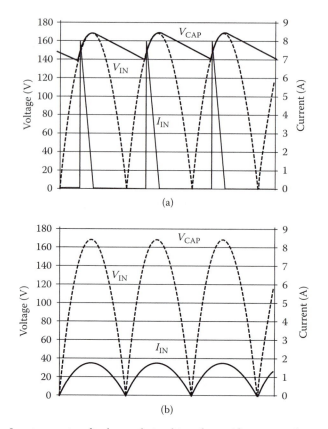

(a)

(b)

FIGURE 7.3 Input current and voltage relationships of a rectifier stage such as in Figure 7.1(a): (a) case of a typical rectifier without PFC; (b) ideal full-wave rectified PFC case with PF = 1.

Figure 7.3(b) illustrates the case of a half-sinusoidal input current waveform with PF = 1. In most cases, PF-corrected designs can achieve PF values of 0.95–0.98, and supplies that are not corrected can have a PF that is significantly less, usually less than 0.65. Given this simple explanation, it is also necessary to appreciate the case of nonsinusoidal rectifier currents, which can generate many harmonics of the line frequency waveform. When a repetitive waveform is not sinusoidal, it generates many harmonics, and the total harmonic distortion is given by

$$V_{THD} = \frac{\sqrt{V_2^2 + V_3^2 + ...V_n^2}}{V_1} \tag{7.3a}$$

where V_1 to V_n are the fundamental and the harmonics of the voltage waveform. V_{THD} is the ratio of all harmonics (added on an rms basis) to the fundamental.

Only current and voltage components with the same frequency can produce non-zero active power. In practice, sometimes it is reasonable to assume that the input voltage waveform is purely sinusoidal, despite any distortions in the current waveform.

Under such an assumption, V, the RMS voltage of the input waveform, is approximately equal to the RMS of the fundamental component, V_1, and we can get the approximate relationship

$$PF = \text{Cos}\phi_1 \cong \frac{V_1 \text{Cos}\phi_1}{VI} = \frac{I_1 \text{Cos}\phi_1}{I} \qquad (7.3b)$$

where I_1 is the in-phase fundamental RMS current and I is the total RMS current. With this discussion, we can appreciate the regulatory bodies defining the limits of harmonics generated by electrical systems connected to the AC utility grid. The EU put into effect EN 61000-3-2 to establish limits on harmonics up to 40th harmonic of the AC line-powered equipment's input current. Amendments in 2001 clearly state that PCs, PC monitors, and TV receivers with power ratings from 75 to 600 W must have PFC power supplies.

In practical cases of off-line SMPS systems, the fundamental idea is to use a power MOSFET switch and an inductor in the series path of the charging capacitor to artificially align the voltage and charging current waveforms. For the case of a boost converter configuration, this can be achieved as shown in Figure 7.4(a), where the smoothing capacitor is now shifted toward the DC-DC converter stage. A power factor control IC that switches the MOSFET at higher frequency smooths out the current waveform and aligns its fundamental component with the input voltage waveform.

Several IC manufacturers, such as Fairchild, Microlinear, and International Rectifier, supply various controller ICs for PFC or PFC/PWM combo operations. Figures 7.4(b) and 7.4(c) are typical configurations in a boost converter topology [8] based on Fairchild parts FAN 752B and FAN 4803. It is important to note that the main smoothing capacitor C_{in} has now moved further toward the DC-DC converter stage. For the case of discontinuous conduction, such as the case of using a FAN 752B-type controller for outputs up to 200 W, the boost converter's MOSFET turns on at zero inductor current and turns off when the current meets the desired input reference voltage. Figure 7.4(d) indicates the waveforms in discontinuous mode. For details of this application case, see Zuk [8]. Sandler et al. [9], Valentine [10], and Chapter 9 of Kularatna [11] provide a practical overview of PFC techniques and application information. Some new PFC techniques are discussed in [12–14].

7.4 Popular Transformer-Isolated Configurations for Off-the-Line Power Supplies and Industry Approaches

Most off-line AC-DC converters offer universal input voltage capability that typically spans 90–265 V AC at 47–63 Hz frequency range. This wide range allows configuration free use anywhere in the world, for portable devices. This concept relieves the designer from the concern that end users may incorrectly set their equipment. However, this condition of universal input range may create problems in selecting

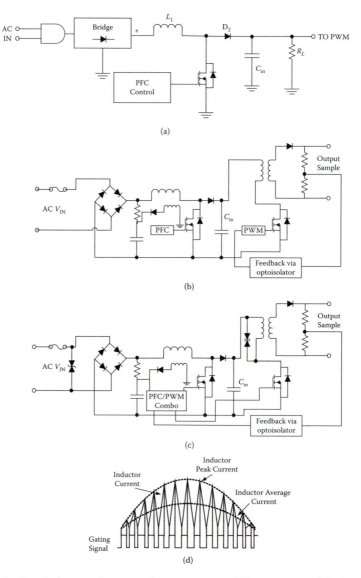

FIGURE 7.4 Practical approach to power factor correction: (a) basic concept of implementation; (b) use of a stand-alone PFC controller IC; (c) use of a PFC/PWM combo IC in a boost topology; (d) waveforms in discontinuous mode. (Courtesy of Zuk, P., *EDN*, September 1, 2006, 67.)

fuses, as at lower input the voltages converter may draw much higher current for the same output power.

Given the above condition, all off-line systems should have one safety-related condition: keeping the output DC loads isolated from the AC mains input. Given the direct rectification-based bulk DC power supplies used in compact systems, this isolation is provided only by a transformer-isolated DC-DC converter topology. In maintaining this isolation criteria, and the total wattage of the output loads, most off-line power supplies are based on transformer-isolated forward-mode converters, and the flyback topologies. For example, in situations such as the silver box of a desktop computer, where most high-current rails are 3.3 or 5 V, mixed with small-capacity 12 V rails, a forward-mode design is used. However, in systems such as set-top boxes and the like where voltage rails of up to 33 V are used, a flyback topology is selected. Figure 7.5 indicates typical cases of forward-mode designs and flyback designs used in common off-line power supplies [15].

For many applications where only 5 V and ±12 V are required by the load, a multirail output AC/DC converter is used, without any intermediate DC-DC converter stages, or POL converters. Some of these units are adhering to Energy Star™ and other similar regulatory guidelines [16]. For compact AC-DC adapters, where weight and the size is important, monolithic off-line switchers are used. An example is TOPSwitch family from Power Integrations [17]. For future-generation GSM cell phones and the like, universal AC wall adaptors with four-star efficiency ratings are expected. The aim is to ensure the mobile industry adopts a common format for mobile-phone chargers for overall reduction of standby consumption up to 50% [18].

Another common industrial requirement is to have battery backup within the off-line power supply. In this case (which is different from an external battery charger [AC adaptor] supplying an internal battery inside a portable product) a DC bus is used with a bidirectional charger based on a charge-discharge algorithm. A typical example is shown in Figure 7.6. For more details, [19] is suggested.

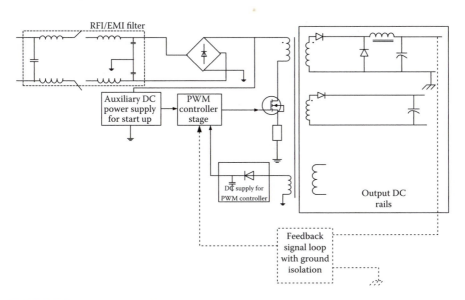

FIGURE 7.5 Simplified block diagrams of off-the-line switch-mode power supplies: (a) a forward-mode case similar to desktop computer power supply; (b) flyback-type design suitable for set-top boxes. (Part (b) adapted from Perzow, J., *EDN*, June 27, 2002, 73–78.)

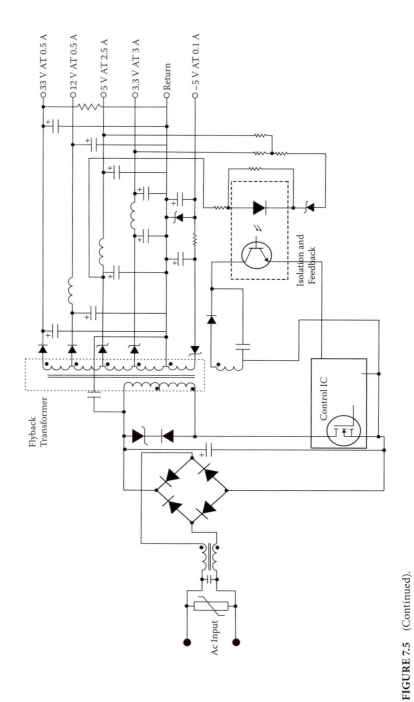

FIGURE 7.5 (Continued).

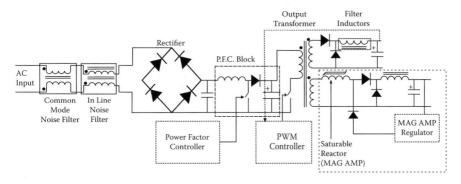

FIGURE 7.6 Typical magnetic component usage in off-line SMPS systems.

7.5 Magnetic Components

7.5.1 Transformers and Inductors

Magnetic elements are the cornerstone of all switching power supply designs but are also the least understood. There are three types of magnetic components inside switching power supplies:

- Forward-mode transformers
- Flyback-mode transformers
- Inductors such as AC filter inductors, DC filter inductors, or magnetic amplifiers

Each has its own design approach. Although the design of each of these magnetic components can be approached in an organized step-by-step fashion, it is beyond the intent of this chapter. For further information, Brown (20) is suggested.

7.5.1.1 Magnetic Materials

Magnetic materials can be divided into three broad categories, each with its own distinct range of characteristics and capabilities:

1. Magnetic metals and alloys, in tape or laminated form (e.g., silicon steel, mumetal, permalloy, amorphous metals)
2. Powdered magnetic metals (e.g., "powdered iron," moly permalloy powder)
3. Magnetic ceramics, or ferrites

7.5.1.1.1 Magnetic Metals

Typically these have a high saturation flux density (7–23 kG) and a high permeability. Electrical conductivity is also high, requiring their use in laminated or tape form to reduce eddy current loss. A large range of shapes is available, limited primarily to "two-dimensional" shapes by the wound tape or laminated construction. The high flux capability makes them most suitable as line frequency (50–400 Hz) transformers and inductors. Eddy current losses limit use at high frequency, although some find application up to several hundred KHz.

7.5.1.1.2 Powdered Metals

These are magnetic metals that are powdered, coated with an insulator, and pressed and sintered to shape. Eddy currents are reduced by the small particle size, extending the useful frequency range to a few megahertz in some applications. The insulated particles create a distributed air gap, giving these materials a low permeability, typically in the range of 8–80, with values to 550 obtainable in moly permalloy powder (MPP). The low permeability and a tolerance for DC current make these materials suitable for filter inductors at medium frequencies (typically 1–100 kHZ).

A relatively stable and well-defined permeability allows them to be used for nonadjustable tuning inductors, with a typical tolerance of a few percent. Saturation flux densities are typically in the range of 5–10 kG, but a very gradual saturation characteristic limits the maximum flux when inductance variations must be minimized. Available core shapes tend to be limited; toroids are the most common, with "E" cores, pot and cup cores, and a few similar shapes also available in some materials.

7.5.1.1.3 Ferrites

Ferrites are dark gray or black ceramic materials. They are very hard, brittle, and chemically inert. Most common ferrite materials such as MnZn and NiZn exhibit good magnetic properties below Curie temperature (T_c). Ferrites are characterized by very high resistivity, making them the most suitable material for transformers and inductors operating at high frequencies (up to tens of MHz) when low loss or high Q is of major concern. Saturation flux densities are low, in the 2–5 kG range, with permeabilities in the middle range of a few hundred to 15,000. Ferrites have the widest range of available core shapes, including any of the shapes for the other materials. Maximum linear dimensions are limited to a few inches by standard fabrication processes, although large cores can be assembled in sections.

The flexibility in shape often makes ferrites the material of choice, even when their other desirable properties are not required. They can easily be magnetized and have a rather high intrinsic resistivity. These materials can be used up to very high frequencies without laminating as is the normal requirement for magnetic metals.

NiZn ferrites have a very high resistivity and are therefore most suitable for frequencies over 1 MHZ, but MnZn ferrites exhibit higher permeabilities (μ_i) and saturation induction levels (B_s). Different kinds of power ferrite material are available from many vendors. Figure 7.6 indicates the typical block diagram of an SMPS with magnetic components. A usage summary is indicated in Chapter 4 of [11].

7.5.1.2 Core Geometry and Common Core Materials

Cores come in many shapes and sizes. The three most common core types are shown in Figure 7.7. There are many more types, but they are all based upon these basic styles. Some of the important considerations when selecting a core type are core material, cost, the output power of the power supply, the physical volume the transformer or inductor must fit within, and the amount of RFI shielding the core must provide. Core material considerations are summarized in Chapter 4 of [11].

A few examples of the commonly used core materials in modern switching power supplies are F, K, N, R, and P materials from Magnetics Inc.; 3C8 and 3C85 from FerroxCube

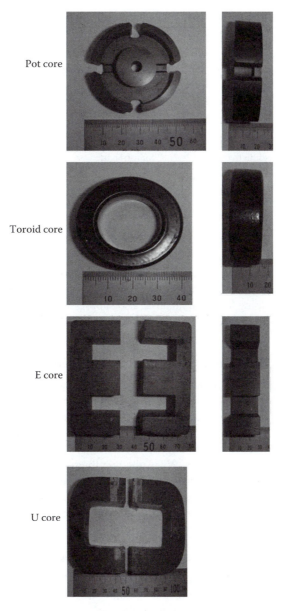

Pot core

Toroid core

E core

U core

FIGURE 7.7   Common core types used in power supplies.

Inc.; and H7C4 and H7C40 materials from TDK. These ferrite materials offer low core losses at operating frequencies between 80 kHz and 2.0 MHz. Table 7.1 indicates the available power ferrite materials from Phillips, their usage, and their characteristics.

Resonant power supplies make it possible to exploit the area beyond 1 MHz, which requires power ferrite transformers with useful properties in the 1–3 MHz frequency

TABLE 7.1   Power Ferrite Materials from Phillips: Their Usage and Frequency Characteristics

Material Grade	Characteristics	Usage	Available Core Types
3B8	Medium frequency (<200 kHz)	Small power transformers and general-purpose transformers	RM, P
3C80	Low frequency (<100 kHz)	Power transformers in TV applications	E, EF, ETD, EC, U
3C10	Low frequency with improved saturation level	Flyback transformers	U
3C85	Medium frequency (<200 kHz)	Industrial use	E, EF, ETD, EC, RMP, EP, Ring core
3F3	High frequency (up to 1 MHz)	Resonant power supplies	E, EF, ETD, RM, P, EP, Ring core
3F4	High frequency (1–3 MHz)	Resonant power supplies	

range. Meeting this need is a newer power transformer core material, 3F4, MnZn ferrite. The chemical composition includes dopants to reduce the hysterisis loss and high-frequency wall damping loss. The material also suppresses electrical conductivity and thus the eddy current losses. For further details, Visser and Shpilman [21] and Bate [22] are suggested.

7.5.1.3 Winding Techniques and Practical Considerations

The design and winding technique used in the magnetic component's design has a great bearing on the reliability of the overall power supply. Two situations arise from a poor transformer design: high-voltage spikes are generated by the rate of transitions in current within the switching supply, and the possibility of core saturation can arise during an abnormal operational mode. Voltage spikes are caused by a physically "loose" winding construction of a transformer. When the windings are physically wound distant from one another, the leakage inductances store and release a portion of the energy supplied to a winding in the form of voltage spikes. It also delays the other windings from seeing the transition in the drive winding. Spikes can cause the semiconductors to enter avalanche breakdown, and the part can instantly fail if enough energy is applied. It can also cause significant RFI problems.

A snubber is usually the solution, but this lowers the efficiency of the power supply. Core saturation occurs when there are too few turns on a transformer or inductor. This causes the flux density to be too high, and at high input voltages or long pulse widths, the core can enter saturation. Saturation is when the core's cross-sectional area can no longer support additional lines of flux. This causes the permeability of the core to drop, and the inductance value to drop drastically.

7.5.1.4 Forward-Mode and Flyback Transformers

For forward-mode operation, transformers may use either tape-wound cores or ferrite cores. Tape-wound cores are used at the lower frequencies because their high flux density minimizes the transformer size. As the frequency increases, high-frequency materials are

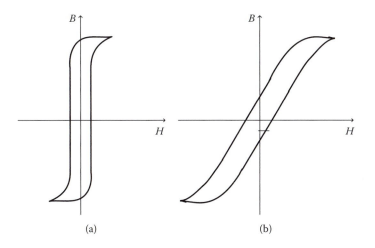

B

H

B

H

(a)   (b)

FIGURE 7.8   B-H loop for a tape-wound core: (a) ungapped; (b) gapped.

required and the tape thickness must be reduced, both of which increase the core cost. This makes the tape-wound core more costly than ferrites at frequencies over 10 kHz.

Tape-wound cores often have to be gapped with a minimal-type air gap. Figure 7.8 shows the hysterisis loop of a typical tape-wound core both with and without an air gap. Because the loop is extremely square, any unbalance in the switching process could cause saturation for some part of the cycle, thus generating extra losses in the circuit and distorting the output. As indicated in Figure 7.8(a), the gapped core will require a large difference in transistor currents before it will saturate the core.

One other method to eliminate this problem of core saturation is to use a composite core. These cores are made with an ungapped tape-wound core surrounded by another tape-wound core containing a small air gap. Composite cores avoid the problems of unbalanced transistors and also reduce or eliminate spikes in the output. In state-of-the-art power supplies running at frequencies over 200 KHz, where the losses could be high, ferrites are the dominant core material because of their lower cost and availability of variety of shapes and sizes.

7.5.1.4.1 Flyback Transformers

The transformer cores for flyback circuits require an air gap in the core so that they do not saturate from the DC current flowing in their windings. As mentioned above, the tape-wound core and ferrite core can both be provided with air gaps. The toroidal permalloy powder core is also used because it offers a distributed-type air gap. The frequency of the unit will dictate the selection of either tape-wound cores or ferrites as described previously, while the powder cores will operate at frequencies up to 50 kHz. Operation of the flyback converter is based on the storage of energy in the core and its air gap during the on time of the switch and discharging the energy into the load during the switch's off time. The magnetic core operates in one quadrant of its B-H curve. For high-energy transfer in a small volume, the core should have the B-H curve shown in Figure 7.9. An ideal core has a large available flux swing (ΔB), low core losses, relatively

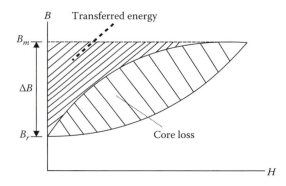

FIGURE 7.9   B-H curve required for a flyback power supply.

low effective permeability, and low cost. MPP cores, gapped ferrites, Kool Mµ, and powdered iron cores are used in this application.

In the case of ferrites, to increase their energy transfer ability ferrites must have an air gap added into the core's structure (Figure 7.10). Although the air gap increases the energy transfer of the ferrite core, it can create EMI problems. Better-shielded core shapes, such as pot cores, PQ cores, RM cores, and EP cores, are used to combat stray magnetic flux emitted from the air gap. The disadvantage of these shapes is higher cost.

7.5.1.5 Inductors

The switching regulator usually includes one or several power inductors. Because of the large DC current through its windings, the core must have a large air gap to keep it from saturating. Cores used here are gapped ferrite cores, permalloy powder cores, and high-flux powder cores. In the current waveform, because there is usually only a small ripple, the AC flux swing in the core is small. Therefore powdered iron cores and silicon laminations may also be used. They are less expensive than the above-mentioned cores but have much higher losses, and therefore care must be taken in the design of such units to avoid overheated inductors and possible damage that may be caused to other components.

Ferrite E cores and pot cores offer the advantages of decreased cost and low core losses at high frequencies. For switching regulators, F or P materials are recommended because of their temperature and DC bias characteristics. By adding air gaps to these ferrite shapes, the cores can be used efficiently while avoiding saturation.

7.5.2 Planar Magnetic

The concept of planar magnetics has existed for many years; however, the idea gained popularity in the late 1990s due to its low-profile configurations. The windings consist of either copper spiral turns on single- or double-sided printed circuit boards or stamped, flat copper foil windings.

The windings are separated from each other and the core by thin sheets of Kapton or Mylar insulation. When necessary, they are separated by special plastic bobbins to maintain safety specifications. These assemblies are available in a fully encapsulated form to meet certain military specifications.

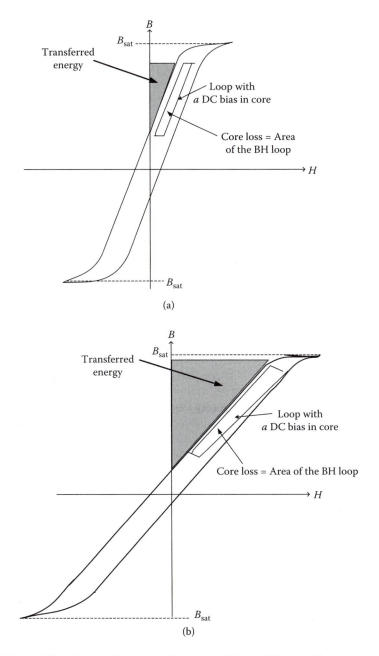

FIGURE 7.10   Effect of gap in ferrite core: (a) ungapped ferrite; (b) gapped ferrite.

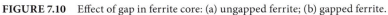

Advantages of the planar construction include low-profile configuration (e.g., high efficiency, lightweight, and uniform construction) and high power density (e.g., low manufacturing times and high frequencies of operation). Some disadvantages include high costs of design and tooling for printed circuit boards and special ferrite cores, inefficient means of terminating windings to the board, and thermal temperature rise of magnetics.

Another approach to planar magnetics is the multilayer circuit board, a more condensed configuration. The primary and secondary windings are stacked vertically on different layers of a circuit board with a thin film of FR4 separating them. The multilayer circuit board offers all the advantages of the planar configuration and provides an even lower-profile transformer. An additional advantage of the multilayer build is that heat is transferred out of the transformer more efficiently so that thermal rise is not as much of a problem compared to single- or double-layer types.

7.5.2.1 Planar Magnetics and New PWM ICs

With the distributed power architectures being widely adopted by the industry, new D/CMOS PWM ICs, combined with integrated PCB transformers, allow simple flyback, and forward converters to operate at frequencies up to 1 MHz, with performance similar to zero-switching topologies. An example of such an IC is Si 9114 from Siliconix Inc. [23].

Planar transformers can be more efficiently used if the subsystem is viewed as an entire unit that includes the power supply and the load. Usually the load is some type of digital circuit and will include a microprocessor and memory, as well as custom-designed ASICs. System complexity has so increased that many designers are now using multilayer printed circuit boards where six and even more layers have become common. These systems can now utilize a "free transformer" directly in the PC board as shown in Figure 7.11. Magnetics Inc. [24, 27], Martin [25], and Horgan [26] provide practical details related to magnetic materials and cores used in switch-mode power supply systems.

7.6 Output Blocks

In general the output section of any switching power supply is comprised of single or multiple DC voltages, which are derived by direct rectification and filtering of the transformer secondary voltages and in some cases further filtering by additional series-pass regulators or LDOs. These outputs are normally low voltage, direct current, and capable of delivering a certain power level to drive electronic components and circuits. The most common output DC voltages are 3.3 V, 5 V, ±12 V, 2.5 V, 1.8 V, 1.2 V, etc., and their power capability may vary from a few watts to thousands of watts. The most common type of secondary voltages that have to be rectified in a switching power supply are high-frequency square waves. These, in turn, require special components, such as Schottky diodes or fast-recovery rectifiers, low-ESR capacitors, and energy storage inductors, in order to produce low noise outputs useful to the majority of electronic components.

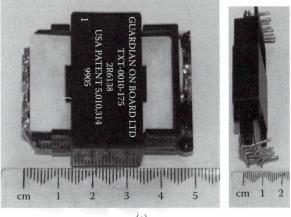

(a)

(b)

FIGURE 7.11 Integrated PC board planar transformer.

7.6.1 Output Rectification and Filtering Schemes

The output rectification and filtering scheme used in a power supply depends on the type of supply topology the designer chooses to use. The conventional flyback converter uses the output scheme shown in Figure 7.12(a). Since the transformer T_1 in the flyback converter also acts as an energy-storing inductor, diode D_1 and capacitor C_1 are the only two elements necessary to produce a DC output. Some practical designs, however, may require the optional insertion of an additional LC filter, shown in Figure 7.12(a), to suppress high-frequency switching spikes. The physical and electrical values of both L_1 and C_2 will be small. For the flyback converter, the rectifier diode D_1 must have a reverse-voltage rating of $[1.2\ V_{in}\ (N_S\ /N_p)]$, minimum.

The output section of a forward converter is shown in Figure 7.12(b). Notice the distinct differences in the scheme compared to the flyback. Here, an extra diode D_2 called the flywheel is added, and also inductor L_1 precedes the smoothing capacitor. Diode D_2 provides current to the output during the off period. Therefore the combination of

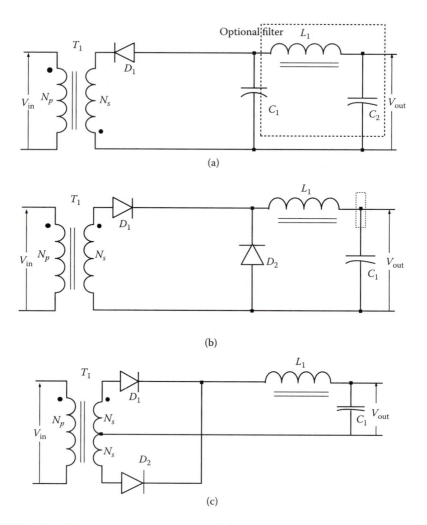

FIGURE 7.12 Output stages: (a) flyback type; (b) forward type; (c) push-pull/half-Bridge/full-bridge.

diodes D_1 and D_2 must be capable of delivering full output current, while their reverse-blocking voltage capabilities will be equal to $[1.2V_{in} (N_s/N_p)]$, minimum.

   The output section that is shown in Figure 7.12(c) is used for push-pull, half-bridge, and full-bridge converters. Since each of the two diodes D_1 and D_2 provides current to the output for approximately half of the cycle, they share the load current equally. An interesting point is that no flywheel diode is needed, because either diode acts as a flywheel when the other one is turned off. Either diode must have a reverse-blocking capability of $[2.4 V_{out} (V_{in,max}/V_{in,min})]$, minimum.

7.6.2 Postregulation Techniques

In many common processor-based load environments, the load may require very tightly regulated outputs together with excellent transient response at the power supply output for fast-changing load currents. The output ripple of transformer-isolated multiple-rail DC-DC converters is typically between 0.5% and 1% of the nominal output, whereas a linear regulator can perform within a millivolt- or even a microvolt-order ripple. The transient response of a 50%–100% (or vice versa) load step can range from 100 to 300 µs for an isolated DC-DC converter, whereas a linear supply can respond within 1–5 µs for a similar load current step. In demanding situations of loads where tight regulation and fast transient response are required, postregulation techniques can be used. This is particularly the case when multiple voltage rails are available in a power supply system. The basic principle behind all postregulation techniques is to use some extra regulator circuits at the output side of the switching power supply (Figure 7.13).

Some of the popular postregulation techniques include linear regulators, added secondary-side DC-DC converters, coupled inductors, magnetic amplifiers (mag amps; sometimes called saturable reactors), and secondary-side postregulators (SSPRs). Figure 7.13 shows these concepts. A linear regulator block used as a secondary-side controller (Figure 7.13[a]) or a postregulator is quite common and provides excellent transient response and minimum ripple.

However, for efficiency reasons these linear regulators should be in the form of LDOs. Adding a second-stage DC-DC converter (Figure 7.13[b]) seems simple but can present design difficulties. Coupled inductors (Figure 7.13[c]) are suitable for secondary-rail situations, where lower regulation is acceptable, in a typical tolerance range of ±5%–±8%. Tight coupling can create unwanted interactions between coupled outputs. Magnetic amplifier techniques are based on a saturable reactor, which can act as a magnetic switch (Figure 7.13[d]), exhibiting high or low impedance toward the output rail to be controlled. Magnetic amplifiers are used for medium and high power requirements. The SSPR technique (Figure 7.13[e]) uses a semiconductor device as a switch, with a switch-mode controller that is synchronized with the main PWM controller. Regulation of the voltage is achieved on the secondary side either by controlling the volt-second product across the output inductor for buck-derived techniques or by controlling the amount of energy for boost and flyback-derived topologies. For details, see Levin [28], Simopoulos [29], and Mammano [30].

7.7 Efficiency Improvements and Synchronous Rectification

In Chapter 2, we briefly discussed the losses associated with a switching power supply. In improving the efficiency of a power supply, every item must be carefully evaluated and optimized. In low-output voltage switching power supplies, losses in the output

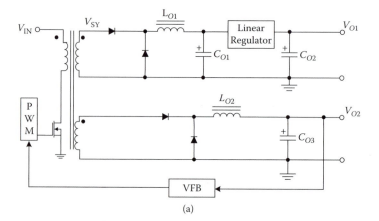

(a)

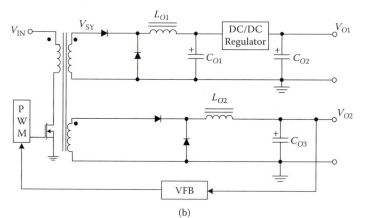

(b)

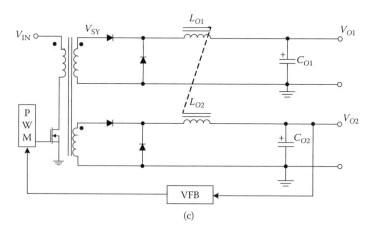

(c)

FIGURE 7.13   Postregulation techniques: (a) linear regulator based; (b) switch-mode controller based; (c) coupled inductor based; (d) magnetic amplifier based; (e) secondary-side postregulator (SSPR) based.

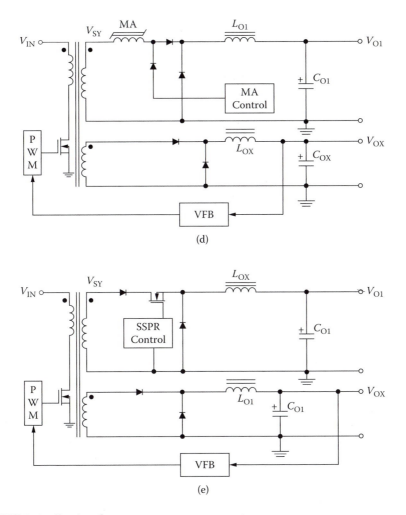

(d)

(e)

FIGURE 7.13 Continued

rectifiers and the switching losses in the transistors contribute a significant share. High-frequency rectifiers usable in the output stage may be of three types: high-efficiency fast-recovery types, high-efficiency very fast rectifiers, and Schottky barrier rectifiers. A discussion of these can be found in [11]. Recently, gallium arsenide (GaAs) and silicon carbide (SiC) devices have been introduced, and some of these could help improve the efficiency of the design.

For the lower DC rail voltage requirements, such as 1.2–3.3 V, of high-performance digital circuits, the high-efficiency requirement comes at very low-output voltages, and the general design approach using a Schottky diode becomes inadequate. In such circumstances, synchronous rectifiers (SRs) configured using low-$R_{DS(on)}$ power MOSFETs can provide much better efficiencies. In all switching topologies, the output rectifiers can

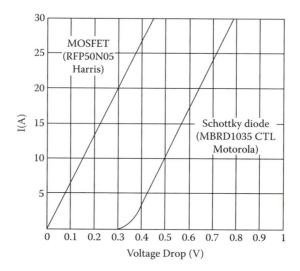

FIGURE 7.14   Voltage drop of a MOSFET compared with a Schottky diode.

be conceptually replaced by power MOSFETs. This basically replaces the DC loss component because of the combination forward voltage ($V_F I_{RMS}$) and the $I^2_{rms} r_D$ (the resistive loss component in the diode) by a single element ($I^2_{rms} R_{DS\ (on)}$ of the MOSFET), providing significant efficiency improvements. Figure 7.14 indicates a comparison of a typical Schottky diode and a MOSFET usable in an SR. For more information, see Sherman and Walters [31], Moore [32, 33], and Christiansen [34].

There are two basic types of SRs: self-driven (SDSR) and control-driven (CDSR). Figure 7.15 shows these types, and Table 7.2 compares their advantages and disadvantages. More design details can be found in How [35]. There are many controllers suitable for SR systems, and details on these advanced techniques can be found in Khasiev [36], Bindra [37, 39], Yee [38], Elbanhawy [40], and Mappus [41].

In off-the-line SMPS similar to the silver box in PCs, there is much room for efficiency improvement. Easily identifiable areas of improvement can be classified into three major categories:

- An appropriate harmonic reduction front end with active PFC
- Architectural-level improvements to eliminate losses
- Component-level improvements and upgrades to reduce losses

More detailed discussion and guidelines can be found in Dalal [42]. Another important design consideration is the reduction of startup current-related losses [43]. Another aspect is to reduce the losses due to stray and leakage inductances [44]. Current-sensing resistors in switchers can also add significant losses. In situations where an inductor current is to be sensed, there are special techniques using a parallel RC network to sense the inductor current and minimize the loss across the series resistor inserted with the inductor [45].

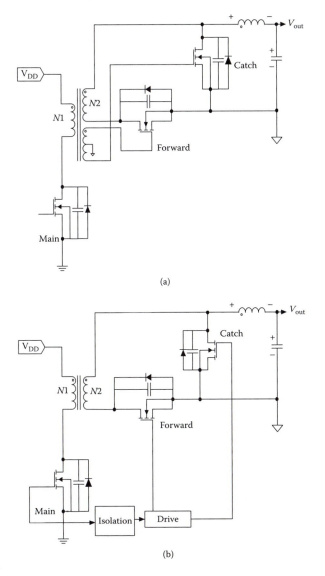

FIGURE 7.15 Different types of synchronous rectifiers: (a) self-driven SR (SDSR); (b) control-driven SR (CDSR).

7.8 EMI Reduction

Because of high-frequency nonsinusoidal waveforms within the circuitry, a switching power supply generates noise that can emerge through conduction or radiation. This noise can be conducted into the load or input source, and radiated components

TABLE 7.2   Comparison of Synchronous Rectification Techniques

Technique	Advantages	Disadvantages
Self-driven SR (SDSR)	Simple	For most topologies, drive signal amplitude varies with the line voltage
		When driven by transformer windings, there is no gate-driving voltage during dead time
		Problems when parallel operated
Control-driven SR (CDSR)	Constant gate signal, irrespective of line voltage and load changes	Complex circuitry
	Constant $R_{Ds(on)}$	Needs accurate timing to prevent cross conduction
	Suitable for wide input voltage ranges	
	Applicable for all topologies	
	Dead time can be kept to a minimum	

Source: Sherman, J., and M. M. Walters, *EDN*, March 14, 1996, 111.

can create annoying situations in portable products such as cellular phones, PDAs, and laptops. U.S. and international standards for EMI-RFI have been established that require manufacturers to minimize the radiated and conducted interference of their equipment to acceptable levels. In the United States, the guiding document is FCC Docket 20780, and internationally the West German VDE- 08XX series is a well-accepted example. These standards generally exclude subassemblies from compliance, but the overall system should strictly adhere to the specifications. The main sources of high-frequency noise are the switching transistors, input and output rectifier stages, protective diodes, and the control ICs. The RFI noise level can vary with the topology used. Flyback converters, because of their near-triangular input current waveforms, generate less conducted RFI than topologies with rectangular input current waveforms. EMI noise reduction is generally accomplished by three means: suppression of the noise source, isolation of the noise coupling path, and filtering and shielding. Some advanced techniques are frequency modulation techniques and slew rate control. Because the total spectrum of high frequencies should be minimized, measurement of conducted EMI noise using a spectrum analyzer is generally carried out, with adequate attention to the resolution bandwidth (RBW) of the spectrum analyzer or a similar test setup [46].

The most common method of conducted noise suppression at the input of an off-line SMPS is the utilization of an LC filter for differential-mode and common-mode RFI suppression. Normally, a coupled inductor is inserted in series with each of the AC input lines, and capacitors are placed between lines (X capacitors) as well as between the lines and the earth terminal (Y capacitors). Figure 7.16 indicates different line filter schemes

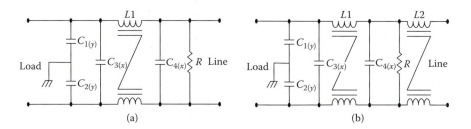

FIGURE 7.16 Input line filters for SMPS systems: (a) basic version (b) an improved version with two filter chokes.

used. The resistor R is for the discharge of the X capacitors, and the values are recommended by the relevant safety specifications under the VDE or IEC series.

Proper component layout and selection are important in controlling EMI. In a typical off-line power supply with common mode filters, the main source of common mode noise is the MOSFET. With the requirement of a heat sink, for example, in the case of a TO-220 package, the capacitance formed between the drain and the ground plane (CP_1 in Figure 7.17[a]) can conduct some common mode noise. In a typical power supply,

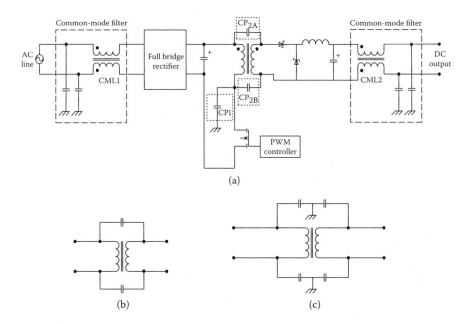

FIGURE 7.17 A simplified example of a typical power supply with common mode filters: (a) a general case indicating parasitic capacitances (b) parasitic capacitances across a winding without a Faraday shield (c) Faraday shield reducing the stray capacitance coupling [Source: [47], Courtesy of EDN]

such as in Figure 7.17, the transformer can also conduct some high-frequency current through the parasitic capacitances formed between the windings. One technique to reduce the effects of these capacitances, such as C_{P2A} and C_{P2B}, is to use a Faraday shield for the windings. The situation is compared in Figures 7.17(b) and 7.17(c).

Another possible source of EMI is gapped cores, such as in flyback transformers. Although the gap increases the stored energy, it can lead to increased EMI problems, and for this reason experienced designers avoid bobbin cores. Details on the selection of transformer cores can be found in Schindler [47]. Another important approach is to pay adequate attention to the layout of the circuit. The most important consideration is the power stage, because it creates the highest circulating currents and acts as the main source of EMI. Next is the drive stage, where currents can be a few hundred milliamps to 10 A or higher. Because of the relatively high currents possible in the drive stage, it should be placed very close to the power stage. In the MOSFET drive stages, if the gate connections are longer than about 5 cm, a rule of thumb is to place a series resistor of 10 Ω near the power MOSFET. Another important consideration is to place the power traces and the returns close to each other to minimize the loop area between the traces and to increase coupling capacitance. An ideal layout design for a multilayer board would have these two traces on adjacent layers, one directly above or below the other. Another important consideration is to have the ground of the controller IC close to the feedback circuit's ground to minimize feedback voltage errors. More details can be found in Rogers [48] and Scolio [49].

Another more recent advancement is the spread spectrum-based controller ICs, where by modulating the PWM frequency, the noise gets spread across the band [50]. Another advanced approach is to introduce dither to the system clock of the DC-DC converter. This approach, with its resulting spread spectrum operation, allows the switching frequency to be modulated by a pseudo-random number (PRN) sequence to eliminate narrowband harmonics [51]. In this approach, a charge pump technique is used. For analysis and spectral characteristics of these techniques, see Tse et al. [52].

In some controller ICs, such as the LT 1533 from Linear Technology Corporation, the slew rate of the switcher voltage and current waveforms are controlled to reduce noise, at the expense of the efficiency [53]. Wittenbreder [54–57] provides a guideline for designing converters for lower EMI. Reducing the ground bounce problems in DC-DC converters is discussed in Barrow [58].

7.9 Power Supply Protection

Safety and reliability should be important aspects of power supply operation. Not only the load but also the power supply and its input source should be safe under all operating conditions. The following are a few important items to consider:

- Thermal design
- Overvoltage and overcurrent protection
- Protection against transients
- Long-term reliability aspects of output and input capacitors
- Age-related aspects

7.9.1 Thermal Design

In any electronic design, semiconductors, as well as the passive parts, have thermal limits for operation. For example, most manufacturers of silicon ICs specify the maximum junction temperature at about 150°C. Similarly, passive parts also have maximum temperature limits. Given such limits, the maximum power dissipation of an IC or a power semiconductor package is given by

$$P_{D,\max} = \frac{T_{J(\max)} - T_{A(\max)}}{R_{\theta JA}} \tag{7.4a}$$

where $T_{J(\max)}$ is the maximum recommended junction temperature, $T_{A(\max)}$ is the worst-case ambient temperature of the application, and $R_{\theta JA}$ is the junction-to-ambient thermal resistance of the package in degrees Celsius per watt (°C/W). A semiconductor's package determines its $R_{\theta JA}$ and quantifies how much the junction temperature will rise for each watt the device dissipates into still air. If a heat sink is used to mount the component, three components contribute to the thermal resistance, as given by

$$R_{\theta JA} = R_{\theta JC} + R_{\theta CS} + R_{\theta SA} \tag{7.4b}$$

where $R_{\theta JC}$, $R_{\theta CS}$, and $R_{\theta SA}$ are thermal resistances of the junction to case, case to heat sink, and heat sink to ambient, respectively. The thermal characteristics of a linear regulator such as CS8121 are shown in Table 7.3 for different possible packages, and it should be noted that the values can have a wide range of variations. Guidelines for designing in a linear regulator IC environment such as LDOs are given in Malley [59]. In compact electronic environments, copper foil of a PCB can be used to remove heat from a device; Figure 7.18 shows the thermal resistance versus PCB foil area.

The designer should be able to calculate the maximum possible power dissipation for a given environment and then translate that value to safe maximum output currents based on approximate relationships for the given circuit and the topology. Another aspect is the transient thermal response, and in this situation an electrical equivalent of an RC element combination can be used with a simulator such as SPICE to generate the thermal response of a power semiconductor [60]. Using a similar RC equivalent approach, when thermal simulation software is not available, or for quick calculations, designers can apply linear superposition to model the thermal performance of systems [61, 62]. Within the last decade, many new cooling systems have been introduced, and

TABLE 7.3 Thermal Characteristics of a Typical Power Package (Example: CS8121 from ON Semiconductor)

Package	$R_{\theta JA}$ (°C/W)	$R_{\theta JC}$ (°C/W)
TO-220	50	3.5
14-lead surface outline (SO)	125	30
8-lead plastic dual inline package	10	52

[a] An example is CS8121 from ON Semiconductor.

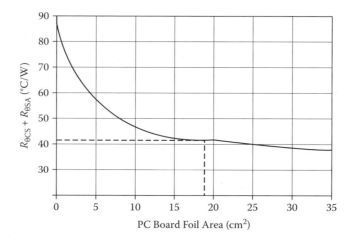

FIGURE 7.18 Thermal resistance from case to ambient with PC-board foil area. (Source: [59], Courtesy of *EDN Magazine*, © 1995, Reed Business Information, a division of Reed Elsevier. All rights reserved.)

designers should consider these systems, such as microchanneled heat sinks [63] and nonlinear fin pattern-based systems [64], in very high power designs.

7.9.2 Overvoltage and Overcurrent Protection

7.9.2.1 Overvoltage Protection

A power supply should not generate any steady or transient overvoltages under all operating conditions, particularly in low-voltage, high-current power supplies. Sometimes an unexpected component failure, such as a shorted power transistor in a buck converter, can generate a disastrous high output voltage. Under a very peculiar and unwarranted load current transient, the power supply may generate a transient overvoltage condition. In most situations, the easiest way to protect the load from overvoltage is to use crowbar circuits, which detect the overvoltage situation quickly and activate a short circuit across a fuse. To enable reliable protection, the overvoltage protection must be independent from the rest of the system's circuits; it must have its own voltage reference source and an independent power source. Figure 7.19(a) indicates a simple implementation of a crowbar circuit [65] in a nonisolated synchronous buck converter. Figure 7.19(b) indicates a case of diode-ORed redundant supplies with independent crowbar circuits.

In isolated power supplies, the transformer provides inherent protection against a switch failure; if any of the components in the feedback path opens accidentally, it can create a dangerous situation. Even in such situations, by using optoisolators, one can design crowbar protection circuits using parts similar to the case in Figure 7.19(a) [65].

7.9.2.2 Overcurrent Protection

One of the most important protection features in the power supply is current limiting. When designing a current limiter, one should think of two main aspects: measure the

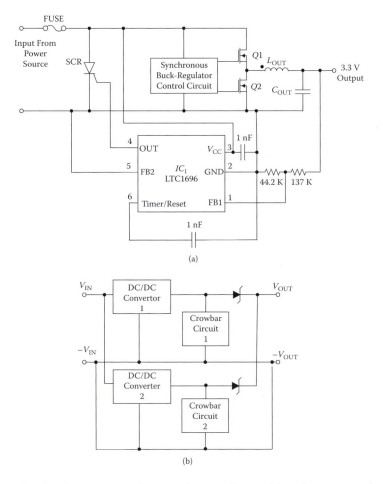

FIGURE 7.19 Crowbar protection for overvoltage conditions: (a) implementation of a simple crowbar circuit using limited components; (b) independent crowbar units in an ORed redundant power supply system. (Source: [65], Courtesy of *EDN Magazine*, © 2007, Reed Business Information, a division of Reed Elsevier. All rights reserved.)

current and develop a limiting circuit using the current signal. Current limiter design necessitates trade-offs among cost, complexity, reliability, and performance. There are several possible current-limiting schemes, as shown in Figure 7.20. In constant current limiting (Figure 7.20[a]), the output voltage drops sharply beyond the limit of the current. In an LDO or a common linear regulator, if such a scheme is applied at the limit, the voltage across the pass element will exceed the normal operation value (from $[V_{in} - V_O]$ to $[V_{in}]$) and the dissipation limit of the transistor and the heat sink can be exceeded, and the designer should allow for such excess dissipation. Figure 7.20(b) shows foldback technique. An advantage in this scheme over the constant current-limiting method is that dissipation within the regulator circuits is minimized. The ratio of I_{SC}/I_{max} is an important parameter for this scheme, where a smaller value means better performance.

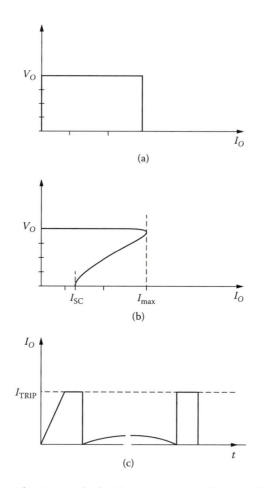

FIGURE 7.20   Current-limiting methods: (a) constant current limiting; (b) foldback limiting; (c) hiccup current limiting.

However, unless the circuit is designed to reset automatically on removal of excess load, this scheme may need manual intervention at overcurrent. Figure 7.20(c) shows the concept of hiccup current limiting. It incorporates overcurrent shutdown but adds an automatic restart mechanism. The power supply shuts down for a limited period of time and automatically restarts after a timeout.

In all these schemes, the traditional approach is to use a low-ohmic-value current sense resistor in series with the load path. To limit dissipation, a lower-value resistor needs to be used, or an alternative such as a PCB track as a resistance is logical. In all these situations, resistance value variation over the operational temperature range creates an uncertainty in the limiting value. When a copper PCB track is used, the approximate trace length can be determined using the following relationship:

$$L_{trace} \approx 40 I R_{sense} \qquad (7.5)$$

where I is the current value and R_{sense} is the expected resistance value in milliohms for a width of 20 mils/A (for 1 oz. type where the thickness is about 1.3 mils [33 μm] at 25°C).

In current-limiting schemes, the sensed current is used to activate the controller shutdown by an appropriate means of additional circuitry. In switching circuits with FETs, with the $R_{DS(on)}$ value for the FET available from the data sheets, one can develop a technique without a special-sense resistor, with knowledge of the converter topology [69]. For the techniques used to activate constant current limiting or foldback current limiting in linear regulators, see Malley [70]. Figure 7.21 shows different concepts to achieve current limiting in a buck regulator [68]. In shunt regulated DC power supplies, techniques can be developed for foldback without resistor based sensing [66, 71].

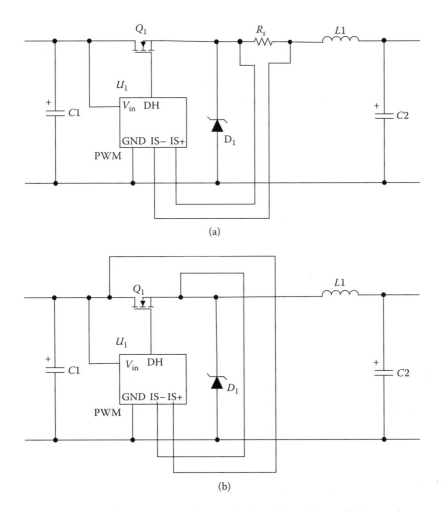

FIGURE 7.21 Current limiting in a synchronous buck regulator: (a) use of PCB trace for sensing; (b) use of FET resistance for sensing. (Source: [68], Courtesy of *Power Electronics Technology*.)

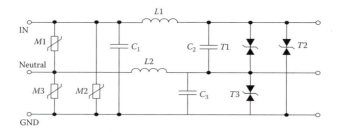

FIGURE 7.22 Transient protection for common- and differential-mode surges: multistage surge suppressor with MOVs (M1 to M3), avalanche diodes (T1 to T3), and passive parts.

7.9.3 Protection against Input Transients

Due to acts of God, such as lightning or inductive load dumps on the AC input power supply, severe surge voltages may occur, and these transients are of very short duration, such as from 50 to 200 μs total. Almost all off-the-line power supplies need to be protected against such events where both common-mode and differential- (or transverse) mode transients can occur. In such situations, nonlinear devices such as metal oxide varistors and avalanche diodes can be combined with small inductors and capacitors, as per the representative schemes in Figure 7.22. It is important for the designers to subject these surge suppressor blocks, as well as the entire prototype of the power supply, to simulated surges, as specified under C62.41 or similar standards that specify surge-testing procedures and waveforms. More details are in Chapter 9.

7.9.4 Reliability of Input/Output Capacitors

One major family of components that can affect the long-term reliability of a switching power supply is the smoothing capacitors, which carry large ripple currents. In these components, such as the output filter capacitors or input filter capacitors, high-frequency ripple currents can flow at the switching frequency, and this can generate heat due to the ESR of the capacitor. Designers tend to select filter capacitors based on the capacitance value rather than the ripple current. This approach can be catastrophic, because ripple current ratings can vary widely among capacitor technologies, manufacturers, and voltage ratings [72]. A typical example [72] is the Nichicon PL series capacitors, with voltage ratings from 6.3 V to 63 V, where the RMS ripple current rating can vary from about 950 mA to about 2.4 A. Power dissipation inside a capacitor is mainly determined by its ESR and is approximately given by

$$P_{cap} = I_{rms}^2 ESR \tag{7.6}$$

The ESR varies widely with the temperature and operating frequency; therefore the operating frequency of a switching supply and the temperature inside the case can have a significant effect on the amount of heat generated within the capacitor [72]. The location of the capacitor within the power supply can also have a significant impact. For example, if output capacitors are placed closer to heat sinks or catch diodes, the heat generated in these external parts can also heat the capacitor. A capacitor's load life specification indicates how many hours the device is likely to operate under severe conditions, and the failure can be catastrophic or parametric. For aluminum electrolytic capacitors, the load life typically ranges between 1000 and 10,000 hr if operated around a maximum of 105°C. If such a capacitor is operated at a lower temperature, the load life is given by

$$L_{actual} = L_{105^0} 2^{\frac{\Delta T_x}{10}} \qquad (7.7)$$

where L_{actual} is the actual lifetime and ΔT_x is the temperature difference between the maximum allowable temperature and the actual temperature. The value of ΔTx is given by

$$\Delta T_x = P_{Cap}/BA \qquad (7.13)$$

where B is the heat transfer coefficient (in W/cm²/°C) and A is the surface area of the capacitor. B may not be available in the data sheet but can be obtained from the manufacturer [72]. One problem arises in the calculation due to nonsinusoidal or DC-type current flowing inside the capacitor in the switching topology, and it varies with each topology. Huffman [72] provides these relationships based on idealized waveforms in each topology. To calculate the ESR at different temperatures, the ESR multiplier in Figure 5.16 can be used. In multiphase buck converters, ripple currents can be improved significantly compared to single-phase versions [73].

7.10 Age-Related Aspects

The reliability of a power supply cannot be solely designed in, tested in, or built in. It takes a team effort, starting with the definition and specification of the product, and does not end with the first shipment to the customer. The following points are key to achieving a high-reliability power supply:

- A rigorous specification review—attention to all aspects of reliability, including packaging, cooling, and connectors
- Proven topologies with proven component sets
- A comprehensive qualification—emphasis on areas of concern such as power line disturbances, parametric variations, lifetime considerations, and extensive testing to specifications
- A rigorous life test, including on/off cycling
- Extensive box testing for electromagnetic compatibility, etc.
- Qualification of the vendor's production
- Analysis of field problems on a continuous basis to understand real failures
- Bathtub curve as applied to reliability of products.

A power supply is a typical product or subsystem that exhibits the typical bathtub curve, as in Figure 7.23. The following tests are typical for achieving high levels of reliability:

- Thermal/shock cycling
- Temperature stress test

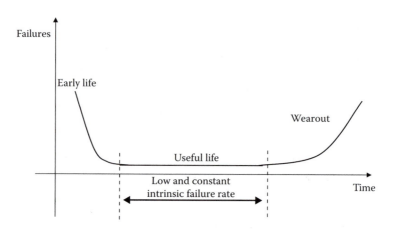

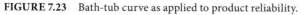

FIGURE 7.23 Bath-tub curve as applied to product reliability.

- Vibration
- EMI
- Thermal imaging (for an overall thermal profile)
- Stress analysis
- Power line disturbances
- Lifetime evaluation
- Test to specification

A discussion on these can be found in Forrester [74] and Pflueger [75].

7.11 Testing of Power Supplies

The power supply is the heart of an electronic system, and it is necessary for designers and test engineers to evaluate a supply's performance. This can be done at different stages, such as while refining the design or releasing the design. The most important basic tests are:

- Load regulation (graph of load current versus load voltage under fixed input voltage)
- Line regulation (graph of load current versus input line voltage under a fixed load current)
- Load transient recovery
- Current limiting
- Startup time
- Output noise (or periodic and random deviations [PARDs])

Other important tests are:

- Inrush current
- Line current
- Efficiency
- VID control (for VRMs)

During the design stage, it may be necessary to measure the loop performance for tuning the power supply's loop behavior.

Essential tests can be easily done using common bench instruments, such as a 5½-digit or better digital multimeter, a digital storage scope, and a reasonable electronic load. Figure 7.24(a) indicates a simple setup for load and line regulation. Load transient recovery, which is very important in high-current power supplies for digital processors, indicates one of the most important behaviors of the system, and Figure 7.24(b) indicates the essential transient parameters to be measured using the scope. In this process, an electronic load, which allows us to program the current step (from high-to-low as well as low-to-high transition), is applied to the power supply being tested. Modern VRMs require very high slew rate capability, which can be in the range of 10–300 A/μs or even

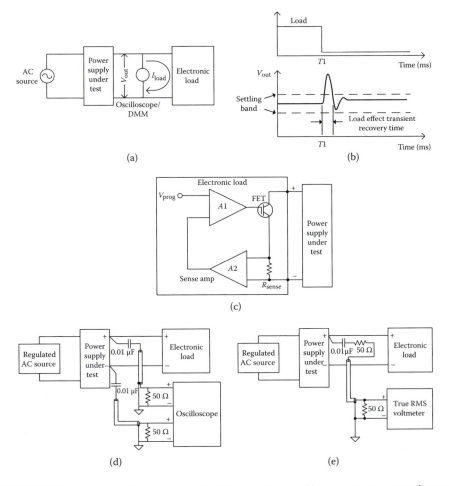

FIGURE 7.24 Power supply measurements: (a) basic test setup; (b) transient parameters for step load changes; (c) concept of an electronic load based on a MOSFET; (d) noise (or PARD) measurement using oscilloscope; (e) noise measurement using true RMS voltmeter. (Source: [76], Courtesy of *Test & Measurement World.*)

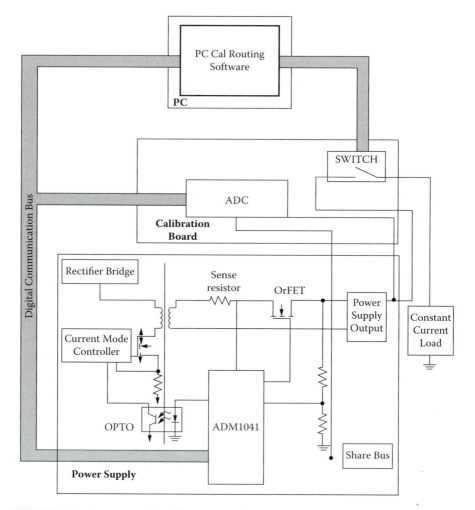

FIGURE 7.25   A power supply calibration set up for a power management and converter system designed around the ADM 1041 controller from Analog Devices, Inc. (Source: [79], Courtesy of *Power Electronics Technology.*)

higher, and this kind of performance cannot be measured without a very high transient capability electronic load. Figure 7.24(c) shows the concept of an electronic load based on a MOSFET and a current sense resistor. (In practice, the electronic load itself may be an expensive piece of equipment.) A few parameters of the electronic load are critical to reliable measurements:

- Response speed of the control loop in the electronic load (T_r)
- Total resistance of the shunt resistor and the $R_{DS(on)}$ of the MOSFET (R)
- Total connector inductance and the inductance of the MOSFET (L)

- Turn-on and turn-off time of the MOSFET (T_f)
- The maximum transient loading level of the step loading (I)

Because the higher value of T_f or T_r determines the rise time of the electronic load (regardless of the load setting), this limits the transient capability of the electronic load. For example, if this parameter is 1 μs, the maximum possible slew rates can be 10 A/μs and 100 A/μs for 10 A and 100 A loads, respectively. Details on these aspects can be found in Romanchik [76] and Lee [77]. Such measurement may allow designers to tune the performance of a DC-DC converter [78]. In POL converters used for processor circuit blocks, transient requirements can be in the range of 300 A/μs or more. Testing of these high-*di/dt* converters is discussed in Callanan [79].

Another important requirement is the measurement of noise or PARD, and this needs either a scope (for peak-to-peak measurement) or a true RF RMS voltmeter for RMS measurement. Figure 7.24(d) shows test setups for PARD measurement. In this process, if the scope inputs are ground referenced, the use of differential probes is necessary.

As discussed in Chapter 5, feedback loop measurements require breaking and injecting a small AC signal, and this may not be easy because monolithic ICs used as controllers do not allow opening up the loop. In such cases, alternative methods may be utilized [80, 81]. Another recent development is that of automatic routines and software that allows the calibration of power supply design parameters. With component tolerances affecting a given design, one needs trimming of various voltage dividers and feedback loop components at different stages. Using PC-based calibration software, a calibration board containing ADCs, and switches, a power supply design can be automatically trimmed to refine the performance. Figure 7.25 shows such a setup for a power supply design based on a controller and a power management IC such as the ADM 1041 from Analog Devices, Inc. [79].

References

1. Smith, K. L. 1985. DC supplies from AC sources—part 3. *Electronics and Wireless World*, February, 24.
2. Smith, K. L. 1985. DC supplies from AC sources—part 4. *Electronics and Wireless World*, May, 67.
3. Williams, K. L. 1982. Mathematical theory of rectifier circuits with capacitor input filters. *Power Conversion International*, October, 42–44.
4. Billings, K. 1989. *Switchmode power supply handbook*. New York: McGraw-Hill.
5. Sadiq, N. 2010. Selecting fuses: Simple procedures to get the right over current protection for DC-DC converters. *Power Electronics Technology*, August, 10–15.
6. Deshayes, R., and J. F. De Palma. 1996. High power semiconductor protection requires the appropriate fuses. *PCIM*, October, 58.
7. Bell, B. 2000. Redesigned input filter limits DC-DC converter inrush current. *PCIM*, March, 66–68.
8. Zuk, P. 2006. Designing offline power supplies using power factor correction. *EDN*, September 1, 67.

9. Sandler, S., C. Hymowitz, and H. Eicher. 2006. Optimizing single stage power factor correction. *Power Electronics Technology*, March, 14.

10. Valentine, M. 2006. PFC controller ICs continue to advance. *Power Electronics Technology*, September, 62–66.

11. Kularatna, N. 1998. *Power electronics design handbook: Low-power components and applications*. London: Elsevier-Newnes.

12. Cuk, S. 2010. Bridgeless PFC converter achieves 98% efficiency, 0.999 power factor—part I. *Power Electronics Technology*, July, 10–18.

13. Cuk, S. 2010. Bridgeless PFC converter achieves 98% efficiency, 0.999 power factor—part II. *Power Electronics Technology*, August, 34–40.

14. Nagaraj, V. S. 2009. Interleaved PFC using digital control lowers cost and boosts efficiency. *Power Electronics Technology*, September, 28–31.

15. Perzow, J. 2002. Point-of-load regulation adds flexibility to set-top-box design. *EDN*, June 27, 73–78.

16. Dodson, S. 2010. Top-down approach simplifies AC-DC power supply selection. *Power Electronics Technology*, May, 29–31.

17. Balakrishnan, B. 2010. Next generation, monolithic off-line switcher improves performance, flexibility. *Power Electronics Technology*, April, 22–24.

18. Davis, S. 2009. Universal AC wall adapters will support future GSM cell phones. *Power Electronics Technology*, September, 39–41.

19. Cleaveland, T. 2007. Digitally managed power circuits. *EDN*, October 27, 59–66.

20. Marty, B. 1990. *Practical switching power supply design*. New York: Academic Press.

21. Visser, E. G., and A. Shpilman. 1991. New power ferrite operates from 1 to 3 MHz. *PCIM*, April, 17–23.

22. Bates, G. 1992. New transformer technologies improve switch mode power supplies. *PCIM*, July, 28–31.

23. Mohandes, B. 1994. Integrated PC board transformers improve high frequency PWM converter performance. *PCIM*, July, 8–17.

24. Magnetics Inc. 1997. Magnetic cores for switching power supplies. Publication-PS-02 2E.

25. Martin, A. W. 1983. Magnetic cores for switching power supplies. Application Note. Magnetics Inc.

26. Horgan, M. W. 1994. "Comparison of Magnetic Materials for Flyback Transformers. *PCIM*, July, 18–24.

27. Magnetics Inc. 1995. Inductor design in switching regulators. Technical Bulletin SR-1A.

28. Levin, G. 1998. Postregulation technique efficiently supplies multiple output voltages, *EDN*. January 15, 133.

29. Simopoulos, A. 2004. Linear postregulators for DC-DC converters. *Power Electronics Technology*, June, 28.

30. Mammano, B. 2001. Isolated power conversion: Making the case for secondary side control, *EDN*, June 7, 123.

31. Sherman, J., and M. M. Walters. 1996. Synchronous rectification: Improving the efficiency of buck converters. *EDN*, March 14, 111.

32. Moore, A. 1995. Synchronous rectification improves the efficiency of a dual output supply. *PCIM*, July, 8, 1995.

33. Moore, B. 1995. Synchronous rectification aids low voltage power supplies. EDN, April 27, 127.

34. Christiansen, B. 1998. Synchronous rectification. *PCIM*, August, 14.

35. How, C. H. 2000. Synchronous rectifiers for DC-DC converters: Better efficiency, but possible problems. *PCIM*, May, 74.

36. Khasiev, V. 2003. Moving forward converters to higher efficiency. *Power Electronics Technology*, May, 38.

37. Bindra, A. 2000. Optimized synchronous rectification drives up DC-DC converter efficiency. *Electronic Design*, January 24, 58.

38. Yee, H. P. 2000. Synchronous rectifier controller IC simplifies and improves isolated DC-DC converter designs. February, 44.

39. Bindra, A. 2003. Synchronous rectifier module eyes isolated power supplies. *Power Electronics Technology*, July, 56.

40. Elbanhawy, A. 2005. Buck converter losses under the microscope. *Power Electronics Technology*, February, 24.

41. Mappus, S. 2003. Predictive control maximizes synchronous converter efficiency. *Power Electronics Technology*, May, 44.

42. Dalal, D. 2005. Boosting power supply efficiency for desk top computers. *Power Electronics Technology*, February, 14.

43. Basso, C. 2004. Reducing wasted startup current in offline supplies. *Power Electronics Technology*, August, 54.

44. Fasching, M. 1996. Losses due to stray inductances in switch mode converters. *EPE Journal* 6:33.

45. Dadafshar, M. 2005. Inductor current sensing boosts regulator efficiency. *Power Electronics Technology*, April, 50.

46. Lin, F. C., and D. Y. Chen. 1994. Reduction of power supply EMI emission by switching frequency modulation. *IEEE Transactions on Power Electronics* 9:132.

47. Schindler, M. 2006. Proper layout and component selection control power supply EMI. *EDN*, October 26, 137.

48. Rogers, P. 1998. Board layout boosts power supply performance. *EDN*, November 5, 175.

49. Scolio, J. 2003. Basic switching regulator layout techniques. *EDN*, November 27, 79.

50. Davis, S. 2003. Spread spectrum ICs cut EMI. *Power Electronics Technology*, February, 70.

51. Armstrong, T. 2003. Alleviating noise concerns in handheld wireless products. *Power Electronics Technology*, October, 26.

52. Tse, K. K., S. H. Chung, S. Y. Hui, and H. O. So. 2000. Analysis and spectral characteristics of a spread spectrum technique for conducted EMI suppression. *IEEE Transactions on Power Electronics* 15:399.

53. Goodenough, F. 1997. Power supply designers trade off efficiency for noise with switcher ICs. *Electronic Design*, August 18, 40.

54. Wittenbreder, E. H. 2003. Power conversion synthesis—part I: Buck converter design. *Power Electronics Technology*, March, 26.

55. Wittenbreder, E. H. 2003. Power conversion synthesis—part II: Zero ripple converters. *Power Electronics Technology*, April, 54.

56. Wittenbreder, E. H. 2003. Power conversion synthesis—part III: Near zero emissions. *Power Electronics Technology*, May, 35.

57. Wittenbreder, E. H. 2003. Power converter synthesis—part IV: Near zero emissions. *Power Electronics Technology*, June, 4.

58. Barrow, J. 2006. Reducing ground bounce in DC/DC converter applications. *EDN*, July 6, 73.

59. Malley, K. 1995. Keep linear regulators in their safe zone. *EDN*, October 26, 137.

60. Walker, L., and G. Dashney. 1996. Spice generates thermal response models of a power semiconductor. *PCIM*, September, 57.

61. Stout, R. 2007. Linear superposition speeds thermal modeling: Part I. *Power Electronics Technology*, January, 20.

62. Stout, R. 2007. Linear superposition speeds thermal modeling: Part II. *Power Electronics Technology*, February, 28.

63. Solovitz, S. A., L. D. Stevanovic, and R. A. Beaupre. 2006. Micro-channel heat sinks take heat sinks to the next level. *Power Electronics Technology*, November, 14.

64. Remsburg, R. 2007. Nonlinear fin patterns keep cold plates cooler. *Power Electronics Technology*, February, 22.

65. Percia, G. 2002. Overvoltage protection circuit saves the day. *EDN*, November 14, 9.

66. Kularatna, N., A Variable shunt regulated power supply, 1978, *Electronic Engineering*, UK, page 21.

67. Malley, K. 2001. Linear regulator protection circuitry. Application Note SR005AN/D, rev. 1. ON Semiconductor, Phoenix, AZ.

68. Rose, D. 1997. Low voltage, high current switching supplies for microprocessors require over current protection. *PCIM*, July, 95.

69. Pelletier, W., and D. Goder. 1998. Current limiting defuses the DC/DC time bomb. *EDN*, April. http://www.edn.com/article/486976-04_09_98_Current_limiting_defuses_the_dc_dc_time_bomb.php

70. Malley, K. 2001. Linear regulator protection circuitry. Application Note SR005AN/D, rev. 1. ON Semiconductor, Phoenix, AZ.

71. Kularatna, N. 1980. Foldback limiter protects high current regulators. Electronics, USA, 98.

72. Huffman, B. 1993. Build reliable power supplies by limiting capacitor dissipation. *EDN*, March 13, 93.

73. Drew, J. 2004. Capacitor ripple current improvements. *Power Electronics Technology*, August, 33.

74. Forrester, S. 1995. International power converter requirements. *PCIM*, December, 8.

75. Pflueger, K. H. 1997. Power supply reliability: A practical improvement guide. *EDN*, March 3, 151.

76. Romanchik, D. 1995. Test verify power supply performance. *Test and Measurement World*, January, 43.

77. Lee, J. 2001. High slew rate electronic load checks new generation voltage regulator modules. *PCIM*, May, 52.

78. Venebale, D. 1995. New signal injection technique simplifies power supply feedback loop measurements. *PCIM*, September, 8.

79. Callanan, S. 2005. Testing high di/dt converters. *Power Electronics Technology*, March, 14.

80. Venable, D. 1997. Testing and stabilizing power supply feedback loops. *PCIM*, September, 8.

81. Daly, B. 2005. Automatic routine speeds power supply calibration. *Power Electronics Technology*, March, 36.

8

Rechargeable Batteries and Their Management

8.1 Introduction

The insatiable demand for smaller lightweight portable electronic equipment has dramatically increased the need for research on battery chemistries as well as the semiconductor components required for the optimal management of the batteries. Estimated worldwide sales for rechargeable batteries, also known as secondary batteries, was around U.S. $36 billion in 2008, and this is expected to grow toward U.S. $51 billion by 2013 [1]. As per market reports, U.S. demand for primary and secondary batteries will increase 2.5% annually to 16.8 billion in 2012, while primary batteries will account for 5.8 billion with a growth rate of 3% [2].

Mature rechargeable battery chemistries are (a) lead acid, (b) nickel cadmium, (c) nickel metal hydride, (d) lithium ion, (e) lithium polymer/lithium metal, and (f) lithium iron phosphate. With the demand growing from electric vehicles and portable consumer products, many organizations spend huge amounts of research money on new battery chemistries. Some significant areas include zinc-based chemistries and silicon as a material for improving some properties of batteries [3]. Higher energy density, superior cycle life, environmental friendliness, and safe operation are among the general design targets of secondary battery manufacturers. Primary or disposable batteries are a reasonably mature market and a product chemistry range, but still there are attempts to increase the energy density, reduce self-discharge rate (to increase the shelf life), and also to improve the usable temperature range. To complement these developments many semiconductor manufacturers continue to introduced new integrated circuit families for battery management.

This chapter describes the characteristics of battery families such as sealed lead acid, NiCd, NiMH, Li-based chemistries, rechargeable alkaline, and zinc-air together with modern techniques used in battery management ICs, without elaborating on the battery chemistries. Concepts and applications related to smart battery systems and related standards, and IEEE 1625/1725 standards for safety of battery-powered systems are also briefly introduced.

8.2 Battery Terminology

8.2.1 Capacity

Battery or cell capacity means an integral of current over a defined period of time.

$$Capacity = \int_0^t i \, dt$$

This equation applies to either charge or discharge; that is, capacity added or capacity removed from a battery or cell. The capacity of a battery or cell is measured in milliampere-hours (mAh) or ampere-hours (Ah).

Although the basic definition is simple, many different forms of capacity are used in the battery industry. The distinctions between them reflect differences in the conditions under which the capacity is measured.

8.2.1.1 Standard Capacity

Standard capacity measures the total capacity that a relatively new, but stabilized production cell or battery can store and discharge under a defined standard set of application conditions. It assumes that the cell or battery is fully formed, that it is charged at standard temperature at the specification rate, and that it is discharged at the same standard temperature at a specified standard discharge rate to a standard end-of-discharge voltage (EODV). The standard EODV is itself subject to variation depending on discharge rate as discussed.

8.2.1.2 Actual Capacity

When the application conditions differ from standard, the capacity of the cell or battery changes. The term *actual capacity* includes all nonstandard conditions that alter the amount of capacity the fully charged new cell or battery is capable of delivering when fully discharged to a standard EODV. Examples of such situations might include subjecting the cell or battery to a cold discharge or a high-rate discharge.

8.2.1.3 Available Capacity

That portion of actual capacity that can be delivered by the fully charged new cell or battery to some nonstandard EODV is called available capacity. Thus if the standard EODV is 1.6 V per cell, the available capacity to an EODV of 1.8 V per cell would be less than the actual capacity.

8.2.1.4 Rated Capacity

Rated capacity is defined as the minimum expected capacity when a new, but fully formed, cell is measured under standard conditions. This is the basis for C rate (defined later) and depends on the standard conditions used, which may vary depending on the manufacturer and the battery type.

8.2.1.5 Retained Capacity

If a battery is stored for a period of time following a full charge, some of its charge will dissipate. The capacity that remains that can be discharged is called retained capacity.

8.2.2 C Rate

The C rate is defined as the rate in amperes or milliamperes numerically equal to the capacity rating of the cell given in ampere-hours (Ah) or milliampere-hours (mAh). For example, a cell with a 1.2 Ah capacity has a C rate of 1.2 A. The C concept simplifies the discussion of charging for a broad range of cell sizes since the cells' responses to charging are similar if the C rate is the same. Normally an 8 Ah cell will respond to a 0.8 A (0.1 C) charge rate in the same manner that a 2 Ah cell will respond to a 0.2 A (also 0.1 C) charge rate. The rate at which current is drawn from a battery affects the amount of energy that can be obtained. At low discharge rates the actual capacity of a battery is greater than at high discharge rates. This relationship is shown in Figure 8.1.

8.2.3 Energy Density

Energy density of a cell is its energy divided by its weight or volume. When weight is used it is called the gravimetric energy density, and volumetric energy density when the volume is used. The terms *energy density* and *specific energy* are sometimes used for volumetric and gravimetric measures respectively.

8.2.4 Power Density of a Battery

Power density is the amount of power a battery can deliver per unit volume at specified state of charge (SOC), usually 20%. It is also called the volumetric power density and usually measured in watts per liter.

8.2.5 Cycle Life

Cycle life is a measure of a battery's ability to withstand repetitive deep discharging and recharging using the manufacturer's cyclic charging recommendations and still provide

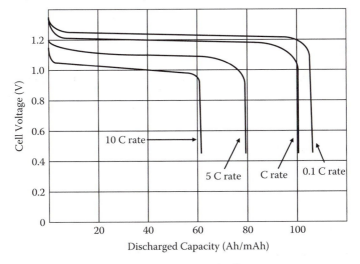

FIGURE 8.1 Capacity versus discharge rate of a typical cell.

minimum required capacity for the application. Cyclic discharge testing can be done at any of various rates and depths of discharge to simulate conditions in the application. It must be recognized, however, that cycle life has an inverse logarithmic relationship to depth of discharge.

8.2.6 Cyclic Energy Density

For purposes of comparison, a better measure of rechargeable battery characteristics is a composite characteristic that considers energy density over the service life of the battery. A composite characteristic, *cyclic energy density*, is defined as the product of energy density and cycle life at that energy density and has the dimensional units, watt-hour-cycles/kilogram (gravimetric) or watt-hour-cycles/liter (volumetric).

8.2.7 Self-Discharge Rate

Self-discharge rate is a measure of how long a battery can be stored and still provide minimum required capacity and be recharged to rated capacity. This commonly is measured by placing batteries on a shelf stand at room (or elevated) temperature and monitoring open-circuit voltage over time. Samples are discharged at periodic intervals to determine remaining capacity and recharged to determine rechargeability.

8.2.8 Charge Acceptance

Charge acceptance is the willingness of a battery or cell to accept charge. This is affected by cell temperature, charge rate, and the state of charge.

8.2.9 Depth of Discharge

Depth of discharge (DOD) is the capacity removed from a battery divided by its actual capacity, expressed as a percentage.

8.2.10 Voltage Plateau

Voltage plateau is the protracted period of very slowly declining voltage that extends from the initial voltage drop at the start of a discharge to the knee of the discharge curve. An example is shown for a lead-acid cell in Figure 8.2.

8.2.11 Midpoint Voltage

Midpoint voltage is the battery voltage when 50% of the actual capacity has been delivered (see Figure 8.2).

8.2.12 Overcharge

Overcharge is defined as continued charging of a cell after it has become fully charged. When a cell is not yet fully charged, the electrical energy of the charge current is

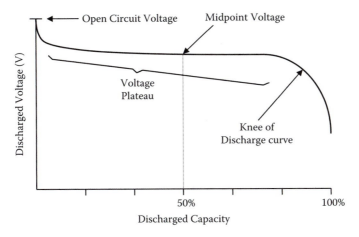

FIGURE 8.2 Midpoint voltage and the voltage plateau.

converted to chemical energy in the cell by the charging reactions. But when all of the available active material has been converted into the charged state, the energy available in the charging current goes to produce gases from the cell or to activate other nonuseful chemical reactions. Usually this results in a temperature rise in the cell.

8.2.13 State of Charge

State of charge (SOC) of a battery pack is defined as the percentage of the remaining charge (or remaining capacity [RC]) divided by its rated capacity (or the full charge capacity [FCC]). In other words it is based on the cumulative sum of the daily charge/discharge of the energy transfers, or simply the remaining percentage of the total ampere hour capacity of the battery.

8.3 Battery Technologies : An Overview

Most commonly available primary batteries (the disposable types) offer significantly greater energy density together with very low self-discharge rates. They are usually cheaper than the rechargeable batteries. Most common primary chemistries include (a) carbon-zinc (Leclanche cells), (b) alkaline-MnO_2, (c) lithium-MnO_2, (d) lithium sulfur dioxide, (e) lithium-iron disulphide, (f) lithium-thionyl chloride ($LiSOCl_2$), (g) silver-oxide, and (h) zinc-air. Mercury primary batteries have gradually gone out of the marketplace due to toxicity and environmental reasons. More details on these chemistries can be found in [4, 5].

Many types of rechargeable chemistry are used in electronic systems. Common rechargeable chemistries are based on variations of lead-acid, nickel-based, and lithium-based systems mainly, while limited zinc-based systems and rechargeable alkaline batteries are also available. The choice of a particular battery technology is limited by size, weight, cycle life, operating temperature range, and cost. A comparison of basic characteristics of major chemistries is depicted in Table 8.1.

TABLE 8.1 Secondary Battery Chemistry Characteristics

Parameter	Units/Conditions	Sealed Lead Acid	NiCd	NiMH	Li-Ion	Li-Polymer	Li iron Phosphate	Rechargeable Alkaline
Average Cell Voltage	Volts	2.0	1.2	1.2	3.6	1.8–3.0	3.2–3.3	1.5
Relative cost	NiCd = 1	0.6	1	1.5–2.0				0.5
Internal resistance		Low	Very low	Moderate	High	High	High	
Self-discharge	%/month	2%–4%	15%–25%	20%–25%	6%–10%	18%–20%	0.3%	
Cycle Life	Cycles to reach 80% of rated capacity	500–2000	500–1000	500–800	1000–1200		1500–2000	<25
Overcharge tolerance		High	Med	Low	Very Low			Med
Energy by volume (volumetric energy density)	Watt-hour/liter	70–110	100–150	200–350	200–330	230–410	200	220
Energy by weight (gravimetric energy density)	Watt-hour/kg	30–45	40–60	60–80	120–160	120–210	100	80

NiCd batteries were used to power most rechargeable consumer appliances before 1990. It is a mature, well-understood technology. However, cadmium is coming under increasing regulatory scrutiny (including mandatory recycling in some jurisdictions), and the maturity of NiCd technology also means most of the capacity and life cycle improvements have already been made. NiMH chemistry, which entered the market in 1989, offers incremental improvements in energy density by both weight and volume over NiCd. Li-ion is better still, offering over twice the watts per liter and per kilogram of NiCd batteries. As always, this higher performance comes at a higher price.

Using liquid lithium, Li-ion cells exhibit a much higher energy density than Ni-based chemistries, with a relatively higher terminal voltage than Ni-based cells, and is the commonly used energy source for most portable systems after 2000. The advantages of Li-ion chemistry, however, also come at the cost of greater electrical fragility. Li-ion particularly is more easily and extensively damaged by less than optimal battery management, so much so that fail-safe circuits are required to disconnect the cells from the load under overcurrent or overtemperature conditions. Therefore these electronic circuits are usually built into the battery pack.

In general, with an increase of the battery energy densities, and any possibilities of quantum jumps in new chemistries, one important concern is the safety and the transportability. Table 8.2 indicates the energy densities of selected fuels and batteries compared with high explosives with transportation capability. A value of 1000 Wh/kg energy density is essentially the practical maximum with its own oxidant for safe transportation [6]. A conventional battery in general is having its own oxidant, similar to this situation, and therefore the maximums shown in the Table 8.2 apply in development options. However, this condition does not apply to metal air batteries, such as zinc-air and the like, since the oxidant is supplied by oxygen from the air.

With Li-based chemistries with high energy densities entering the market, and with some fires and explosions occurring in the past, Li battery manufacturers are not

TABLE 8.2   Energy Densities of Selected Fuels and Batteries

Fuel or Battery	Volumetric Energy Density (Wh/kg)
Diesel fuel	10,000
Methanol	5,000
High explosive	1,000
Primary battery (maximum estimated around 1995)	500
Rechargeable battery (maximum estimated around 1995)	200
$LiSO_2$ battery	175
Alkaline battery	80
Ni-Cd battery	40
Li-ion battery	120

Source: From Hamlen, R. P., H. A. Christopher, and S. Gilman, *IEEE AES Systems*, June 1995.

supposed to sell them to uncertified battery pack assemblers (BPAs), and strict charge and discharge control practices are in place.

During the last decade, newer lithium-based chemistries entered the market. Important ones are the Li-polymer (sometimes known as Li-ion polymer) and Li-iron phosphate chemistries, which have better performance than the Li-ion [7, 8]. Note that Table 8.1 is only a guide; for more precise details, individual manufacturer data sheets need to be consulted.

Lead-acid batteries are most familiar in automobiles because they are the most economical chemistry for delivering large currents. Lead acid also has a long trickle life and therefore serves well for classic "floating" applications. While flooded lead-acid technology is popular for automobile and similar applications, sealed lead-acid batteries serve the electronic engineering environments. On the downside, lead acid has the least capacity by volume and weight.

8.4 Lead-Acid Batteries

Lead-acid batteries are the oldest and most widely used rechargeable battery systems today, mostly due to the automotive industry's attachment to the lead-acid chemistry. There are basically two types of lead-acid batteries, namely the (a) flooded lead-acid type and (b) sealed lead-acid type.

8.4.1 Flooded Lead-Acid Batteries

The flooded lead-acid battery today basically uses the design developed by Faure in 1881. It consists of a container with multiple plates immersed in a pool of dilute sulfuric acid. Recombination is minimal, so water is consumed through the battery life and the batteries can emit corrosive and explosive gases when experiencing overcharge. So-called "maintenance-free" forms of flooded batteries provide excess electrolyte to accommodate water loss through a normal life cycle. Most industrial applications for flooded batteries are found in motive power, engine starting, and large-system power backup. Today, other forms of battery have largely supplanted flooded batteries in small- and medium-capacity applications, but in larger sizes flooded lead-acid batteries continue to dominate. By far, the biggest application for flooded batteries is starting, lighting, and ignition (SLI) service on automobiles and trucks. Large flooded lead-acid batteries also provide motive power for equipment ranging from forklifts to submarines and provide emergency power backup for many electrical applications, most notably the telecommunications network.

8.4.2 Sealed Lead-Acid Batteries

Sealed lead-acid batteries first appeared in commercial use in the early 1970s. Although the governing reactions of the sealed cell are the same as other forms of lead-acid batteries, the key difference is the recombination process that occurs in the sealed cell as it reaches full charge. In conventional flooded lead-acid systems, the excess energy from overcharge goes into electrolysis of water in the electrolyte with the resulting gases being vented. This occurs because the excess electrolyte prevents the gases from diffusing to the opposite plate and possibly recombining. Thus electrolyte is lost on overcharge with the resulting need for replenishment. The sealed lead-acid cell, like the sealed nickel-cadmium, uses recombination to reduce or eliminate this electrolyte loss.

Sealed lead-acid batteries for electronics applications are somewhat different from the type commonly found in the automobile. There are two types of sealed lead-acid batteries: the original gelled electrolyte and retained (or absorbed) system. The gelled electrolyte system is obtained by blending silica gel with an electrolyte, causing it to set up in gelatin form. The retained system employs a fine glass fiber separator to absorb and retain liquid electrolyte. Sometimes the retained system is named absorbed glass mat (AGM). AGM is also known in the industry as "starved design." Starved refers to the absorption limits of the glass separator creating a limitation to the AGM design relating to diffusion properties of the separator. In certain cases the AGM battery must be racked and trayed in a specific position for optimum performance. Both these types, gelled and AGM, are called valve-regulated lead-acid (VRLA) systems. Today, sealed lead-acid cells are operating effectively in many markets previously closed to lead-acid batteries. For a detailed account of lead-acid cells, [9–12] are suggested.

Meanwhile, some manufacturers have introduced special versions of sealed lead-acid batteries with higher volumetric energy density [12,13].

8.4.2.1 Discharge Performance of Sealed Lead-Acid Cells

The general shape of the discharge curve, voltage as a function of capacity (if the current is uniform) is shown in Figure 8.2. The discharge voltage of the starved-electrolyte sealed lead-acid battery typically remains relatively constant until most of its capacity is discharged. It then drops off sharply. The flatness and the length of the voltage plateau relative to the length of the discharge are major features of sealed lead-acid cells and batteries. The point at which the voltage leaves the plateau and begins to decline rapidly is often identified as the knee of the curve.

Starved-electrolyte sealed lead-acid batteries may be discharged over a wide range of temperatures. They maintain adequate performance in cold environments and may produce actual capacities higher than their standard capacity when used in hot environments. Figure 8.3 indicates the relationships between capacity and cell temperature. Actual capacity is expressed as a percentage of rated capacity as measured at 23°C.

8.4.2.2 Capacity during Battery Life

The initial actual capacity of sealed lead-acid batteries is almost always lower than the battery's rated or standard capacity. However, during the battery's early life the actual capacity increases until it reaches a stabilized value, which is usually above the rated capacity. The number of charge-discharge cycles or length of time on float charge required to develop a battery's capacity depends on the specific regime employed. Alternatively, if the battery is on charge at 0.1 C, it is usually stabilized after receiving 300% (of rated capacity) overcharge. The process may be accelerated by charging and discharging at low rates.

Under normal operating conditions the battery's capacity will remain at or near its stabilized value for most of its useful life. Batteries will then begin to suffer some capacity

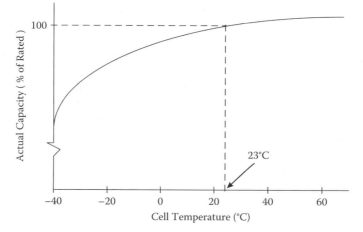

FIGURE 8.3   Typical discharge capacity as a function of cell temperature.

degradation due to their age and the duty to which they have been subjected. This permanent loss usually increases slowly with age until the capacity drops below 80% of its rated capacity, which is often defined as the end of useful battery life. Figure 8.4 shows a representation of the capacity variation with cycle life that can be expected from sealed lead-acid batteries.

8.4.2.3 Effect of Pulse Discharge on Capacity

In some applications, the battery is not called upon to deliver a current continuously. Rather, energy is drawn from the battery in pulses. By allowing the battery to "rest" between these pulses, the total capacity available from the battery is increased. Figure 8.5 represents typical curves representing the voltage delivered as a function of discharged capacity for pulsed and constant discharge at the same rate.

For the pulsed curve, the upper row of dots represents the open-circuit voltage, and the lower sawtooth represents the voltages during the periods when the load is connected.

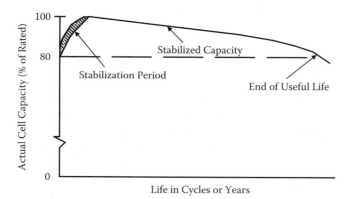

FIGURE 8.4   Typical cell capacity during its life.

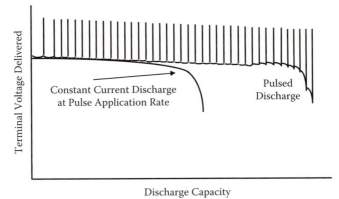

FIGURE 8.5 Typical pulsed discharge curve.

The use of discharged capacity as the abscissa eliminates the rest periods and shows only the periods of useful discharge.

8.4.3 Charging

In general, experience with sealed lead-acid chemistry indicates that application problems are more likely to be caused by undercharging than by overcharging. Since the starved-electrolyte cell is relatively resistant to damage from overcharge, designers may want to ensure that the batteries are fully charged, even at the expense of some degree of overcharge. Obviously, excessive overcharge, in either magnitude or duration, should still be avoided.

The charge acceptance of sealed lead-acid batteries in most situations is quite high, typically greater than 90%. A 90% charge acceptance means that for every amp-hour of charge introduced into the cell, the cell will be able to deliver 0.9 Ah to a load. Charge acceptance is affected by a number of factors including cell temperature, charge rate, cell state of charge, age of the cell, and the method of charging.

The state of charge of the cell will dictate to some extent the efficiency with which the cell will accept charge. When the cell is fully discharged, the charge acceptance initially is quite low. As the cell becomes only slightly charged, it accepts current more readily and the charge acceptance jumps quickly, approaching 98% in some situations. The charge acceptance stays at a high level until the cell approaches full charge.

As mentioned earlier, when the cell becomes fully charged some of the electrical energy goes into generating gas, which represents a loss in charge acceptance. When the cell is fully charged, essentially all the charging energy goes to generate gas except for the very small current that makes up for the internal losses that otherwise would be manifested as self-discharge. A generalized curve representing these phenomena is shown in Figure 8.6(a). As with most chemical reactions, temperature has a positive effect upon the charging reactions in the sealed lead-acid cell. Charging at higher temperatures is more efficient than it is at lower temperatures, all other parameters being equal, as shown in Figure 8.6(b).

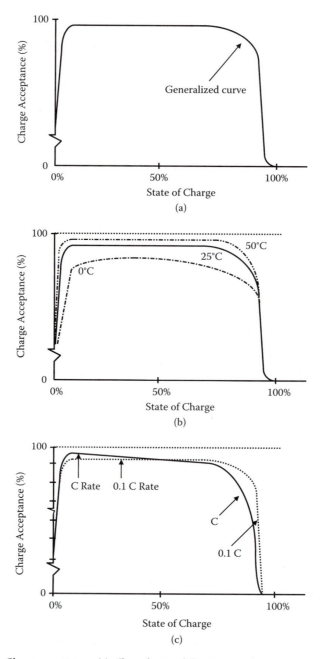

FIGURE 8.6 Charge acceptance: (a) effect of state of charge upon charge acceptance; (b) charge acceptance at various temperatures; (c) charge acceptance at various charge rates.

The starved-electrolyte sealed lead-acid cell charges very efficiently at most charging rates. The cell can accept charge at accelerated rates (up to the C rate) as long as the state of charge is not so high that excessive gassing occurs. And the cell can be charged at low rates with excellent charge acceptance.

Figure 8.6(c) shows the generalized curve of charge acceptance now further defined by charging rates. When examining these curves, one can see that at high states of charge, low charge rates provide better charge acceptance.

8.5 Nickel Cadmium (NiCd) Batteries

Sealed NiCd batteries are well suited to applications where a self-contained power source increases the versatility or reliability of the end product. Among the significant advantages of NiCd families are: higher energy density and discharge rates, fast recharge capability, long operating and storage life. This used to place NiCd families at the top of usage in the portable products. In addition, the NiCds are capable of operating over a wide temperature range and in any orientation with reasonable continuous overcharge capability.

8.5.1 Construction

NiCd secondary batteries operate at 1.2 V, using nickel oxyhydroxide for the active material in the positive electrode. The active material of the negative electrode consists of cadmium, while an alkali solution acts as the electrolyte. In NiCd batteries, a reaction at the negative electrode consumes the oxygen gas that generates at the positive electrode during overcharge. The design prevents the negative electrode from generating hydrogen gas, permitting a sealed structure. NiCd batteries mainly adopt cylindrical or prismatic type configurations. Because NiCd batteries contain cadmium, an environmentally hazardous substance, their disposal has become controversial. This has spurred research into other alternative chemistries.

8.5.2 Discharge Characteristics

The discharge voltage of a sealed NiCd cell typically remains relatively constant until most of its capacity is discharged. It then drops off rather sharply. The flatness and length of the voltage plateau relative to the length of discharge are major features of sealed NiCd cells and batteries. The discharge curve, when scaled by considering the effects of all the application variables, provides a complete description of the output of a battery. Differences in design, internal construction, and conditions of actual use of cell affect the performance characteristics. For example, Figure 8.7 illustrates the typical effect of discharge rate.

8.5.3 Charge Characteristics

Nickel-based batteries are easily charged by applying a controlled current. The charging current can be pure direct current (DC) or it may contain a significant ripple component such as half-wave or full-wave rectified current.

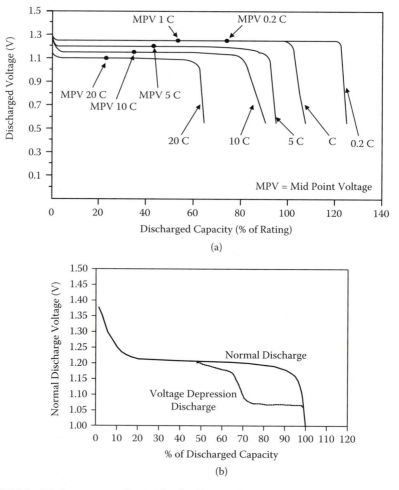

FIGURE 8.7 Discharge curves for NiCd cells: (a) typical curves at 23°C; (b) voltage depression effect.

This section on charging sealed nickel-cadmium batteries refers to charging rates as multiples (or fractions) of the C rate. These C rate charging currents can also be categorized in descriptive terms, such as standard-charge, quick-charge, fast-charge, or trickle-charge as shown in Table 8.3. When a nickel-cadmium battery is charged, not all of the energy input is converting the active material to a usable (chargeable) form. Charge energy also goes to converting active material into an unusable form, generating gas, or is lost in parasitic side reactions.

Figure 8.8 shows the charge acceptance of NiCd cells. The ideal cell, with no charge acceptance losses, would be 100% efficient. All the charge delivered to the cell could be retrieved on discharge. But nickel-cadmium cells typically accept charge at different levels of efficiency depending upon the state of charge of the cell, as shown by the bottom curve of Figure 8.8.

TABLE 8.3   Definition of Rates for Charging an NiCd Cell

| Method of Charging | Charge Rate | | Charge Control |
	Multiples of C-Rate	Recharge Time (Hours)	
Standard	0.05	36–48	Not required
	0.1	16–20	
Quick	0.2	7–9	Not required
	0.25	5–7	
	0.33	4–5	
Fast	1	1.2	Required
	2	0.6	
	4	0.3	
Trickle	0.02–0.1	Used for maintaining charge of a fully charged battery	

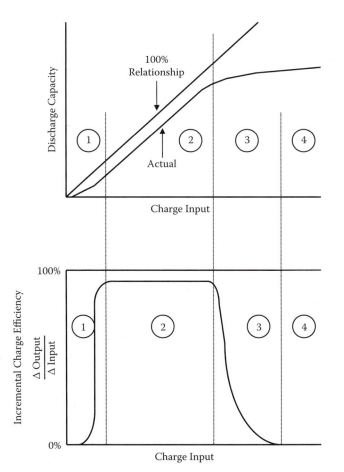

FIGURE 8.8   Charge acceptance of a sealed NiCd cell at 0.1°C and 23°C.

Figure 8.8 describes this performance for successive types of charging behavior (zones 1, 2, 3, and 4). Each zone reflects a distinct set of chemical mechanisms responsible for loss of charge input energy.

In zone 1, a significant portion of the charge input converts some of the active material mass into a nonusable form; that is, charged material not readily accessible during medium- or high-rate discharges, particularly in the first few cycles. In zone 2, the charging efficiency is only slightly less than 100%; small amounts of internal gassing and parasitic side reactions are all that prevent the charge from being totally efficient. Zone 3 is the transition region.

As the cell approaches full charge, the current input shifts from charging positive active material to generating oxygen gas. In the overcharge region, zone 4, all of the current coming into the cell generates gas. In this zone the charging efficiency is practically none.

The boundaries between zones 1, 2, 3, and 4 are indistinct and quite variable depending upon cell temperature, cell construction, and charge rate. The level of charge acceptance in zones 1, 2, and 3 is also influenced by cell temperature and charge rate. For details, [9] is suggested.

8.5.4 Voltage Depression Effect

When some NiCd batteries are subjected to numerous partial discharge cycles and overcharging, cell voltage decreases below 1.05 V/cell before 80% of the capacity is consumed. This is called the *voltage depression effect*, and the resultant lower voltage may be below the minimum voltage required for proper system operation, giving the impression that the battery has worn out. See Figure 8.7(b). When cells are exposed to overcharge, particularly at higher temperatures, this is quite common, and the voltage may be about 150 mV lower than the normal cell voltage. Voltage depression is an electrically reversible condition and disappears when the cell is completely discharged and charged. This process is sometimes called conditioning. This effect is sometimes erroneously called the "memory effect."

8.6 Nickel Metal Hydride Batteries

While NiCd battery performance was rapidly advancing after 1980, nickel metal hydride (NiMH) cells introduced in early 1990s have shown nearly a 170% increase in energy density. By 2000 volumetric energy density of NiMH cells had increased over 300 Wh/liter [13]. These extensions of the Ni-based chemistries have become popular with product applications such as notebook computers, cellular phones, and the like. The first practical NiMH batteries entered the market in the early 1990s. In these cells the environmentally unfriendly cadmium negative electrode was replaced by an alloy that could reversibly absorb and desorb hydrogen. This chemistry was the choice for electric vehicles till about 2005 [14, 15].

8.6.1 Construction

In many ways, nickel-metal hydride (NiMH) batteries are the same as NiCd types, but they use nickel for the positive electrode, and a recently developed material known as a hydrogen-absorbing alloy for the negative electrode. When an NiMH cell is charged, hydrogen generated by reaction with cell electrolyte is stored in the metal alloy (M) in the negative electrode. Meanwhile at positive electrode, which consists of nickel oxyhydroxide loaded into a nickel foam substrate, a hydrogen ion is ejected and nickel is oxidized [15]. With an operating voltage of 1.2 V, they provide high-capacity, large-energy-density characteristics comparable to those of NiCd models.

8.6.2 A Comparison between NiCd and NiMH Batteries

The NiCd cell is more tolerant of fast recharging and overcharging than NiMH cells. NiCd cells hold their charge longer than do NiMH cells. NiCd cells will withstand between 500 and 2000 charge/discharge cycles compared to about 500–800 cycles for NiMH cells. Further, NiCd cells will withstand a wider temperature range than NiMH cells. On the other hand, NiMH cells do not exhibit the notorious "memory effect" that NiCd cells sometimes do. As with any new technology at the time of entering the market, NiMH's prices were higher than those of NiCds [16,17].

Voltage profile of NiMH cells during discharging is very similar to that of the NiCd cells. NiMH cells' open-circuit voltage is 1.3–1.4 V. At moderate discharge rates, NiMH cells' output voltage is 1.2 V. Both NiCd and NiMH cells have relatively constant output voltage during their useful service. Figure 8.9 is a typical graph from a battery company comparing the output voltage of 700 mAh NiCd and 1100 mAh NiMH AA cells while under load. Note that the NiMH cell's greater capacity results in approximately 50% longer service life.

NiMH chemistry advances helped introduce batteries suitable for electric vehicles and aerospace applications. These bipolar NiMH batteries were meeting the volumetric and gravimetric energy density needs of these applications during the mid-1990s to mid-2005 [18, 19].

Figure 8.10 is another typical battery-company graph showing that NiCd and NiMH batteries and cells charge in similar fashions as well. However, the little bumps at the end

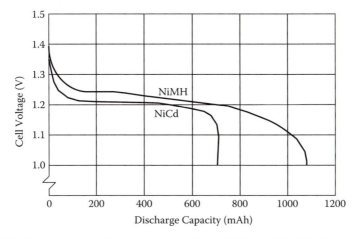

FIGURE 8.9 Comparison of discharge characteristics of NiCd and NiMH batteries.

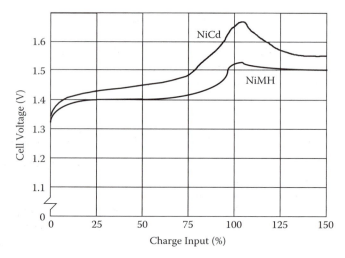

FIGURE 8.10   Battery voltage at the achievement of 100% charge.

of the two cells' charge curves bear closer examination. You will always see these nega-tive excursions even though absolute cell voltages vary significantly with temperature.

The negative excursions signal a fully charged cell more or less independently of tem-perature, a useful quirk that sophisticated battery chargers exploit. Note that the NiCd cell's negative-going voltage excursion after reaching full charge is more pronounced than the NiMH cell's.

8.7 Lithium-Based Rechargeable Batteries

Toward the latter part of the 1990s, the demand for portable systems was increasing at a dramatic rate. Meeting these goals required improvements in battery technolo-gies beyond traditional NiCd and NiMH systems. Newer lithium-based rechargeable battery systems have overcome the safety and environmental obstacles posted by early efforts and are, in general, the most efficient rechargeable battery packs avail-able. With an energy density by weight about twice that of nickel-based chemistries (Table 8.1), Li-ion batteries can deliver lighter-weight packs of acceptable capac-ity. Li-Ion also has about three times the cell voltage of NiCd and NiMH batter-ies; therefore fewer cells are needed for a given voltage requirement. Li-ion batteries have become the choice in recent versions of notebook PCs and many other portable systems because of their high energy density, declining costs, and readily available management circuits.

Rechargeable lithium cells come in several different chemistries, namely (a) lithium-ion, (b) lithium-polymer, (c) lithium-metal, and (d) lithium-iron phosphate. Most of these chemistries came into commercial use after 1992 onwards [20–23]. Figure 8.11(a) indicates the progress of Li-ion chemistry based on the commonly used 18650-type cells, which is the most mature of all the lithium cells. Figure 8.11(b) indicates the rela-tive capabilities of the four common rechargeable chemistries.

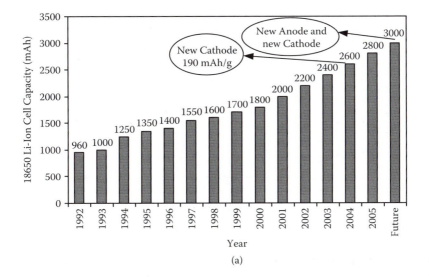

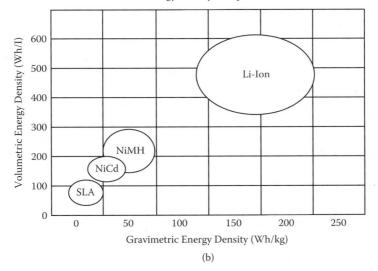

FIGURE 8.11 Progress of the Li-ion cells and comparison with other chemistries: (a) milliam-pere-hour capacity increase over the period from 1992 for 18650 cells; (b) comparison of Li-ion with other chemistries. (Part [a[courtesy of Morrison, David, 2006. *Power Electronics Technology*, January 2006, 50–52.)

The first noticeable difference between Li-based chemistries and nickel-based chemistries is the higher internal impedance of the lithium-based batteries. Figure 8.12 shows this by graphing the actual discharge capacity of an Li-ion cell at different discharge currents compared to an NiCd cell. At a 2A discharge rate (2 C), less than 80% of the rated capacity is available for Li-ion compared to nearly 95% of the rated capacity for NiCd [24].

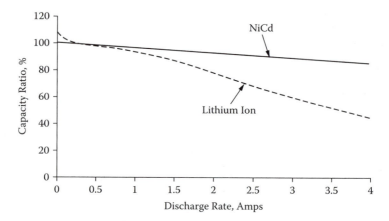

FIGURE 8.12 Li-Ion and NiCd capacity versus discharge current.

For systems with discharge currents greater than 1 A, the capacity realized from the Li-ion battery may be less than expected. Parallel battery stack configurations are often used in Li-ion battery packs to help reduce the severity of this problem. Due to the nature of the lithium chemistries, lithium batteries cannot tolerate overcharge and overdischarge.

Given the comparison of energy densities of fuel, batteries, and explosives (Table 8.2), the most important factor in using lithium chemistries in a portable consumer product is safety. Fortunately, battery safety is comprehensively addressed by cell and battery pack assemblers (BPAs) in concert with the semiconductor manufacturers. For details, [25] is suggested. Commercially available Li-ion packs have an internal protection circuit that limits the cell voltage during charge to between 4.1 and 4.3 V per cell, depending on the manufacturer. Voltages higher than this rating could permanently damage the cell. A discharge limit of between 2.0 and 3.0 V (depending on the manufacturer) is necessary to avoid reducing the cycle life of the battery and damaging the battery.

8.7.1 Construction

The anode, or negative electrode, in an Li-ion cell is comprised of a material capable of acting as a reversible Li-ion reservoir. This material is usually a form of carbon, such as coke, graphite, or pyrolytic carbon. The cathode, or positive electrode, is also a material that can act as a reversible lithium ion reservoir. Due to lithium ions shuttling back and forth between these two reservoirs, these batteries are sometimes called the rocking-chair cell [26].

Preferred cathode materials are currently $LiCoO_2$, $LiNiO_2$, or $LiMn_2O_4$ because of their high oxidation potentials of about 4 V versus lithium metal. Commercially available Li-ion cells use a liquid electrolyte made up of mixtures that are predominantly organic carbonates containing one or more dissolved lithium salts [27]. While $LiCoO_2$ was the preferred cathode material at the early stages of development, in newer systems $LiMn_2O_4$ cathodes are used. Chemical process details are summarized in [27]. In the latest generation of Li-ion cells, a new cathode material known as Nickel oxide-based New Platform (NNP) is used by Panasonic batteries [21], extending the commonly used 18650 series to 2.9 Ah.

In lithium polymer batteries, the liquid electrolyte is replaced with a polymer in gel or solid form. The polymer electrolyte provides the required electrode stack pressure, so the metal can is no longer required and it becomes possible to package the cells easily. A laminate of aluminium foil and plastic, the pouch occupies less space and weighs less than the metal can. For these reasons lithium polymer cells came in compact prismatic cell formats around early 2000 [28]. Also, they are considered safer than liquid Li-ion, as they don't leak when they are punctured. As a result simplifications are possible for in-pack protection circuits [29].

LiFePO$_4$ batteries, also known as LFP battery systems, were introduced around the late 1990s, and recently they have gone into mass production, due to demand from EVs, hybrid EVs, electric bikes, and power tools. In these batteries' cathode material, LiFePO$_4$ is environmentally benign, inexpensive, and relatively abundant [30].

8.7.2 Charge and Discharge Characteristics

Today, the predominant Li-ion technologies use coke or graphite for an anode material. Figure 8.13 illustrates the differences in the two types of cells during discharge. The graphite anode discharge voltage is relatively flat during a majority of the discharge cycle, while the coke anode discharge voltage is more sloped [28].

The energy available from the graphite anode cell is higher for a given capacity due to the higher average discharge voltage. This may be useful in systems that need the maximum watt-hour capacity for a given battery size. Also, the charge and discharge cut-off voltages between the two Li-ion systems vary among manufacturers.

Figure 8.14 shows the typical charge profile for Li-ion batteries. The charge cycle begins with a constant-current limit, transitioning to a constant-voltage limit, typically specified between 4.1 V and 4.3 V ±1%, depending on manufacturer's recommendation.

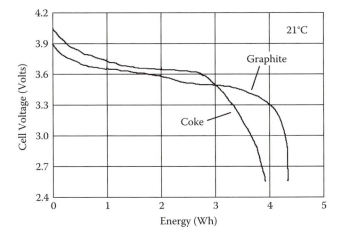

FIGURE 8.13 Li-ion discharge profile for different electrodes. (Adapted from Juzkow, M. W., and C. St. Louis, 1996. *Portable by Design Conference Proceedings*, 1996, 13–22. Source Moli Energy Limited, USA).

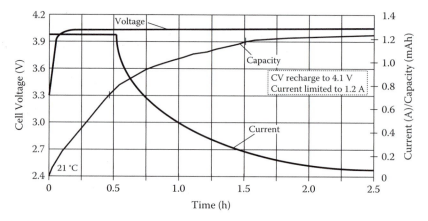

FIGURE 8.14   Li-ion charge profile at constant potential charging at 4.1 V and current limited to 1.2 A. (Courtesy Moli Energy Ltd, USA.)

This allows maximum charge capacity without cell damage. Charging to a lower voltage limit does not damage the cell, but the discharge capacity will be reduced. A 100 mV difference could change the discharge capacity by more than 7%.

8.8 Reusable Alkaline Batteries

Alkaline technology has been used in primary batteries for several years. With the development of the reusable alkaline manganese technology, secondary alkaline cells have quickly made their way into many consumer and industrial applications. In many applications, reusable alkaline cells can be recharged from 75 to over 500 times, and initially have three times the capacity of a fully charged NiCd battery. These cells do not compete with NiCds in high-power applications, however.

Intensive research and development activities carried out at Battery Technologies Inc. (BTI), Canada, and at the Technical University in Graz, Austria, in the late 1980s and early 1990s resulted in the successful commercialization of the rechargeable alkaline manganese dioxide zinc (RAM™) system. BTI has chosen to sell licenses and production equipment, where necessary, for the manufacturing and worldwide marketing rights of its proprietary RAM technology. For example, Rayovac Corporation, one of the licensees, launched its line of reusable alkaline products under the name RENEWAL™ in the United States, Pure Energy Battery Corporation in Canada (PURE ENERGY™), and Young Poong Corporation in South Korea (ALCAVA™). For details, [31–33] are suggested.

The chemistry behind the reusable approach depends on limiting the zinc anode to prevent overdischarge of the MnO_2 cathode. Additives are also incorporated to control hydrogen generation and other adverse effects on charge. Rated cycle life is around 25 cycles to 50% of initial capacity. Longer cycle life is possible depending on drain rate and depth of discharge. To take advantage of the reusable alkaline cell and increase its life, a special "smart charger" is required.

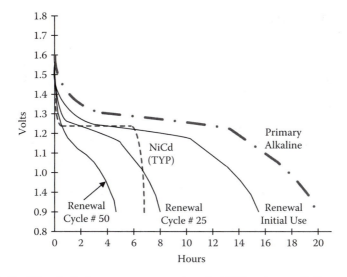

FIGURE 8.15 100 mA discharge curve comparison for NiCd, primary alkaline, and reusable alkaline. (Courtesy Benchmarq Microelectronics/BTI Technologies.)

8.8.1 Cumulative Capacity

Using reusable alkaline cells can drive down the total battery cost to the consumer. This cost saving can be determined by looking at the cumulative capacity of a reusable cell versus the one-time use of a primary alkaline cell. Figure 8.15 illustrates the capacity of AA cells being discharged down to 0.9 V at 100 mA. It shows that although the initial use of the reusable alkaline is almost that of primary alkaline, the reusable one can be recharged for continued use. Table 8.4 shows the increase in cumulative capacity by limiting the DOD and achieving more cycles.

Overcharging also affects the cycle life of reusable alkaline. Reusable alkaline is not tolerant of overcharge and high continuous-charge currents, and may be damaged if high current is forced into them after they have reached a partially recharged state. Proper charging schemes should be used to prevent an overcharged condition.

TABLE 8.4 Capacity of "AA" Cells at Various Depths of Discharge (DOD) (Values in mAh)

Condition	125 mA to 0.9 V		
	100% DOD	30% DOD	10% DOD
Cycle 1	1,500	450	150
Cycle 50	400		
Cumulative 50	33,000	22,000	7,000
Cumulative 100		44,000	15,000
Cumulative 500			73,000

Source: Nossaman, P., and J. Parvereshi. *Proceedings of HFPC Conference (USA)*, 1995.

8.9 Zn-Air Batteries

Primary Zn-air batteries have been in existence for over 50 years with applications such as hearing aids and harbor buoys. The light weight and high energy content in Zn-air technology has promoted research on Zn-air rechargeable chemistry by companies such as AER Energy Resources, USA, in the late 1990s, and around 2009 by companies such as RWE Innogy (Germany) and ReVolt Technology AS [34]. The focus is on electric vehicles and portable appliances. Another application is solar-powered rural telecom systems.

Rechargeable Zn-air technology is an air-breathing technology where the oxygen in ambient air is used to convert zinc into zinc oxide in a reversible process. Cells use air-breathing carbon cathode to introduce oxygen from air into potassium-hydroxide electrolyte. Cathode is multilayered with a hydrophilic layer, and the anode is comprised of metallic zinc.

The characteristic voltage of zinc-air systems is nominally around 1 V. For example, during discharge, they will operate within a voltage between 1.2 and 0.75 V. The current and power capability of the system is proportional to the surface area of the air-breathing cathode. For more current and power, a larger-surface-area cell is required. For less current and power, a smaller cell may be used. Compared to other rechargeable chemistries, Zn-air needs an air manager for an intake and exhaust of air to allow the chemical process.

Figure 8.16 compares the performance of recent developments on zinc-air chemistry [34] with other common rechargeable chemistries. This clearly indicates that Zn-air

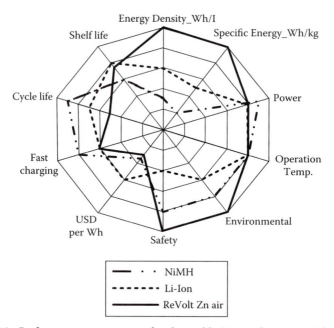

FIGURE 8.16 Performance comparison of rechargeable Zn-air chemistry with other chemistries. (Adapted from Green Car Congress, January 16, 2009, http://www.greencarcongress.com/2009/09/revolt-20090901.html.)

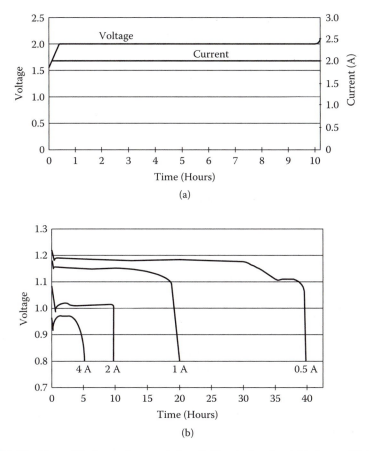

FIGURE 8.17   Charge/discharge characteristics of a Zn-air chemistry: (a) charge; (b) discharge. (Courtesy AER Energy Technologies.)

batteries require less weight and volume. Discharge and charge characteristics of Zn-air batteries by AER Energy Resources are shown in Figure 8.17.

The cells exhibit a flat voltage profile over the discharge cycle. Typical charge voltage is 2 V per cell using a constant voltage/current taper approach. Life cycle varies between 50 and 400, depending on the depth of discharge. Cost per watt-hour is apparently the lowest compared to Ni- and Li-based chemistries. For details, [35] is suggested.

8.10 Battery Management

Two decades ago, battery management consisted of having a reliable, fast, and safe charging methodology to be selected for a battery bank, together with the monitoring facilities for detecting the discharged condition of the battery pack. With modern battery technologies emerging, the demands from the cost-sensitive portable product market, as well as the medium-power range products such as UPS and telecom power units, attributes of a modern battery management system may include:

- Battery modeling
- Battery-charging methods and charge control
- Determination of state of charge (SoC)/end of discharge
- Gas gauging
- Monitoring battery health issues
- Communication with the host system/or power management subsystems
- Battery safety

The following sections provide some concepts and techniques related to modeling of batteries for best performance extraction, managing a given chemistry for best run time and the longest life, the safety of battery packs, and prognostics for health management of battery packs.

8.10.1 Modeling of a Cell to Reflect Its Electrochemistry

As engineers we tend to work with device data sheets, leaving the device design and its behavior to be decided by the device expert, and a team of electrochemists to give us a mass-producible cell. In an oversimplified circuit equivalent, we tend to simplify a battery to a voltage source (with open circuit voltage) and a series resistance. However, voltage profiles of batteries during discharge and charge do not exactly reflect the accuracy of this simplified assumption. Figure 8.18 indicates the chemical process within a battery as it discharges its stored chemical energy to a load.

As we have seen in Figure 8.1, battery capacity decreases as the discharge rate increases. Also, as per Figure 8.5 we see that a cell can have high pulse discharge, if we allow resting between pulses. These kinds of behavior are simply explained by Figure 8.18. Figure 8.18(a) shows the case of a fully charged cell with maximum concentration of active species. When a load is connected, the load current causes this active species to be consumed at the electrode surface, Figure 8.18(b), and replenished by the diffusion process from the bulk of the electrolyte. A higher load current causes a higher concentration gradient, and thus a lower concentration of active species near the electrode surface. When this concentration falls below a particular threshold, a voltage cutoff occurs. However, the unused charge is not physically lost but is simply unavailable at the electrode surface due to lag between reaction and diffusion rates. Decreasing the discharge rate effectively reduces this lag as well as the concentration gradient in the vicinity of the electrode. When the recovery occurs (Figure 8.18[c]) a high current discharge is possible until the active species are nearly fully consumed as in Figure 8.18(d). For more details and related modeling of batteries, [36] is suggested.

This simplified discussion on the process of charge transfer, diffusion process, etc., leads to curiosity toward more complex chemical behavior of a cell, which is the subject of an electro-chemist. Such complex chemical processes inside a cell lead us to develop a more accurate circuit model suitable for modeling a battery based on simplified engineering information extracted from a data sheet. Based on recent research [37–39],

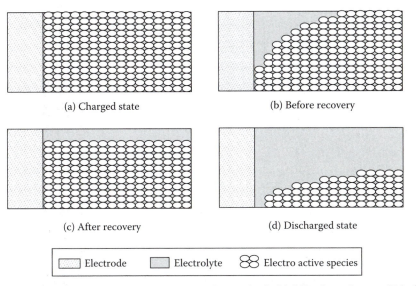

(a) Charged state

(b) Before recovery

(c) After recovery

(d) Discharged state

Electrode   Electrolyte   Electro active species

FIGURE 8.18 Behavior of a symmetric electrochemical cell: (a) fully charged state; (b) before recovery; (c) after recovery; (d) discharged state. (From Rao, R., S. Vrudhala, and D. N. Rakhmatov, 2003. *IEEE Computer* 36, 12, 2003, 77–87. With permission.)

NiMH or Li-ion cells can be modeled using modern finite element methods [39] for temperature behavior, while automated test systems can be used to extract information for simplified models [38] to reflect the cell behavior. For much more simplified requirements, extracting data from cell manufacturer data sheets is also a possibility [37].

8.10.2 Battery Equivalent Circuits

Battery equivalent circuits are useful for modeling the behavior of a battery pack, to predict its short-term behavior; its long-term performance with regard to discharge rates, temperature changes, and life cycle prediction; as well as its pulse discharge performance. There are many different models available, ranging from a simple voltage source with internal resistance to equivalent circuits modeling the effects of charge state, temperature, pressure, aging, and many more aspects of the battery pack.

To predict the performance of a battery in the short term, simplified equivalents such as are shown in Figure 8.19(a) are possible. This model is valid for time scales ranging from milliseconds up to a few seconds at a constant state of charge/discharge and temperature.

In Figure 8.19(a):

- U_B and I_B are the terminal voltage and current.
- U_o is the open-circuit voltage.
- R_i is the constant part of the internal resistance (connector, electrodes, electrolyte).
- R_D and C_D are the effects on the surface of the electrode (double-layer capacitance effect).
- R_k and C_k reflect the effect of diffusion process in the electrolyte.

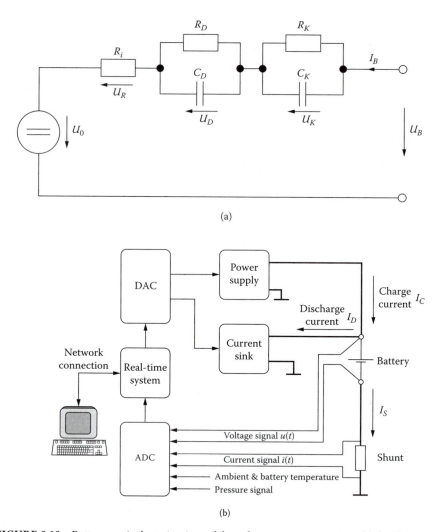

(a)

(b)

FIGURE 8.19 Battery equivalent circuit models and parameter extraction: (a) for short-term behavior and thermal performance; (b) test setup for extracting parameters of model. (Reproduced from Schweighofer, B., K. M. Raab, and G. Brasseur, *IEEE Transactions on Instrumentation and Measurement* 52, 4, 2003, 1087–91. With permission.)

To deal with longer prediction times or different temperature situations, the values of components have to be adapted according the changes inside the battery [40–42]. Reference [38] discusses the details of a test setup that can be used to extract the parameters for the equivalent circuit model shown in Figure 8.19(a). Figure 8.19(b) shows some details of such a test setup. Table 8.5 indicates the component values related to a 9 Ah NiMH battery with 270 A allowed current. Given the time constants for the surface

TABLE 8.5 Model Parameters for a Typical 9 Ah NiMH Cell

Components of Cell Internal Resistances		Capacitances	
Ri	1 mΩ		
RD	0.35 mΩ	CD	171 F
RK	1.6 mΩ	CK	16,000 F

Source: Adapted from Schweighofer, B., K. M. Raab, and G. Brasseur, *IEEE Transactions on Instrumentation and Measurement*, 52, 4, August 2003.

effects of the electrodes, R_D and C_D and the time constants related to the diffusion process, R_K and C_K, one should be able to predict the battery behavior to match the experimental data available in data sheets. For details, [38] is suggested.

If a more simplified model is useful, public information available on battery data sheets can be used to arrive at the model [37] shown in Figure 8.20. In this case the battery voltage curves based on different discharge rates, measured in multiples or fractional C rates, can be used using simplified graphical and calculative approaches. This three-component model is based on:

- Open-circuit voltage (equilibrium potential), E
- An internal resistance R_{int} having two components, R_1 and R_2
- An effective capacitance that characterizes the transient response of charge double layers in porous electrodes

More details related to this procedure are discussed in [37]. For design of battery packs for portable systems, an accurate prediction of battery models for run time and

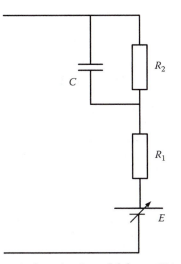

FIGURE 8.20 A simplified equivalent circuit model for an Li-ion cell usable for data sheet parameter extraction.

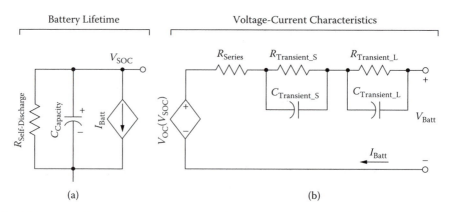

FIGURE 8.21 A battery model suitable for Cadence simulation: (a) model suitable for capacity, SOC, and run-time estimation; (b) model suitable for transient response estimation. (Reproduced from Chen, M., and G. A. Rincon-Mora, *IEEE Transactions on Energy Conversion* 21, 2, 2006, 504–11. With permission.)

V-I performance Cadence simulation can be practically used. More details and various equivalent circuits suitable for this purpose are discussed in [43], from a design engineer's viewpoint. A set of battery models suitable for Cadence-type simulation work as proposed in [44] is shown in Figure 8.21. This uses two sets of parameters where Figure 8.21(a) indicates a capacitor and a controlled current source inherited from run time-based models. An RC network-based model in Figure 8.21(b) based on Thevenin models simulates the transient response [44].

Another important condition in a battery is the battery temperature and the ambient temperature of the battery pack. By using finite element models (FEMs), it is possible to analyze and predict the thermal behavior of a battery pack or a cell using modern FEM software combined with thermal models of the battery [39].

8.10.3 Temperature Variations and Battery Performance

In military and industrial circumstances, batteries are expected to perform well despite any wide temperature variations. However, due to intrinsic chemical reasons and packaging issues the cells perform differently at different temperatures. For example, Figure 8.22(a) indicates the voltage droop and reduced capacity of Li-ion chemistry. Figure 8.22(b) indicates the impact of temperature on the self-discharge of Li-ion cells. In situations where rechargeable lithium chemistries don't allow wide temperature range, lithium primary cells could be one of the limited choices, despite their voltage delay effect at lower temperatures. Figure 8.22(c) indicates this situation in two different lithium primary cell types. This situation is related to the growth of a cell's passivation layer when stored at higher temperature, and then the battery is to deliver current at a lower temperature [43]. These situations force the battery pack designers to consider multiple

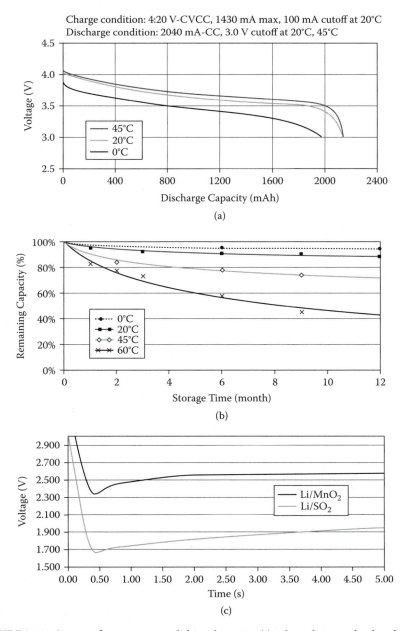

FIGURE 8.22 Impact of temperature on lithium batteries: (a) voltage droop and reduced capacity at lower temperatures; (b) impact of temperature on self-discharge rate; (c) voltage delay in lithium primary chemistries at lower temperatures. (Courtesy of VanZwol, J., *Power Electronics Technology*, July 2006, 40–44.)

temperature sensors within battery packs. If a pack will be used in high-temperature environments, several specific design principals must be applied [43]:

- In high-temperature charge/discharge conditions, to monitor associated temperature rises a thermal sensor needs be used to disconnect at specified temperatures to avoid thermal breakdown.
- Placement of components within the pack needs be critically reviewed, particularly when using heat-generating components such as FETs.
- Packs need be designed with vent holes to dissipate the generated heat or exhaust the vented gases.

A major aim of using commercial FEM packages is to predict the dependence of SOC of a battery on temperature and the instantaneous discharge rate. In this process, a very accurate knowledge of the U_B in Figure 8.19(a) is necessary in addition to the estimation of the heat generation and transfer within the battery pack. To calculate temperature rise, the following heat transfer processes need to be accounted for:

- Heat conduction
- Heat transfer by convection based on boundary conditions
- Estimating the heat sources inside the cell

A detailed discussion on these aspects of the thermal problem and an FEM-based solution is available in [39]. Electrochemical-thermal modeling of Li-polymer cells is discussed in [48].

8.10.4 Prognostics in Battery Health Management

Prognostics attempt to estimate the remaining component lifetime when an abnormal condition has been detected. The key to useful prognostics is not only an accurate remaining lifetime estimate but also an assessment of confidence of the uncertainty estimate. The phrase "battery health monitoring" has a wide variety of connotations, ranging from estimating the approximate voltage and other parameters to fully automated online monitoring of various measurements and estimated battery parameters. Prognostics of battery health could be very critical in aerospace and space exploration applications. For example, a catastrophic battery failure occurred in NASA's Mars Global Surveyor, which stopped operating in November 2006. It was revealed later that when the spacecraft was commanded to go into safe mode, the radiator of the batteries was positioned toward the sun, which created a temperature issue in the battery system [45].

Prognostics and health management (PHM) in batteries for EVs and HEVs is a growing topic [37, 45]. PHM dynamic models use sophisticated reasoning schemes applied with the goal of estimating the SOC, state of health (SOH), and state of life (SOL). However, it remains a difficult task to predict the end of life for a battery pack [45]. In estimating battery SOC, SOH, and SOL, monitoring the variation of battery impedance becomes very useful, as this parameter has a significant variation as the battery pack ages.

8.10.5 Battery Impedance and Its Time Variation Estimation as a Prognostic Parameter

Impedance of a battery is one of the key parameters vary as the battery ages. Due to various electrochemical reasons within a cell, as the cell ages impedance keeps increasing in general. Figure 8.23 indicates this situation for NiMH and Li-ion cells. Figure 8.23(a)

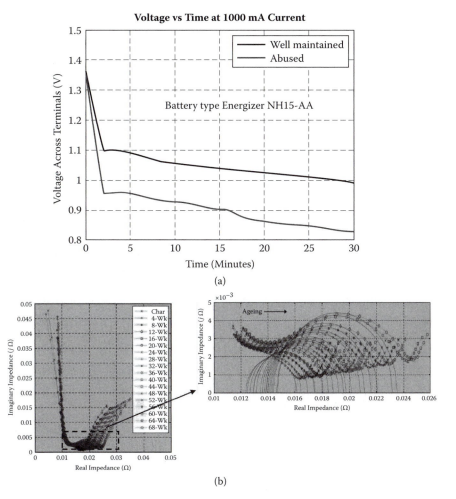

(a)

(b)

FIGURE 8.23 Battery impedance variation with age and abuse: (a) DC impedance comparison of a well-used NiMH cell and an abused one; (b) electrochemical impedance spectrometry (EIS) data for a 18650-type Li-ion cell with aging; (c) typical variation of Li-ion cells with life cycle at different frequencies. (Part [b] reproduced from Goebel, K., B. Saha, A. Saxena, J. R. Celaya, and J. P. Christopherson, 2008. *IEEE I & M*, August 2008, 33–40. With permission. Part [c] courtesy of Karfthoefer, B., *Power Electronics Technology*, January 2005, 30–38.)

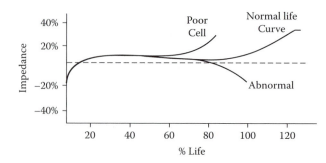

FIGURE 8.24 VRLA battery impedance behavior over lifetime. (Adapted from Langnan, P. E. 2000. *Power Quality Assurance*, September/October 2000, 24–29.)

is a simple DC resistance variation for a well-used versus a badly used (overcharged repeatedly) NiMH cell. Figure 8.23(b) indicates the results of measuring the impedance of a 18650-type Li-ion cell in a PHM exercise [45], and Figure 8.23(c) indicates the variation of the Li-ion cell impedance versus frequency after different numbers of discharge cycles [46]. In Figure 8.23(c) it clearly shows that at lower frequencies below 10 Hz, as the discharge cycles are increased the impedance gradually increases. There are new battery management ICs that monitor this behavior in gas gauging (discussed later) [46].

In large standby telecommunication battery banks or UPS installations, large numbers of series- and parallel-connected VRLA cells (sealed lead-acid cells) are used, and any single or multiple misbehaving cells could in the longer run create expensive replacement issues. In these cases of large and expensive battery banks, battery impedance testing is used as a prognostic technique. By automatic or manual measurement of individual cell voltages over the lifetime of the bank, test engineers can easily predict the potential failure of individual cells. In these situations, typical VRLA cells behave as per Figure 8.24 in general [47]. A detailed discussion on the impedance variations and terminal voltages is available in [47].

8.10.6 Energy-Aware Battery-Modeling Concepts for Best Run Time

With the availability of techniques for computationally feasible mathematical models of cells, "energy-aware system design" is another new concept currently developing to get the best out of a battery pack, with multiple cells or batteries connected to a given electronic system. Given the chemical structure and the processes inside a battery, as depicted in Figure 8.18, if a battery pack can contain multiple cells, where each one of them can be connected to the load selectively depending on the best utilization model for the total energy within the pack, run time of a portable product can be improved significantly [36]. This is particularly the case with systems based on CMOS logic where the power supply source voltage V_{DD} and threshold voltage V_{th} of the FETs are the key variables for best power consumption. If these parameters can be optimized with a battery-efficiency model, an interleaved power supply system (with multiple batteries selectively supplying the load conditions) can provide the best run time, given the case that a

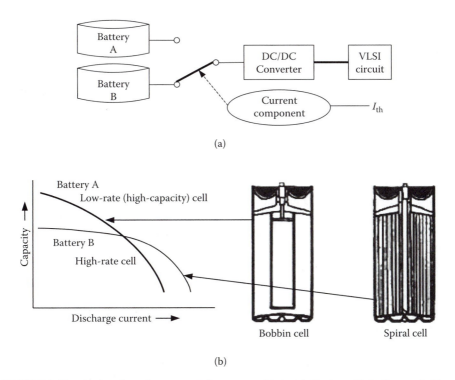

(a)

(b)

FIGURE 8.25 An interleaved power supply system with two batteries with battery-switching scheme for best run time: (a) basic concept; (b) an example of implementation with a high-rate spiral cell and a low-rate bobbin cell. (Reproduced from Karfthoefer, B., *Power Electronics Technology*, January 2005, 30–38. With permission.)

battery pack has only limited energy after an optimal charging process. This concept is shown in a simplified case in Figure 8.25, with two batteries that can be selectively connected to a suitable DC-DC converter, which supplies the VLSI circuit.

In this example, where an interleaved dual battery (IDB) concept is used as in Figure 8.25(b), the working principle is simple. When the discharge current is lower than a threshold value, I_{th}, the system uses battery A as the energy source and vice versa with battery B. Battery A is a low-rate/high-capacity cell, while battery B is a high-rate/low-capacity cell. In this process of switching based on suitable algorithms, a suitable battery model is taken into account for optimum discharge behavior [49]. This kind of dynamic power management of battery-powered electronics can be used for battery life optimization as well [50]. Energy-efficient communication is another application of these concepts with electrochemical battery models [51].

8.10.7 Fast-Charging Nickel-Based Batteries

While four major battery families can accept either a standard (16–24 hr) or a fast (2–4 hr) charge, most of the discussion here is limited to fast-charging methods. Slower-charging schemes tend to be found in simpler, price-sensitive applications, which do not need (or cannot afford) much beyond a simple charger and a low-battery indicator.

The objective of fast-charging a battery is to cram as much energy as it takes to bring the battery back to fully charged state in the shortest possible time without damaging the battery or permanently affecting its long-term performance. Since current is proportional to charge divided by time, the charging current should be as high as the battery systems will reasonably allow.

Nickel-based chemistries prefer constant-current-type charging, while lead-acid and lithium cells prefer constant-voltage (current-limited) charging. For the constant-current cells (NiCd and NiMH), a 1 C charge rate will typically return more than 90% of the battery's useable discharge capacity within the first hour of charging. The constant-potential cells (lead-acid and lithium-ion) are a bit slower to reach the 90% mark, but can generally be completely recharged within five hours. See Table 8.6 for representative and typical charging recommendations for different chemistries. The most appropriate method should be selected in consultation with the manufacturer.

Fast charging has compelling benefits but places certain demands upon the battery system. A properly performed fast charge, coordinated to the specifications of a battery rated for such charging, will deliver a long cycle life. The high charging rates involved, however, cause rapid electrochemical reactions within the cells of the battery. After the battery goes into overcharge, these reactions cause a sharp increase in internal cell pressure and temperature.

Uncontrolled high-rate overcharge quickly causes irreversible battery damage. Thus as the battery approaches full charge, the charging current must be reduced to a lower "top-off" level, or curtailed entirely.

TABLE 8.6   Representative Charging Recommendations

	NiCd or NiMH	Sealed Lead-Acid	Li-Ion
Charging current	1 C	1.5 C	1 C
Voltage per cell (volts)	1.80	2.5	4.20 ± 0.05
Charging time (hours)	~3	~3	2.5–5.0
Method for optimum fast-charge termination point	See Table 5.4		
		Current cutoff	Typically a timer[a]
Backup charge termination method			
See Table 5.4			
Timer			

"Top off" rate	0.1 C	0.002 (trickle)	
Temperature range			
(°C)	10°–40° (NiCd)		
15°–30° (NiMH)	0–30°	0–40°	

[a] Depending on manufacturer's recommendation.

8.10.7.1 Charge Termination Methods

If a rapid charge is applied to a battery pack, it is necessary to select a reliable method to terminate charging at the fully charged position. Two practical approaches for charge termination are temperature termination method and voltage termination method.

8.10.7.1.1 Voltage Termination Methods

Four commonly available voltage termination methods are maximum voltage (V_{max}), negative delta voltage (–DV), zero slope, and inflection point (d^2V/dt^2). The maximum-voltage method senses the increase in battery voltage as the battery approaches full charge. However, this is accurate only on a highly individualized basis. It is necessary to know the exact value at the voltage peak, otherwise the batteries may be over- or undercharged.

Temperature compensation is also required because of the negative temperature coefficient of battery voltage. The maximum voltage will increase if the batteries are cold, causing an undercharge because the charging voltage will reach the maximum voltage trip point early. If the batteries are hot, the maximum voltage may never be reached and the batteries may be discharge. Therefore the V_{max} method is generally not recommended for fast charge rates.

The negative delta voltage is the most popular of the fast-charge termination schemes. It relies on the characteristic drop in cell voltage that occurs when a battery enters overcharge, as shown in Figure 8.26. With most NiCd cells, the voltage drop is a very consistent indicator, and the –DV method is fine for charge rates up to 1 C. An inherent problem with this method is that the batteries must be driven into the overcharge region to cause the voltage decrease. Pressures and temperatures rise very rapidly at fast charge rates beyond 1 C. In cyclic applications the battery must be able to endure that continual abuse.

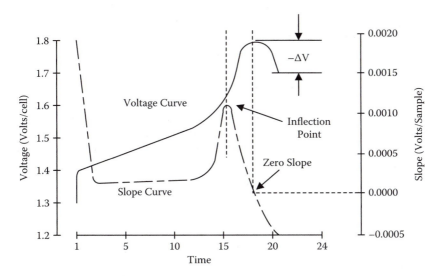

FIGURE 8.26 Termination methods based on changes in voltage and voltage slope.

Another concern is that cells such as NiMH types do not always have the characteristic decrease in voltage, compared to NiCds as shown in Figure 8.10. This creates a problem of forward compatibility when moving from NiCd to NiMH cells. Most manufacturers of NiMH cells do not advocate the –DV method of charge termination.

The zero-slope method monitors the point where the slope of the battery voltage reaches zero. This method is reliable for rapid charge rates up to 4 C, and is less susceptible to noise on the voltage sense lines. However, a few types of batteries such as the NiCd button cells may have a voltage slope that never quite reaches zero. Therefore the zero-slope method is better suited as a backup method.

In the inflection point (d^2V/dt^2) method, the system monitors the change in voltage over time and is the most sensitive indicator for preventing overcharge. The inflection point method relies on the changes in the voltage slope, shown in Figure 8.26, which occurs during charge and is an excellent primary termination method for up to 4 C charge rates.

The change in the voltage slope is an extremely reliable and repeatable indication of charge. It does not rely on the decrease in voltage, which may not always occur. Instead, this method looks for the flattening of the voltage profile as the battery reaches full charge. By monitoring the relative change in the steepness of the voltage slope, this method avoids having to use absolute numbers.

8.10.7.1.2 *Temperature Termination Method*

Temperature is the main cause of failure in a rechargeable cell, so it makes sense to monitor the cell temperature to determine when to shut off charge to a battery. Three methods of charge termination, based on temperature, are common: maximum temperature cutoff (MTC), temperature difference (DT), and temperature slope (dT/dt). The maximum temperature cutoff system is the easiest and cheapest to implement, but is the least reliable. Using a bimetallic thermal switch or a positive temperature coefficient thermistor, a simple, low-cost circuit can shut down a charging current at an appropriate temperature.

The DT method measures the difference in ambient and cell temperatures to compensate for a cool environment. It requires monitoring two temperature sensors, one for the battery temperature and one for the ambient. This method may be unsuitable if the difference between cell and ambient temperature is very large.

The DT method can become unreliable with a quickly changing ambient temperature unless an equal thermal mass is attached to the ambient sensor. This means that the DT method is suitable for a primary charge termination at lower charge rates up to C/5 if the ambient temperature is not going to change often. The DT method also provides an excellent backup charge termination scheme.

The temperature slope method (dT/dt), a more sophisticated temperature termination scheme, measures the change in temperature over time. This method uses the slope of the battery temperature, and therefore is less dependent on changes in ambient temperatures or in large differences between ambient and battery temperatures. Accurately adjusted to a particular pack, and with careful attention paid to the type and placement of the temperature sensor, the dT/dt method works very well. This dT/dt method is suitable for charge rates up to 1 C and provides an excellent backup method.

8.10.7.2 NiCd and NiMH Fast-Charge Methods

Nickel-based batteries such as NiCd and NiMH types are mature chemistries. Although it is not correct to consider the NiCd and NiMH electrochemistries or charging regimens as being interchangeable, they are similar enough that they can be discussed together.

There is no one best way to fast-charge a NiCd or NiMH battery. Variables introduced by the allowable cost and size of the end application, the choice of charge termination method(s), and the specific battery vendor's recommendations will all influence the final choice of charging technique.

Figure 8.27(a) shows the voltage, pressure, and temperature characteristics of an NiCd cell being charged at 1 C rate. Figure 8.26(b) shows similar data for an NiMH cell.

These curves illustrate the need for a reliable termination of the high-current portion of the charge cycle, and assist in understanding the various fast-charge termination methods outlined in Table 8.7. For both electrochemistries, the ideal fast-charge termination point is at 100%–110% of returned charge. The charging current is then reduced to the top-off value for one to two hours, to bring the cell into a state of slight overcharge. This compensates for the inefficiencies of the charging process (e.g., heat generation). If the specific application will have the battery on standby for more than several weeks, or at high temperatures, the top-off charge is followed by a continuous, low-level "trickle" charge to counter the self-discharge characteristics of NiCd and NiMH cells.

Under certain conditions, particularly following intervals of storage, an NiMH battery may give an erroneous voltage peak as charging commences. For this reason, the charger should deliberately disable any voltage-based charge termination technique for the first five minutes of the charging interval. Table 8.7 summarizes the fast-charge termination methods for NiCd and NiMH cells.

8.10.8 Charging Sealed Lead-Acid Batteries

Unlike nickel based batteries, sealed lead-acid (SLA) batteries are charged using a constant-potential (CP) regimen. CP charging employs a voltage source with a deliberately imposed current limit (a current-limited voltage regulator). A significantly discharged battery undergoing CP charging will initially attempt to draw a high current from the charger. The current-limiting function of the CP regimen serves to keep the peak charging current within the battery's ratings.

Following the current-limited phase of the charging profile, an SLA battery exposed to a constant voltage will exhibit a tapering current profile as shown in Figure 8.28. When returned charge reaches 110%–115% of rated capacity, allowing a dischargeable capacity of 100% of nominal, the charge cycle is complete.

The specifics of fast-charging SLA batteries are more vendor-dependent than those of NiCd or NiMH units. The information in Table 8.8 uses data from GS Battery (USA) Inc. The primary termination method, "current cutoff," looks at the absolute value of the average charging current flowing into the battery.

When that current drops below 0.01 C, the battery is fully charged. If it will be in standby for a month or more, a trickle current of 0.002 C should be maintained. The

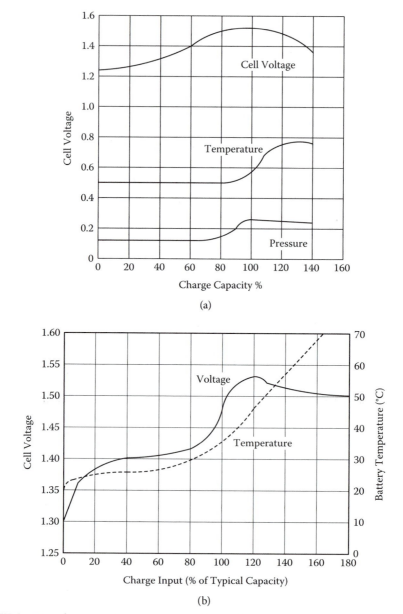

FIGURE 8.27 Charging indications for nickel-based batteries at 1 C charge rate: (a) cell voltage, temperature, and pressure for a typical NiCd cell at 1 C rate charging; (b) NiMH voltage and temperature characteristics.

TABLE 8.7   Fast-Charge Termination Methods for NiCd and NiMH Batteries

Charging Technique	Description
Negative DV(–DV)	Looks for the downward slope in cell voltage that a cell exhibits (30–50 mV for NiCd, 5–15 mV for NiMH) upon entering overcharge. Very common in NiCd applications due to its simplicity and reliability.
Zero DV	Waits for the time when the voltage of the cell under charge stops rising, and is "at the top of the curve" prior to the downslope seen in overcharge. Sometimes preferred over $-\Delta V$ for NiMH, due to NiMH's relatively small downward voltage slope.
Voltage slope (dV/dt)	Looks for an increasing slope in cell voltage (positive dV/dt), which occurs somewhat before the cell reaches 100% returned charge (prior to the zero DV point).
Inflection point cutoff (d^2V/dt^2,IPCO)	As a cell approaches full charge, the rate of its voltage rise begins to level off. This method looks for a zero or, more commonly, slightly negative value of the second derivative of cell voltage with respect to time.
Absolute temperature cutoff (TCO)	Uses the cell's case temperature (which will undergo a rapid rise as the cell enters high-rate overcharge) to determine when to terminate high-rate charging. A good backup method, but too susceptible to variations in ambient temperature conditions to make a reliable primary cutoff technique.
Incremental temperature cutoff (DTCO)	Uses a specified increase in the cell's case temperature, relative to the ambient temperature, to determine when to terminate high-rate charging. A popular, relatively inexpensive and reliable cutoff method.
Delta temperature/ delta time (DT/Dt)	Uses the rate of increase of a cell's case temperature to determine the point at which to terminate the high-rate charge. This technique is inexpensive and reliable once the cell and its housing have been properly characterized.

backup termination method, according to the vendor's recommendations, should be a 180-minute time-out on the charging cycle [52].

To satisfy more stringent charge control recommendations, where the battery temperature, voltage, and current need to be sampled, many dedicated charge controller ICs are available on the market. The Bq2031 lead-acid fast-charge IC from Texas/Benchmarq and UC3906 (sealed lead-acid charger) from Texas/Unitrode Integrated Circuits are examples. The UC3906 battery charger controller contains all of the necessary circuitry to optimally control the charge and hold cycle for sealed lead-acid batteries. These integrated circuits monitor and control both the output voltage and current of the charger through three separate charge states: a high-current bulk-charge state, a controlled overcharge, and a precision float-charge or standby state. Reference [53] and Application Note U-104 [54] provide details about sealed lead-acid charge control using UC3906.

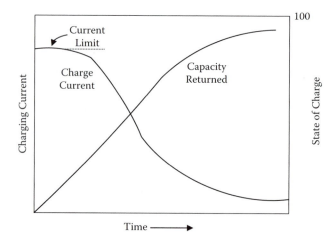

FIGURE 8.28 Typical current and capacity-returned Vs charge time for CP charging in sealed lead-acid batteries.

TABLE 8.8 Comparison of Recharge Requirements

Parameter	SLA	NiCd	NiMH	Li-Ion (Coke Electrode)	Li-Ion (Graphite Electrode)
Standard Charge					
Current (C rate)	0.25	0.1	0.1	0.1	0.1
Voltage per cell (V)	2.27	1.5	1.5	4.1 ± 50 mV	4.1 ± 50 mV
Time (hours)	24	16	16	16	16
Temperature range (°C)	0–45	5–40	5–40	5–40	5–40
Termination	None	None	Timer	None	None
Fast Charge					
Current (C rate)	1.5	1	1	1	1
Voltage per cell (V)	2.45	1.5	1.5	4.1 ± 50mV	4.1 ± 50 mV
Time (hours)	1.5	3	3	2.5	2.5
Temperature range (°C)	0–30	15–40	15–40	10–40	10–40
Primary termination method	I_{min},[a] ΔTCO	dT/dt, −ΔV	Zero dv/dt, −ΔV, d^2V/dt, ΔTCO	I_{min} + timer, dT/dt, dT/dt	I_{min} + timer, dT/dt, dT/dt
Secondary termination method	Timer, ΔTCO	TCO, timer	TCO, timer	TCO, timer	TCO, timer

[a] Minimum current termination threshold.

Source: Adapted from Israelsohn, J., *EDN*, January 18, 2001.

8.10.9 Li-Ion Chargers

Lithium-ion batteries require a constant-potential charging regimen, very similar to that used for lead-acid batteries. Because the voltage of an Li-ion cell under constant-current charge could continue to rise to the point of cell destruction, absolute voltage limits are required [55]. Typical recommendations for Li-ion fast-charge are indicated in Table 8.8. As with lead-acid batteries, an Li-ion cell under charge will reduce its current draw as it approaches full charge. Figure 8.29 indicates this condition. However, if the charger has poor regulation on the output voltage and creates a lower output voltage, it will decrease the capacity as depicted in Figure 8.29(b). The area between the graphs of current indicates the lost charge.

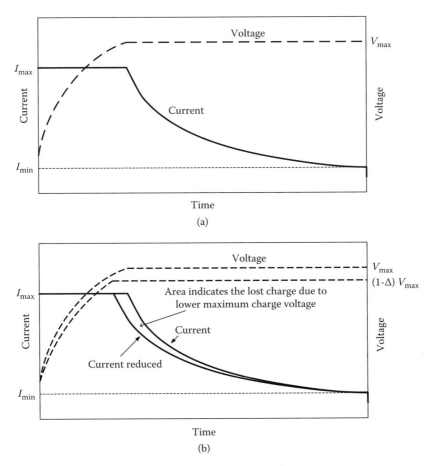

FIGURE 8.29 Voltage-regulated charging profile for Li batteries: (a) current-limited voltage-regulated charging profile; (b) charge dependence on voltage regulation. (Adapted from Bentley, W. F., and D. K. Heacock, *Proceedings of the 11th Annual Battery Conference*, Long Beach, CA, January 1996.)

If the cell vendor's recommendation for charging voltage (generally 4.10 V ±50 mV at 23°C) is followed, the cells will be able to completely recharge from any "normal" level of discharge within five hours. At the end of that time, the charging voltage should be removed. Trickle current is not recommended. If the voltage on an Li-ion cell falls below 1.0 V (or to an abnormal level), recharging of that cell should not be attempted. If the voltage is between 1.0 V and the manufacturer's nominal minimum voltage (typically 2.5–2.7 V), it may be possible to salvage the cell by charging it with a 0.1 C current limit until the voltage across the cell reaches the nominal minimum, followed by a fast charge.

Due to special characteristics of Li-ion batteries, most Li-ion manufacturers incorporate custom protection circuits into their battery packs to monitor the voltage across each cell within the battery and to provide protection against overcharge, battery reversal, and other major faults. These circuits are not to be confused with charging circuits. For example, the MC 33347 protection circuit is such an IC from Motorola [56].

8.10.10 Portable Chargers and Comparison of Recharge Requirements for Different Chemistries

With the availability of charge controllers, protectors, and other battery management ICs today, a designer can find many commercial ICs and supporting reference designs [57]. Choosing the best chemistry requires a deliberative comparison between battery attributes and the required power source specifications in general. As summarized above, the four common chemistries require different recharge algorithms and give different indications when they have completed the discharge cycle. Table 8.8 provides a summary, as adapted from [57].

Most secondary batteries tolerate trickle charging for long periods. The simplest charging strategy, therefore, uses a simple linear regulator IC or a pulse-width modulator IC in conjunction with a series transistor and a currents sense resistor. Circuits such as these are available from many vendors and with a range of auxiliary features for charging single cells or nickel chemistry stacks [57]. They sometimes use an adaptive method that adjusts the charger's behavior according to the battery's SOC. A charger of this type may start by testing the battery for deep discharge, which is determined by comparing the battery's terminal voltage with a threshold value. If the battery is deeply discharged, the charger enters a precharge mode in which it limits its current to some fraction, k, of resistor-programmed maximum charge current, I_{PGM}. Figure 8.30 gives a representative options of charging algorithms. Figure 8.30(a) is a constant-current/constant-voltage charging algorithm while Figure 8.30(b) is a current foldback mode.

8.10.11 End-of-Discharge Determination

Determination of the point at which a battery has delivered all of its usefully dischargeable energy is important to the longevity of the cells that form that battery. Discharging a single cell too far will often cause irreversible physical damage inside the cell.

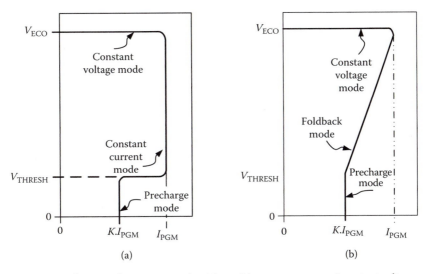

FIGURE 8.30 Charger voltage-current algorithms: (a) constant-current/constant-voltage case; (b) current foldback case. (From Israelsohn, J., Battery management included. *EDN*, January 18, 2001, 65–74. With permission.)

If multiple cells are placed in series, unavoidable imbalances in their capacities can cause the phenomenon known as "cell reversal," in which the higher-capacity cells force a backward current through the lowest-capacity cell. Knowing the end-of-discharge (EOD) point provides a "zero capacity" reference for coulometric gas gauging (discussed later).

The actual determination of the EOD point is typically done by monitoring the cell voltage. For the most accurate determination of EOD when the load is varying, correction factors for load current and the battery's state of charge should be applied, especially to SLA and Li-ion batteries. The essentially flat discharge profiles of NiCd and NiMH make these corrections a matter of user discretion for most load profiles. Table 8.9 shows voltages commonly used to indicate the EOD point for the four battery types.

TABLE 8.9 Typical End-of-Discharge Voltages

Cell Type	EODV (Volts)	Comments
Lead Acid	1.35–1.9 V (1.8 V typical)	Dependent upon loading, state of charge, cell construction, and manufacturer
NiCd	0.9	Essentially constant
NiMH	0.9	Essentially constant within recommended range of discharge rates
Li-Ion	2.50–2.70	Dependent upon manufacturer, loading, and state of charge

8.10.12 Gas Gauging

The "gas-gauging" "or fuel-gauging" concept discussed here does not refer to the gases that may be evolved by the battery reactions, but rather to the concept of the battery as a fuel tank powering the product. Gas gauging, therefore, involves real-time determination of a battery's state of charge, relative to the battery's nominal capacity when fully charged.

It is possible to make an inexpensive and moderately useful state of charge measurement from a simple voltage reading, if the battery being used has a sloping voltage profile. Hence, Li-ion batteries and, to a lesser extent, SLA batteries should be amenable to such an approach. In practice, the results are less than optimal: cell voltages are dependent upon loading, internal impedance, cell temperature, and other variables. This reduces the attractiveness of the simple voltage method of gas gauging; the fact that it is not suitable for NiCd or NiMH batteries, due to their essentially flat voltage profile, makes it commercially untenable. A clever and effective alternative is the "coulometric" method.

Coulometric gas gauging, as its name implies, measures the actual charge ($\int Idt$) going into and out of the battery. By integrating the difference of current in and current out, it is possible to determine the charge status of the battery at any given time. There are of course real-world details that must be observed in the actual implementation of such a gas gauge; some of the most important of these are:

1. It is necessary to have an accurate starting point for the integrator, corresponding to a known state of charge in the battery. This is often resolved by zeroing the integrator when the battery reaches its EODV.
2. Temperature must be compensated for. The actual capacity of lead-acid batteries increases with temperature; that of nickel-based batteries decreases as battery temperature rises.
3. Appropriate conversion factors should be applied for the particular charge regimen and discharge profile used. Under conditions of highly variable battery loading, dynamic compensation may be advisable.

8.10.13 Battery Health

SOC is a widely used concept to represent the remaining capacity of a battery in a charge/discharge cycle. However, due to cycle-aging phenomenon, using SOC alone may result in large errors because the full charge capacity (FCC) may be significantly less than the design capacity (DC) that denotes the FCC of a new battery.

Considering the aging effect, SOH of a battery is defined as the FCC of a cycle-aged battery to its DC. Analytical modeling of batteries can be used to get theoretically accurate values for SOC and SOH [58]. With battery health defined as a battery's actual capacity relative to its rated capacity, the health of the battery can be determined and maintained in three steps:

1. Discharge the battery to the EOD point, preferably into a known load.
2. Execute a complete charge cycle, while gas gauging the battery.
3. Compare the battery's measured capacity to its rated capacity.

This sequence will simultaneously "condition" the battery (which means to overcome the so-called "memory effect" of capacity of NiCd batteries, for example), and will indicate the capacity of the battery after conditioning. The information so obtained can be used to ascertain whether the battery is in good shape, or is approaching the end of its useful life. Online SOC estimation with monitoring system is discussed in [59].

8.11 Battery Communication and Related Standards

In the early 1990s power management and battery management were rarely used in portable products. More-advanced processors were introduced around the mid-1990s, with very high power consumption and relatively high load currents of tens of amperes, which placed a burden on the available rechargeable battery chemistries due to rapid increase of discharge current. To meet system run times expected by the users, innovative power management concepts had to be employed. For this reason, around the mid-1990s new industry standards were proposed to standardize the battery and power management subsystems within portable products, where concepts such as Advanced Configuration and Power Interface (ACPI) by Intel were utilized. Following were the standards proposed:

System Management Bus Specification (SMBus)
Smart Battery Data Specification
Smart Battery Charger Specification
Smart Battery Selector Specification

The above specifications form the Smart Battery Systems (SBS) Specification. This specification presents a solution for many of the issues related to batteries used in portable equipment such as laptop computer systems, cellular telephones, and video cameras. Fundamental to the SBS system is the concept that the battery contains all of the necessary components to determine the battery's state of charge, predict time to full and time to empty, specify the charging voltage and current, and determine when the battery is fully charged to fully discharged.

The concept of a smart battery system is shown in Figure 8.31(a). Figure 8.31(b) indicates details. It consists of an AC-DC converter (unregulated), power switch, system power supply, smart battery charger, and smart battery selector, all of which communicate with the system host and the system elements themselves via SMBus. In this case, smart battery A powers the system while battery B is getting conditioned and/or charged.

The SMBus is a two-wire interface through which simple power-related chips can communicate with the rest of the system. It uses I^2C as its backbone [60]. A system using SMBus passes messages to and from devices instead of tripping individual control lines. Removing the individual control lines reduces pin count. Accepting messages ensures future expandability. With SMBus, a device can provide manufacturer information, tell the system what its mode/part number is, save its state for a suspend event, report different types of errors, accept control parameters, and return its status. The SMBus may share the same host device and physical bus, provided that an appropriate electrical bridge is provided between the respective devices. For details, [62–66] are suggested.

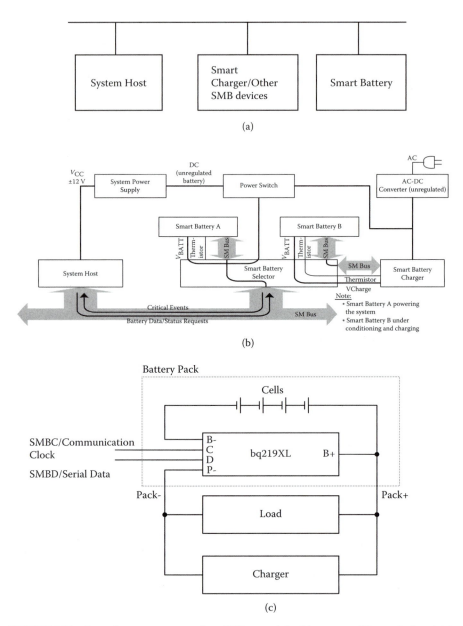

FIGURE 8.31 Smart battery system and an SMBus module: (a) concept; (b) a typical multiple smart battery system; (c) typical control IC (bq 219XL) and its connections.

8.12 Battery Safety

With the introduction of Li-based batteries, which have high gravimetric energy density, safety of the battery powered portable products has become a concern during the last decade. In the early 2000s, some laptop fires scared the public and promoted widespread recalls of Li-ion cells. Since these events, individual companies have taken many steps to ensure that such disasters do not recur, and the industry has joined forces to unify the standards. By 2004, IEEE had published two standards relevant to the design and manufacture of portable systems: the IEEE 1625 Standard for Rechargeable Batteries for Portable Computing, and a related IEEE 1725 Standard for Rechargeable Batteries for Cellular Phones.

The standards cover design approaches that ensure reliable operation and that minimize the occurrence of faults leading to hazards in portable computing devices and other rechargeable battery-powered systems. The standards guide the system and subsystem design through five major areas:

• Systems integration
• Cell
• Pack
• Host device
• Total system reliability

Also covered are the critical operational parameters and how they change with time and environment, the effect of extremes in temperature, and the management of component failure. Figure 8.32 indicates the case of a portable computing device broken into subsystems and within the environment.

In general, portable equipment battery packs usually contain an analog front end (AFE), which is monitoring the conditions of the cells, and it usually provides a first level of protection to the pack. A battery management unit (BMU) (similar to the example in Figure 8.32[c]) usually communicates with the cell stack to monitor and communicate with the host system via SMBus. Referring to Figure 8.32(b), in the 1625 standard it proposes to move the second level of protection into the AFE, which communicates directly with the BMU [66]. An industry-wide methodology such as design failure modes and effects analysis (DFMEA), which can be used to highlight and prioritize the possible roots of faults and hazards, is to be used with the 1625 conditions in order to get the benefits proposed in the standard. Figure 8.32(c) indicates the block diagram of a battery pack compliant with the SBS 1.1 with four series Li-ion cells [66]. This change in architecture helps clarify the DFMEA analysis. Compared to the older architecture, in the new architecture (Figure 8.32[b]) BMU and AFE monitor each other as well as the cell stack. Theoretically this allows the second-level protection to be integrated with the first-level protection device.

As in Figure 8.32(c) the new architecture consists of a BMU, an AFE integrated circuit, current sense resistor, safety FET, and a chemical fuse. BMU could be a 3.3 V microcontroller with measurement capabilities and a programmable flash. The AFE provides a high-voltage interface to the cell stack for voltage measurements, along with cell-balancing control, overcurrent protection, and a low-dropout regulator to power the

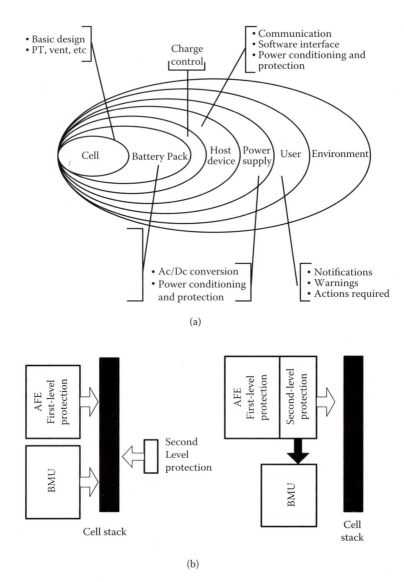

(a)

(b)

FIGURE 8.32 A portable computing device and its environment within the case of IEEE 1625: (a) portable computing device broken out into subsystems; (b) proposed level of safety compared with older battery packs; (c) implementation of IEEE 1625 concepts.

BMU. Based on possible problem levels, for discrete components BMU and AFE interaction DFMEA tables can be prepared similar to the case in Table 8.10, and each item and score can be evaluated according to the following criteria:

- Severity of the fault (SEV)
- Probability of occurrence (OCC)
- Difficulty of detection (DET)

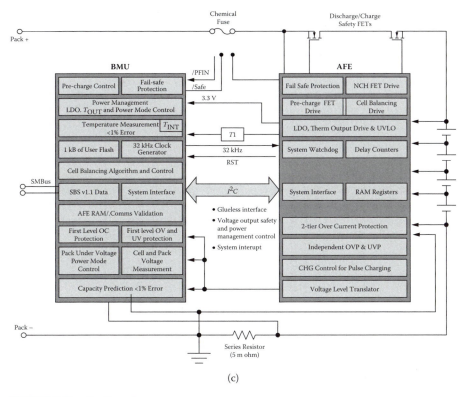

(c)

FIGURE 8.32 Continued

Typically, evaluation and scoring are based on experience and statistical reliability data. However, this data may not be always available, and guidance in the 1625 standard may be helpful. For more details, [66] is suggested.

8.13 Future

With the consumer electronics, EV, and the portable-product markets growing, newer battery technologies may enter the market. Major growth areas for batteries are electric vehicles and bicycles. Battery packs can challenge the designer because they are no longer a simple configuration of cells, but could contain many safety features, much intelligence, energy-aware models, selective batteries feeding the host product, serial data communication, and even recycling suggestions.

In all these situations, a simple and useful concept that can be remembered by designers is that "batteries are like human beings," and they need care and useful intelligence.

TABLE 8.10 An Example of a DFMEA Table for Discrete Components

Problem	Level of the Problem	Severity (SEV)	Issue	Probability of Occurrence (OCC)	Protection Features	Difficulty of Detection (DET)	Risk Product Number (RPN)
AFE reset output latched high	Severe customer experience	4	BMU is held in reset mode.	1	As BMU no longer functions, the AFE turns off the FET(and blows fuse as an option).	1	4
AFE timeout (TOUT) output latched high	Low customer experience	1	Thermistor will be always powered.	1	AFE will consume extra current, but no safety issue.	7	7
AFE TOUT output latched low	Severe safety	9	Thermistor will never be powered.	1	BMU will measure out-of-range temperature and create fuse blow action. All current and voltage protection methods will be functional.	1	9
BMU sense of cell latched high	Moderate safety	7	BMU voltage measurements will be 0 V.	2	First-level protection is not valid, but second-level OV protection is fully operational (as it is an independent circuit).	2	28

Source: Adapted from Elder, G., *Power Electronics Technology*, April 2004, 34–43. With permission.

References

1. Marketresearch.com. 2009. Advanced rechargeable battery market: Emerging technologies and trends worldwide, March 1.
2. Marketresearch.com. 2009. US demand for batteries to reach $16.8 billion in 2012. *Power Electronics Technology*, March, 9.
3. Marketresearch.com. 2009. Silicon-air battery touts unlimited shelf life. *EE Times Asia*, November 25, 2009. http://www.eetasia.com/login.do?fromWhere=/ART_8800590556_765245_NT_17302c65.HTM.
4. Quinnell, R. A. 1991. The business of finding the best battery. *EDN*, December 5, 162–66.
5. Schimpf, M. 1996. Choosing lithium primary-cell types. *Electronic Design*, January 8, 141–44.
6. Hamlen, R. P., H. A. Christopher, and S. Gilman. 1995. US Army battery needs—present and the future. *IEEE AES Systems*, June, 30–33.
7. Kim, S. 2001. Lithium-ion polymer batteries promise improved size, safety, energy density. *PCIM*, January, 30–39.
8. Chu, B. 2009. LiFePO$_4$ batteries help consumer devices come to life. *Power Electronics Technology*, October, 10–15.
9. Gates Energy Products Inc. 1992. *Rechargeable batteries applications handbook*. Oxford: Butterworth-Heinemann.
10. Hirai, T. 1990. Sealed lead-acid batteries find electronic applications. *PCIM*, January, 47–51.
11. Moore, M. R. 1993. Valve regulated lead acid vs flooded cell. *Power Quality Proceedings*, October, 825–27.
12. Moneypenny, G. A., and F. Wehmeyer. 1994. Thinline battery technology for portable electronics. *HFPC Conference Proceedings*, April, 263–69.
13. Nelson, B. Pulse discharge and ultrafast recharge capabilities of thin-metal film technology. *Proceedings of Portable by Design Conference (USA)*, 13–18.
14. Powers, A. R. 2000. Sealed nickel cadmium and nickel metal hydride cell advances. *IEEE AES Systems*, December, 15–18.
15. Stempel, R. C., S. R. Ovshinsky, P. R. Gifford, and D. A. Corrigan. 1998. Nickel metal hydride: Ready to serve. *Spectrum, IEEE* 35 (11, November): 29–34.
16. Small, C. H. 1992. Nickel-hydride cells avert environmental headaches. *EDN*, December 10, 156–61.
17. Briggs, A. 1994. NiMH technology overview. *Portable by Design Conference Proceedings*, BT-42–BT-45.
18. Reisner, D .E., and M. Klein. 1994. Bipolar nickel-metal hydride battery for hybrid vehicles. *IEEE Aerospace and Electronic Systems* 9 (5, May): 24–28.
19. Cole, J. H., M. Eskra, and M. Klein. 2000. Bipolar nickel-metal hydride batteries for aerospace applications. *IEEE Aerospace and Electronic Systems* 15 (1, January): 39–45.
20. Dan, P. 1997. Recent advances in rechargeable batteries. *Electronic Design*, February 3, 112–16.
21. Morrison, David. 2006. New materials extend Li-ion performance. *Power Electronics Technology*, January, 50–52.

22. Morrison, David. 2006. Li-ion cells build better batteries for power tools. *Power Electronics Technology*, February, 52–54.

23. Morrison, David. 2007. Cathode modeling builds better batteries for power tools. *Power Electronics Technology*, March, 52.

24. Freeman, D., and D. Heacock. 1995. Lithium-ion battery capacity monitoring within portable systems. *HFPC Conference Proceedings*, 1–8.

25. Bennett, P. D., and G. W. Brawn. 1997. Introduction to applying Li-ion batteries. *Proceedings of Portable by Design Conference (USA)*, 125–34.

26. Fuller, T. F, M. Doyle, and J. Newman. 1994. Simulation and optimization of the dual lithium ion insertion cell. *Journal of the Electrochemical Society* 141 (1, January): 1–10.

27. Levy, S. C. 1995. Recent advances in lithium ion technology. *Portable by Design Conference Proceedings*, 316–23.

28. Juzkow, M. W., and C. St. Louis. 1996. Designing lithium-ion batteries into today's portable products. *Portable by Design Conference Proceedings*, 13–22.

29. Morrison, D. 2000. Thinner Li-ion batteries power next generation portable devices. *Electronic Design*, February 7, 95–106.

30. Jiayuan, W., S. Zechang, and W. Xuezhe. 2009. Performance and characteristic research in LiFePO4 battery for electric vehicle applications. *Vehicle Power and Propulsion Conference Proceedings*. Washington, DC: IEEE, 1657–61.

31. Nossaman, P., and J. Parvereshi. 1995. In systems charging of reusable alkaline batteries. *Proceedings of HFPC Conference (USA)*.

32. Sengupta, U. 1995. Reusable alkaline™ battery technology: Applications and system design issues for portable electronic equipment. *Proceedings of Portable by Design Conference*, 562–70.

33. Ivad, J. D. and K. Kordesch. 1997. In-application use of rechargeable alkaline manganese dioxide/zinc (RAM⁻) batteries. *Proceedings of Portable by Design Conference (USA)*, 119–24.

34. RWE Innogy invests €5.5M in ReVolt; rechargeable zinc-air storage systems. 2009. January 16. http://www.greencarcongress.com/2009/09/revolt-20090901.html.

35. Cutler, Tim. 1997. Rechargeable zinc-air design options for portable devices. *Proceedings of Portable by Design Conference (USA)*, 112–18.

36. Rao, R., S. Vrudhala, and D. N. Rakhmatov. 2003. Battery modeling for energy aware system design. *IEEE Computer* 36 (12, December): 77–87.

37. Gao, L., S. Liu, and R. A. Dougal. 2002. Dynamic lithium-ion battery model for system simulation. *IEEE Transactions on Components and Packaging Technologies* 25 (3, September): 495–505.

38. Schweighofer, B., K. M. Raab, and G. Brasseur. 2003. Modeling of high power automotive batteries by the use of an automated test system. *IEEE Transactions on Instrumentation and Measurement* 52 (4, August): 1087–91.

39. Renhart, W., C. Magele, and B. Schweighofer. 2008. FEM based thermal analysis of NiMH batteries for hybrid vehicles. *IEEE Transactions on Magnetics* 44 (6, June): 802–05.

40. Bernardi, D. M., and M. K. Carpenter. 1995. A mathematical model of the oxygen recombination lead acid cell, *Journal of the Electrochemical Society* 142 (8): 2631–42.

41. Gu, W. B., C. Y. Wang, and B. Y. Liaw. 1995. Numerical modelling of coupled electrochemical and transport processes in lead-acid batteries. *Journal of the Electrochemical Society* 144 (6): 2053–61.

42. Landfors, J., D. Simonsson, and A. Sokirko. 1995. Mathematical modelling of a lead-acid cell with immobilized electrolyte. *Journal of Power Sources* 55:217–30.

43. VanZwol, J. 2006. Designing battery packs for thermal extremes. *Power Electronics Technology*, July, 40–44.

44. Chen, M., and G. A. Rincon-Mora. 2006. Accurate electrical battery model capable of predicting run time and I-V performance. *IEEE Transactions on Energy Conversion* 21 (2, June): 504–11.

45. Goebel, K., B. Saha, A. Saxena, J. R. Celaya, and J. P. Christopherson. 2008. Prognostics in battery health management. *IEEE Instrumentation & Measurement* 11 (4, August): 33–40.

46. Karfthoefer, B. 2005. Measure battery capacity precisely in medical design. *Power Electronics Technology*, January, 30–38.

47. Langnan, P. E. 2000. VRLA battery impedance analysis: Cell evaluation via changes in the slope of the impedance curve. *Power Quality Assurance*, September/October, 24–29.

48. Song, L., and J. W. Evans. 2000. Electrochemical-thermal model of lithium polymer batteries. *Journal of the Electrochemical Society* 147 (6): 2086–95.

49. Wu, Q., Q. Qiu, Q., and M. Pedram. 2000. An interleaved dual-battery power supply for battery operated electronics. *Proceedings of Asia and Pacific Design Automation Conference*, 387–90.

50. Benini, L., D. Bruni, A. Macii, E. Macii, and M. Poncino. 2003. Discharge current steering for battery life optimization. *IEEE Transactions on Computers* 92 (8, August): 985–95

51. Chiasserini, C., and R. R. Rao. 2001. Energy efficient battery management. *IEEE Journal on Selected Areas of Communications* 19 (7, July): 1235–45.

52. Schwartz, P. 1995. Battery management. *Portable by Design Conference Proceedings*, 525–47.

53. Sacarisen, P. S., and J. Parvereshi. 1995. Lead acid fast charge controller with improved battery management techniques. South Conference, March.

54. Unitrode Inc. Improved charging methods for lead-acid batteries using the UC 3906. Application Note U-104.

55. Bentley, W. F., and D. K. Heacock. 1996. Battery management considerations for multichemistry systems. *Proceedings of the 11th Annual Battery Conference*, Long Beach, CA, January.

56. Alberkrack, J. 1996. A programmable in-pack rechargeable lithium cell protection circuit. *HFPC Conference Proceedings*, September, 230–37.

57. Israelsohn, J. 2001. Battery management included. *EDN*, January 18, 65–74.

58. Rong, P., and M. Pedram. 2006. An analytical model for predicting the remaining capacity of lithium-ion batteries. *IEEE Transactions on VLSI Systems* 14 (5, May): 441–51.

59. Kutluay, K., Y. Cadirici, Y. S. Ozkazanc, and I. Cadirici. 2005. A new on line state of charge estimation and monitoring system for sealed lead-acid batteries in tele-communication power supplies. *IEEE Transactions on Industrial Electronics* 52 (5, October): 1315–27.

60. Phillips Semiconductors. 1995. The I²C bus and how to use it. Document no. 98-8080-575–01.

61. Benchmarq Microelectronics Inc., Duracell Inc., Energizer Power Systems, Intel Corporation, et al. 1995. Smart battery system specifications—system management bus specifications. Revision 1.0, February 15.

62. Benchmarq Microelectronics Inc., Duracell Inc., Energizer Power Systems, Intel Corporation, et al. 1995. Smart battery system specifications—smart battery data specifications. Revision 1.0, February 15.

63. Benchmarq Microelectronics Inc., Duracell Inc., Energizer Power Systems, Intel Corporation, et al. 1996. Smart battery system specifications—smart battery charger specifications. Revision 1.0, June 27.

64. Benchmarq Microelectronics Inc., Duracell Inc., Energizer Power Systems, Intel Corporation, et al. 1996. Smart battery system specifications—smart battery selec-tor specifications. Revision 1.0, September 5.

65. Heacock, D. 1998. Enabling smart batteries for portable devices. *Proceedings of Global Forum on Mobile Handsets (London)*, June.

66. Elder, G. 2004. IEEE 1625 helps promote safety and reliability. *Power Electronics Technology*, April, 34–43.

9

Protection of Systems from Surges and Transients

9.1 Introduction

At the turn of the century, around 75% of the power generated was processed by power electronics. In the modern scenario of electronics, ultra-large-scale integrated (ULSI) circuits, which progress toward system-on-a-chip (SoC) concepts, are powered by DC power rails as low as 1.2–0.7 V. These two scenarios have made the demands within the power conversion interface complex. Compact ultra-low-DC voltage sources with energy backup, fast transient response, and power management have become mandatory in powering modern electronics. Uninterruptible power supplies and advanced power conditioning with surge protection have become essential in providing clean and reliable AC power.

Modern electronic systems are uniquely vulnerable to power line disturbances because they bring together the high-energy power line and sensitive low-power integrated circuits controlling power semiconductors in the AC-to-AC conversion interfaces as well as the DC-DC converters. The term *power conditioning* is used to describe a broad class of products designed to improve or assure the quality of the AC voltage connected to sensitive electronic systems.

Utilities realize that different types of customers require different levels of reliability, and make every effort to supply disturbance-free power. However, normal occurrences, most of which are beyond control and are acts of God, make it impossible to provide disturbance-free power 100% of the time. In addition to these external disturbances, sources within buildings, such as switching of heavy inductive equipment loads, poor wiring, overloaded circuits, and inadequate grounding, can cause electrical disturbances. Many of these power disturbances, particularly the transient surges, can be harmful to sensitive electronic equipment with low-voltage DC power supplies. Power disturbances can cause altered or lost data and sometimes equipment damage, which may, in turn, result in loss of production, scheduling conflicts, lost orders, and accounting problems.

There are methods and devices available to prevent these disasters from happening. Protective systems range from those providing minimal protection to those that construct a new power source for critical loads, converting the standard "utility-grade power," which may be adequate for most equipment, into the "electronic-grade power" required by many critical loads. This chapter discusses the protection against transients and surges in power conversion systems required for electronic systems demanding 99.999% or higher reliability. Treatment will be on a low-voltage electronic circuit protection basis only, and hence a discussion on power-engineering aspects is not treated in the chapter.

9.2 Types of Disturbances and Power Quality Issues

In a single- or three-phase utility power supply, commercial power companies are expected to supply AC power at a nominal RMS voltage with a percentage tolerance such as ±6% with limited amounts of harmonics as per applicable standards. However, the practical utility grade power supply carries many unwanted RMS voltage disturbances, harmonics, noise, and transients. Figures 9.1(a) and 9.1 (b) depict the RMS voltage disturbances and transients and noise. Figure 9.1(c) indicates the case of a 50 Hz, 230 V RMS power supply and its corresponding frequency spectrum.

9.2.1 RMS Voltage Fluctuations

9.2.1.1 Voltage Surges

Voltage surges are voltage increases that typically last from about 15 milliseconds to one-half of a second. Surges are commonly caused by the switching of heavy loads and power network switching. Surges don't reach the magnitudes of sharp spikes, but generally exceed the nominal line voltage by about 20%. This kind of deviation can cause common computer data loss, equipment damage, and erroneous readings in monitoring systems. A surge that lasts more than two seconds is typically referred to as an *overvoltage*.

9.2.1.2 Voltage Sags

Voltage sags are undervoltage conditions, which also last from 15 milliseconds to one-half of a second. Sags often fall to 20% below nominal voltage and are caused when large loads are connected to the power line. Sags can cause computer data loss, alteration of data in progress, and equipment shutdown. A sag that lasts for more than two seconds is typically referred to as an *undervoltage*.

9.2.2 Transients

As depicted in Figure 9.1(b) voltage transients are sharp, very brief increases in the supply waveform. These spikes are commonly caused by the on and off switching of heavy loads such as air conditioners, electric power tools, machinery, and elevators. Lightning can cause even larger spikes. Although they usually last less than 200 microseconds, spikes can be dangerous to unprotected equipment, with amplitudes ranging from about 180% of the peak value to over a few thousands of volts, positive or negative, and sometimes as high as 6000 V or even larger. This high magnitude of sudden voltage variations can wipe out stored data, alter data in progress, and cause electronic-hardware damage.

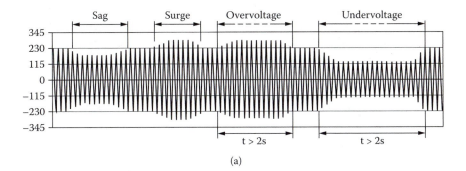

(a)

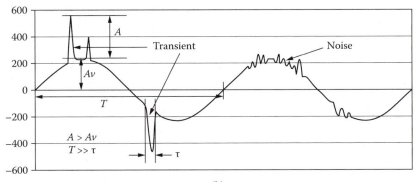

(b)

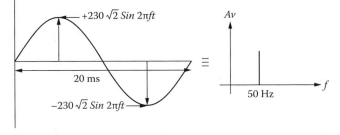

(c)

FIGURE 9.1 Examples of disturbances on a 230 V RMS, 50 Hz utility AC power supply: (a) RMS voltage fluctuations; (b) transients and noise superimposed on the waveform; (c) ideal waveform and its frequency spectrum.

9.2.3 Electrical Noise

Electrical noise is a high-frequency interference in the frequency spectrum of 7000 Hz to over 50 MHz. Noise can be transmitted and picked up by a power cord acting as an antenna, or it can be carried through the power line. These disturbances can be generated by radio frequency interference (RFI) such as radio, TV, cellular, and microwave transmission, radar, arc welding, and distant lightning. Noise can also be caused by electromagnetic interference (EMI) produced by heaters, air conditioners, white goods, and other thermostat-controlled or motor-operated devices.

Although generally nondestructive, electrical noise can sometimes pass through a power supply as if it were a signal and wipe out stored data or cause erroneous data output. Problems result when microelectronic circuitry is invaded by transient, high-frequency voltages collectively called "line noise," which can be grouped into one of two categories: *normal mode* or *common mode* (described later).

9.2.4 Harmonics

Harmonic distortions are usually caused by the use of nonlinear loads by the end users of electricity. Nonlinear loads, a vast majority of which are loads with power electronic devices, draw current in a nonsinusoidal manner. With the increased use of such devices in consumer loads, the presence of distortions in current and voltage waveforms has become a frequent occurrence today. For instance, virtually all modern computer power supplies use a direct rectification mode of AC input to derive a bulk DC power supply for conversion into lower DC rails using switch-mode techniques. A simple example is illustrated in Figure 9.2. Since the rectifiers can conduct only when the instantaneous line voltage is higher than the smoothing capacitor voltage, the charging of capacitors takes place only during a small time period, resulting in narrow input current pulses as shown in Figure 9.2(b). Referring to Figure 9.2(c), if any one or several of the loads connected to the distribution bus draws a nonlinear current, terminal voltage at the consumer terminal will be a distorted sinewave. It is quite common to have a flattened-top sinewave due to non-linear loads such as computer power supplies, UPS rectifiers, or even energy-saving lamps. The ultimate result is that the terminal voltage at the consumer end will be a nonsinusoidal waveform with lot of harmonics.

In view of the widespread use of power electronic equipment connected to utility systems, various international agencies and national agencies have proposed limits on harmonic content injected into the system. More details are available in [1].

9.2.5 Modes of Transients and Noise

The noise and transients sources superimposed on the utility voltage can come in two different forms, namely common mode and differential mode. Differential mode is sometimes also referred to as normal mode or transverse mode. In tackling surge protection in power electronic systems, it is very important to deal with both cases. Figure 9.3 indicates the two cases.

9.2.5.1 Differential Mode (Normal Mode or Transverse Mode)

As the name implies, differential-mode signals appear between the live wire and the accompanying neutral for a case of AC supply input. Similar to a DC power supply, it means the positive or negative rail and its return current path. These two lines represent the normal path of power through the electric circuits, which gives any normal-mode signal a route into sensitive components.

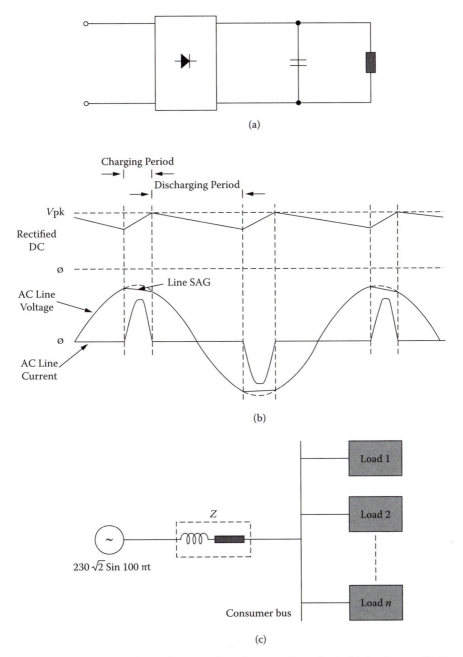

FIGURE 9.2 Distortion of the voltage waveform due to nonlinear loads: (a) simple case of bridge rectifier; (b) rectifier input current in relation to input voltage waveform; (c) simplified case of distribution bus, with a sinusoidal generator, and linear bus impedance before the point of common coupling (PCC).

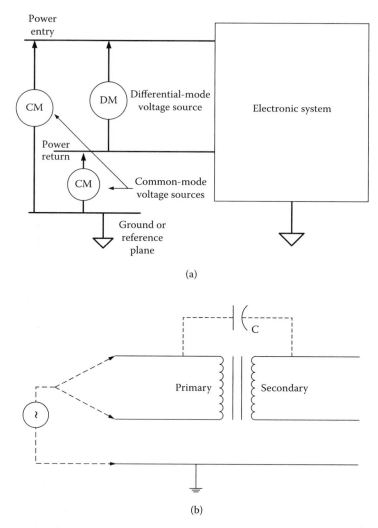

FIGURE 9.3   Different modes of signal inputs and the effect of isolation transformer: (a) differential and common modes; (b) common-mode signal coupling toward secondary side due to coupling capacitance.

9.2.5.2 Common-Mode Signals

Common-mode signals are voltage differentials that appear between the ground and either of the two supply lines. Common-mode (CM) transients are most often the cause of disruption, because digital logic or analog signals are either directly or capacitively tied to the safety ground as a zero-voltage data reference point for semiconductors. As a result, transient CM voltage differences as small as 0.5 V can cause that reference point to shift, momentarily " confusing" the semiconductor chips. Figure 9.3(b) indicates the possibility of transferring common-mode signals due to interwinding capacitances in a transformer with electrical isolation. With the transformer winding process focused on reducing interwinding capacitances, the better an isolation transformer allows you to reduce noise and transients coupled toward load. However, manufacturing isolation transformers with very high common-mode isolation properties could be expensive.

9.2.6 Development of Integrated Circuits, Processors, and Their Power Supply Requirements

After the invention of the transistor in 1947, followed by the development of integrated circuit concepts in the mid-1950s, an exponential growth of integrated circuits occurred following Moore's law [2]. While silicon and GaAs progressed on similar paths, newer compound semiconductor materials such as SiGe were also gradually introduced to cater to high-frequency requirements of commercial electronic systems.

With integrated circuits gradually progressing toward system-on-chip (SoC) concepts with a massively increased number of transistors, the feature size of the transistors was gradually dropping toward less than 0.1 μm. For example, companies such as Intel are planning to progress into 22 nm semiconductor processing as early as 2012. Figure 9.4(a) indicates the Moore's law-based general progress of integrated circuits, while the DC rail voltages are dropping toward sub-1.0 volt levels. With the equivalent noise levels increasing within the complex ICs, dropping of the logic levels makes the scenario even much more complex [3]. Figure 9.4(b) indicates the development of processors similar to the Intel family, and their power supply requirements. Figure 9.4(c) indicates the scenario in terms of clock speed, power consumption, and, most importantly, the equivalent impedance of the equivalent processor load. With the processor equivalent impedance dropping below 1 milliohm, while the DC rail voltages were dropping toward sub-1 V levels, if a transient surge voltage appears on the power supply, it could create disastrous consequences.

In summary, this situation is creating a signal integrity crisis, and the chip designers, as well as the power supply designers, are required to pay adequate attention to these situations for reliable product design. Protection against transients and surges therefore becomes a paramount interest.

9.2.7 Development of DC Power Supplies

During the 1970s and early 1980s, most DC supplies were linear types as depicted in Figure 9.5(a). These simple linear power supplies were based on a step-down transformer, which had the advantage of suppressing high-voltage differential- or common-mode transients. Particularly dangerous common-mode transients were easily suppressed by a reasonably well-designed input transformer with galvanic isolation between the windings and with very low interwinding capacitances. If the step-down transformer is designed to have very low interwinding capacitances, the probability of transferring dangerous levels of common-mode transients toward the rectifier stage and the linear regulator chips is pretty low.

Figure 9.5(b) depicts the situation of modern SMPS systems designed based on power semiconductors such as BJTs, MOSFETS, or IGBTs where higher VA-rated units carry a power factor correction block as well. In these units essential (regulatory) condition

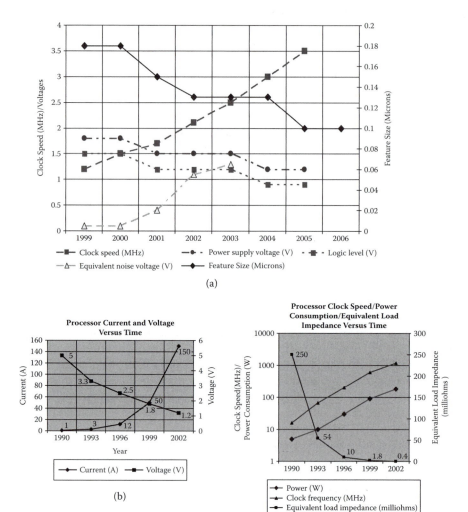

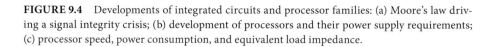

FIGURE 9.4   Developments of integrated circuits and processor families: (a) Moore's law driving a signal integrity crisis; (b) development of processors and their power supply requirements; (c) processor speed, power consumption, and equivalent load impedance.

of ground isolation (for minimizing safety issues) is maintained only by the high-frequency switching transformer, which becomes an integral part of the DC-DC converter stage [4]. It is important to note that the input bulk DC rail is created by direct rectification of 230 V/50 Hz or 120 V/60 Hz input source creating an unregulated DC rail of 320 V DC or 165 V DC respectively. Given the complexity of the modern off-the-line SMPS systems, where all circuit blocks from the AC input up to the DC-DC converter stage, there is a disadvantage to having any common-mode signals directly coupled toward the semiconductors. In practice, these systems are much more vulnerable to high-voltage

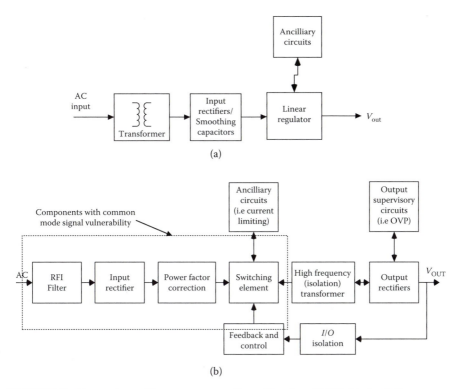

FIGURE 9.5 Comparison of linear power supplies and modern SMPS: (a) simple linear power supply with an isolation transformer; (b) modern off-the-line switching power supply and its components vulnerable to common-mode surges.

transients, and well-designed surge protection circuits are necessary to protect them. This is particularly true for the protection of components within the dotted lines of Figure 9.5(b).

9.3 Principles of Surge Protection Techniques

Almost all transients and surges are unpredictable and statistical in nature. It is not possible to stop them totally, but in planning engineering facilities every effort is made to minimize their effects. However, there is some part of the transients either induced on circuits or directly conducted into the circuits, which can cause disastrous effects on various end user circuits. Protecting a circuit against transients is based on two simple principles: (1) limiting the amplitude of the surge at each component to lower values, and (2) diverting the surge currents through protection-specific components. In practice a designer is expected to protect a system against superimposed surges as shown in Figure 9.6. The case is shown for a system where the input supply is 230 V/50 Hz. The voltage spike shown in the figure, which resembles a power surge or a superimposed lightning transient, usually lasts only a few hundred microseconds, compared to the 20

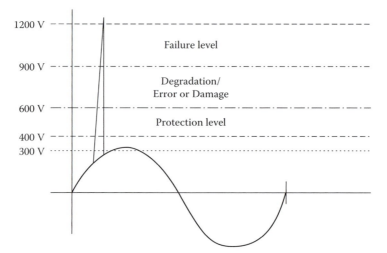

FIGURE 9.6   Transient superimposed on AC-input waveforms with their effects.

ms-long AC cycle in this case. In the design of the basic power conversion system, the designers disregard the possibility of these superimposed signals and assume that the input utility source is a sinewave as per specifications considered in the project. In general, the possibility of surge damage is related to the height of the spike and the duration of the surge. As indicated in Figure 9.6, if the surge peak is between 400 and 600 V, the basic circuit blocks are still considered to be safe. When the spike amplitude is within 600–900 V, degradation of performance or temporary errors could be expected, while values above this can cause permanent faults.

9.3.1 Practical Devices Used in Surge Protection Circuits

In developing surge protection circuits, based on the simple concepts discussed above the designer has to look after two important aspects, namely: (1) surge arrestor stages inserted should not alter the normal operation of the power conversion circuit; (2) components used for surge absorption, diversion, or attenuation should withstand the surges safely. Reliability of the surge protection system is another important concern, since in practical circumstances if a very high level surge enters the system, it might kill the surge arrestor itself.

In designing practical surge arrestor circuits, nonlinear devices such as metal oxide varistors (MOVs), and semiconductor devices such as thyristors or avalanche-type zeners are combined with inductors and capacitors. Gas discharge tubes (GDTs) are also used in some systems.

9.3.1.1 Gas Discharge Tubes

Gas discharge tubes (GDTs, or simply gas tubes) are devices that employ an internal inert gas that ionizes and conducts during a transient event. Because the internal gas requires time to ionize, gas tubes can take several microseconds to turn on or "fire." In fact, the reaction time and firing voltage are dependent on the rising slope of the transient front. A circuit protected by a gas tube arrestor will typically see overshoot voltages ranging from a few hundred volts to several thousand volts. These devices are available for protection of AC line input as well as signal and communication circuits.

9.3.1.2 Metal Oxide Varistors

Metal oxide varistors (MOVs) are devices composed of ceramic-like material usually formed into a disk shape. They are typically constructed from zinc oxide plus a suitable additive. Each intergranular boundary displays a rectifying action and presents a specific voltage barrier. When these conduct, they form a low ohmic path to absorb surge energy [5]. A varistor's voltage-current behavior is defined by the relationship:

$$I = KV^{\alpha}$$

where K and α are device constants. K is dependent on the device geometry, and α defines the degree of nonlinearity in the resistance characteristic and can be controlled by selection of materials and the manufacturing process. High α implies a better clamp, and zinc-oxide technology comes with α values in the range of 15–30. High transient energy absorbent capability is achieved by increasing the size of the disc. Typical sizes range from 3 to 20 mm in diameter. MOVs turn on in a few nanoseconds and have high clamping voltages, ranging from approximately 30 V to 1.5 kV. Figure 9.7 compares the behavior of an MOV with an ideal clamping device and with a solid-state device like a transient voltage suppressor (TVS) zener diode. For more implementation details on MOVs, [6, 7] are suggested. For basic theory of operation of ZnO varistors, [22, 23] are suggested.

Subjecting an MOV to continuous abnormal voltage conditions rather than short-duration transients may cause the MOV to go into thermal runaway, resulting in overheating, smoke, and even potential fire. To prevent this condition, many modern MOVs include an internal thermal fuse or a thermal cutoff device. Some even extend this capability with an internal indicator that provides a logic output if the device's thermal protection is engaged [12]. An overview comparison of thermally protected MOVs (TPMOVs) with traditional MOVs is provided in [13]. These TPMOVs are useful in occasions where continuous AC line overvoltages could occur.

9.3.1.3 Solid-State Devices

Two basic types of solid-state devices for surge absorption are available. TVS diodes are solid-state p-n junction devices. Both these types are available as back-to-back devices (bidirectional devices) suitable for protection against positive or negative surges appearing at power or signal entry inputs.

Compared to a zener, which is designed to regulate a steady-state voltage, the TVS diode is designed to clamp a transient surge voltage. A large cross-sectional area is employed for a TVS diode junction, allowing it to conduct high-transient currents. These diodes respond almost instantaneously to transient events. Their clamping voltage (which is the voltage measured across the device terminals when the device is fired, and a peak pulse current is flowing through the device) ranges from a few volts to a few hundred volts, depending on the breakdown voltage of the device. The fast response time of the TVS

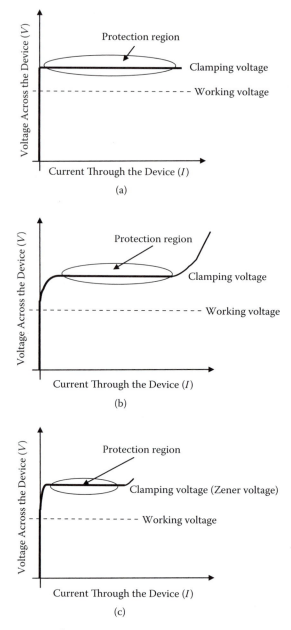

FIGURE 9.7   Comparison of transient protection devices: (a) ideal case; (b) zinc-oxide varistor; (c) TVS zener diode.

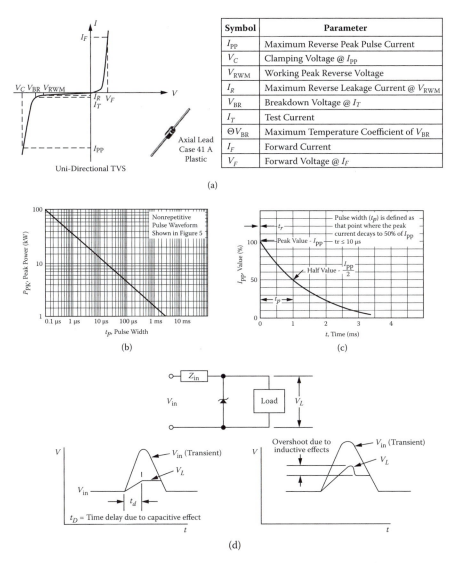

The following table appears in part (a) of the figure:

Symbol	Parameter
I_{PP}	Maximum Reverse Peak Pulse Current
V_C	Clamping Voltage @ I_{PP}
V_{RWM}	Working Peak Reverse Voltage
I_R	Maximum Reverse Leakage Current @ V_{RWM}
V_{BR}	Breakdown Voltage @ I_T
I_T	Test Current
ΘV_{BR}	Maximum Temperature Coefficient of V_{BR}
I_F	Forward Current
V_F	Forward Voltage @ I_F

FIGURE 9.8 An example of a TVS diode: (a) device behavior and data sheet parameters; (b) pulse power absorption capability versus pulse width; (c) test pulse used; (d) typical application circuit and voltage at protected load due to device capacitance and lead inductance. (Adapted from ON Semiconductor, Data Sheet 1N6267A series.)

diodes means that any voltage overshoot is primarily due to lead inductance. Figure 9.8 provides an example of a commercial device from ON Semiconductor [8]. Data sheet terminology for silicon TVS devices, and design calculations are discussed in [14].

TVS thyristors are solid-state devices constructed with four alternating layers of p-type and n-type material. The resulting device is capable of handling very high pulse

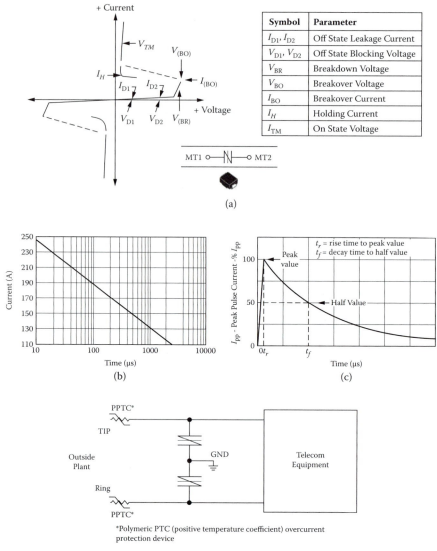

FIGURE 9.9 An example of a thyristor breakover device: (a) device behavior and data sheet parameters; (b) pulse power absorption capability versus pulse width; (c) test pulse used; (d) typical application circuit in telecommunication plant with series current-limiting device. (Adapted from ON Semiconductor Data Sheet, rev. 3, August 2005.)

currents. These devices respond in nanoseconds and have operating voltages that start at about 28 V and go up. Figure 9.9 provides an example of a commercial device from ON Semiconductor, with a typical telecommunication outside plant protection circuit [9].

The ability of TVS diodes to dissipate surge voltages that contribute to the early failure of semiconductors can be evaluated using SPICE macromodels [10].

TABLE 9.1   Comparison of TVS Devices

Suppression Element	Advantages	Disadvantages	Expected Life
Gas tube	Very high current-handling capability Low capacitance High insulation resistance	Very high firing voltage Finite life cycle Slow response times Nonrestoring under DC	Limited
MOV	High current-handling capability Broad current spectrum Broad voltage spectrum	Gradual degradation High clamping voltage High capacitance	Degrades
TVS diodes	Low clamping voltage Does not degrade Broad voltage spectrum Extremely fast response time	Limited surge current rating High capacitance for low-voltage types	Long limited
TVS thyristors	Does not degrade Fast response time High current-handling capability	Nonrestoring under DC Narrow voltage range Turn-off delay time	Long

Source: Adapted from Russell, W. *PCIM*, April 1996, 66–71.

Table 9.1 compares the characteristics of the most widely used TVS devices. References [15, 16] provide more details.

9.3.2 Levels of Surge Protection

TVS device protection levels may be divided into three categories: (a) primary protection, (b) secondary protection, and (c) board-level protection.

9.3.2.1 Primary Protection

Primary protection applies to power lines and data lines exposed to an outdoor environment, service entry, and AC distribution panels. Transient currents can range from tens to hundreds of kiloamps at these sources.

9.3.2.2 Secondary Protection

Secondary protection is for equipment inputs, including power from long-branch circuits, internal data lines, PBX, wall sockets, and lines that have primary protection at a significant distance from the equipment. Transient voltages can exceed several kilovolts with transient currents ranging from several hundred to several thousand amps.

9.3.2.3 Board-Level Protection

Board-level protection is usually internal to the equipment; it is for protection against residual transient from earlier stages of protection, system-generated transients, and electrostatic discharge (ESD). Transients at this level range from tens of volts to several thousand volts with currents usually in the tens of amps.

9.3.3 Circuit Concepts Used for Surge Protection

In general practical circuits developed for surge protection are based on few simple concepts: (1) design the protection circuit separately as an add-on block to the base circuit; (2) attenuate the incoming transient using passive series impedances, or bypass the surge currents via passive low-impedance circuits (they act as filters for high-frequency components of the surge); (3) use nonlinear devices such as GDTs, MOVs, or TVS to divert the surge currents and absorb the transient energy. Figure 9.10 depicts four different kinds of TVSS circuits.

In Figure 9.10(a), which is designed to protect against both common and differential mode transients, M_1 to M_3 are MOVs that will enter their firing or conduction mode when the transient exceeds the threshold voltage limit. Inductors L_1 and L_2 act as high impedances for the transient signal, since its spectrum contains many higher-frequency components, compared to the impedance of the inductors at the line frequency. Similarly, the capacitors indicated as C_{nx} and C_{ny} act as low-impedance paths to the transient signal, which helps divert the transient energy from reaching the critical load end. The overall effect of these circuitry is to minimize the transfer of the transient voltage toward the critical load side. However, due to clamping voltages of the MOVs, some share of the transient energy reaches the load side. The TVS diodes (or thyristors) are used to absorb the remaining transient energy to minimize the possible damage. Figure 9.10(b) indicates a communication line protection scheme, and Figure 9.10(c) is an example of line driver/receiver protection. Figure 9.10(d) is an example of generic IC protection.

Figure 9.11(a) indicates the typical V-I characteristics of an MOV. At low currents the V-I curve approaches a linear ohmic relationship and shows a significant temperature dependence [11]. Under this condition the MOV is in a very high resistance mode, approaching about 1 GΩ or higher. Under this near-open-circuit condition, the nonlinear resistance R_X in Figure 9.11(b) can be ignored, as R_{OFF} value in parallel will dominate. Figure 9.11(c) indicates equivalent circuits under off and conduction states, as applicable in a design calculation. R_{OFF} value is dependent on the temperature, but remains in the range of 10–1000 MΩ. It also depends on the frequency in an inversely proportional manner. Under conduction mode the value of R_X in Figure 9.11(c) becomes many orders of magnitude less than the R_{OFF} value. The effect of this MOV set in Figure 9.10(a) is to absorb much of the transient energy into the MOVs if a common- or differential-mode transient occurs.

9.4 Surge Protection Standards and Practices

Transients that could induce on the equipment inputs could be due to several different categories such as (a) inductive switching, (b) lightning, (c) electrostatic discharge, and (d) electromagnetic pulses (EMPs). All these are very unpredictable random occurrences. Electrostatic discharge may occur due to buildup of static charge on human body up to voltages as high as 20,000 V. These can generate transients with rise times as high as 2 kV/ns. Electromagnetic pulses due to nuclear activities where gamma rays can be released could cause transients with rise times with 5 kV/ns, while lightning activity related transients can generate signals with rise times around 600 V/ns [17]. Due to the

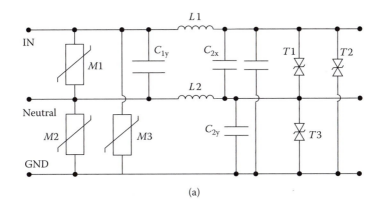

(a)

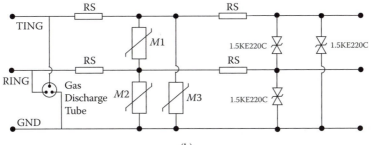

(b)

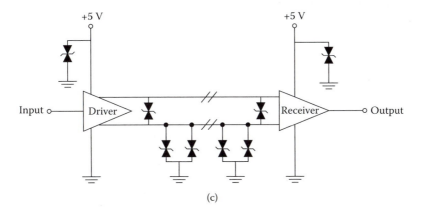

(c)

FIGURE 9.10 Different types of surge protection circuits: (a) multistage surge suppressor (SS) using a combination of devices; (b) communication or signal line protection using multistage SS devices including a GDT; (c) line driver-receiver protection using back-to-back thyristor elements. (d) generic IC protection.

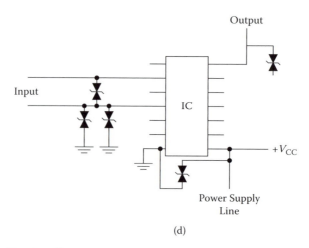

(d)

FIGURE 9.10   (Continued).

statistical nature of the transient voltages superimposed, many useful guidelines are provided by the standards and practices available for the designers. Transient activity can be divided into external sources and internal sources.

9.4.1 External Sources of Transient Activity

Power companies have no control over transients induced by lightning or high-power switching at substation levels. Figure 9.12 illustrates the enormous energy content of a typical lightning waveform. Currents from a direct or indirect strike may enter conductors of a suspended cable or enter a buried cable by ground currents. Either way, the surge will propagate in the form of a traveling wave bidirectionally on the cable from the point of origin. Severity of impact to the end user is directly proportional to the proximity of the lightning strike. If the facility is at a 10–20 pole distance from the strike, little harm will occur, since the surge current will have been dissipated by the utility ground system. Such is not the case where the strike is much closer. In this case, the residual current can migrate through the facility's service equipment and cause severe damage. Other externally generated transients result from switching in nearby industrial complexes, which can send transients back into the power line, causing damage to equipment. EMPs are another rare case of external transient activity.

9.4.2 Internal Sources of Transient Activity

Internally generated transients result from switching within the facility. Any time the flow of inductive currents is altered, such as in the simple act of turning a motor or when a fluorescent light with a magnetic choke is on and off, transient activity can result in an "inductive kick." An excellent example of this phenomenon in day-to-day life is the transients generated by turning off a fluorescent light with a magnetic ballast.

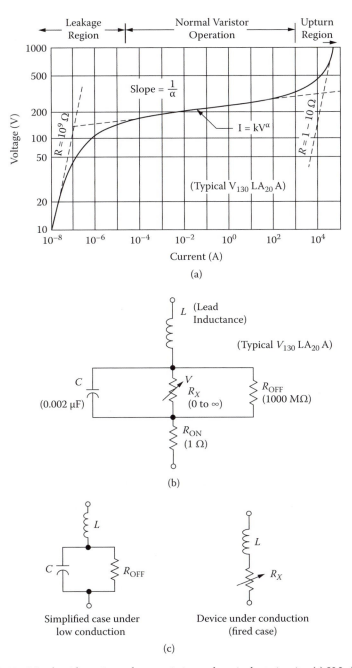

FIGURE 9.11 Metal oxide varistor characteristics and equivalent circuits: (a) V-I characteristics; (b) generalized equivalent circuit; (c) simplified cases under low conduction and fired cases. (From Littelfuse, USA.)

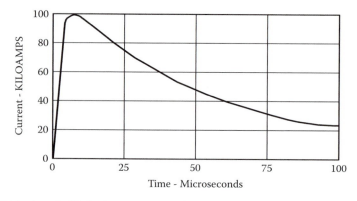

FIGURE 9.12 A typical lightning current waveform.

9.4.3 Transient Energy

The energy of a transient waveform may be readily calculated for transients that are internal to the circuit, such as those caused by inductive switching. Transients external to the circuit are more difficult to quantify. The energy absorbed by the suppression element may be approximated by:

$$E = \int_0^\tau V_c I_c dt$$

where, E = energy in joules; V_c = clamping voltage in volts; I_p = peak pulse current in amperes; and τ = impulse duration.

9.4.4 Transient Protection Standards

Committees such as ANSI, IEEE, and IEC have defined standards for transient wave shapes based on the threat environment. European or IEC transient standards include:

- IEC 61340-3-1 Electrostatics: Methods for simulation of electrostatic effects— human body model (HBM) electrostatic discharge test waveforms
- IEC 61340-3-2 Electrostatics: Methods for simulation of electrostatic effects — Machine model (MM) electrostatic discharge test waveforms
- IEC 61000-4-4 Electromagnetic compatibility (EMC): Testing and measurement techniques - Electrical fast transient/burst immunity test
- IEC 61643 series – For low voltage surge protection

U.S. transient standards include:

- ANSI/IEEE C62.4x series for power line transients
- FCC Part 68 for telecommunication lines
- UL 1449 and various military standards

In all these standards, the transient wave shape is usually defined as an exponentially decaying pulse or a ring wave (damped sinusoidal) for electrical surges. The

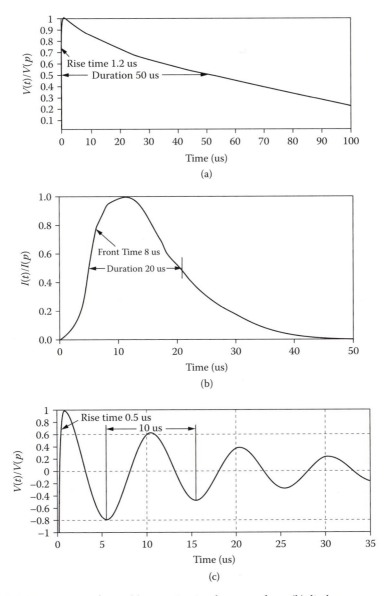

FIGURE 9.13 Surge waveforms: (a) open-circuit voltage waveform; (b) discharge current waveform; (c) ring waveform for testing long branch circuits.

exponentially decaying open-circuit voltage and short-circuit current waveforms are shown in Figures 9.13(a) and 9.13(b), respectively. These impulse waveforms are defined by their rise times and half-amplitude duration. For example, an 8 × 20 μsec impulse current would have an 8 μsec rise time from 10% of the peak current to 90% of the peak current. A 20 μsec time is measured between half-amplitude points. The ring wave results from the effect of a fast rise time transient encountering the impedance of a wiring system.

The waveform shown in Figure 9.13(c) is a 100 kHz ring wave with an initial rise time of 0.5 μsec. The wave shape is defined as rising from 10% to 90% of its peak amplitude in 0.5 μsec and decaying while oscillating at 100 kHz. Each peak is 60% of the amplitude of the preceding one. Most surge protection component data sheets define the surge capability of a suppression device using an 8 × 20 μsec or 10 × 1000 μsec current waveform.

9.4.5 IEEE C62.41 : "IEEE Recommended Practice on Characterization Surge in Low-Voltage (1000 V) and less AC Power Circuits"

IEEE C62.41 identifies location categories within a building, described such as A1, A2, A3, B1, B2, and B3 for surge locations. The "A" and "B" location prefixes represent wiring run distances within a building; the "1," "2," and "3" suffixes represent surge severity. "A" category locations receive their power after more than 60 ft. of wiring run from the main power service entrance, with frequent exposure to comparatively low-energy surges.

External surges will be of a lesser threat in "A" locations than "B" locations due to the impedance protection provided by the inductance of the building wiring in "A" locations. IEEE specifies a low-energy "ringwave" surge waveform for the "A" locations. "B" category locations are within a building, close to the power service entrance, with greater exposure to infrequent, high-energy surges originating outside the building. The "B" location category surges can be caused by lightning, power outages due to storms, and normal utility switching functions. A "combination wave" with high surge energy is specified by IEEE for these "B" locations.

"1," "2," and "3" denote low, medium, and high exposures respectively, in terms of the number and severity of surges, with "1" being the least severe and "3" being the most severe. A "B3" location, therefore, would have the highest exposure to surge energy, while an "A1" location would have the lowest incidence of surge energy.

Table 9.2 shows possible annual surge magnitudes, frequency of occurrences, and surge waveform as extracted from the IEEE C62.41 Standard. Figure 9.14 depicts the

TABLE 9.2   C62.41 Location Categories, Frequency of Occurrences, and Surge Waveforms

IEEE LOCATION CATEGORY	IEEE EXPOSURE	2000 Volt, 70 Ampere Ringwave Surges (0.63 Joules)	4000 Volt, 130 Ampere Ringwave Surges (2.34 Joules)	6000 Volt, 200 Ampere Ringwave Surges (5.4 Joules)
A1	Low	0	0	0
A2	Medium	50	5	1
A3	High	1000	300	90
		2000 Volt, 1000 Ampere Combination Wave Surges (9 Joules)	4000 Volt, 2000 Ampere Combination Wave Surges (36 Joules)	6000 Volt, 3000 Ampere Combination Wave Surges (81 Joules)
B1	Low	0	0	0
B2	Medium	50	5	1
B3	High	1000	300	90

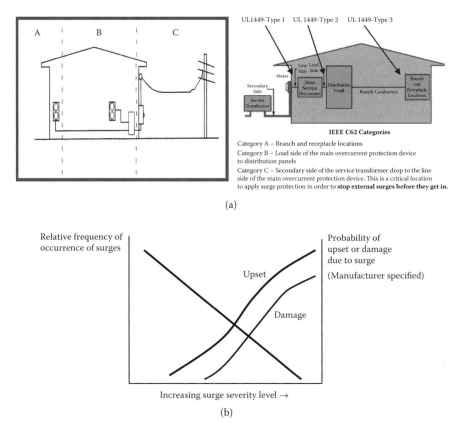

(a)

(b)

FIGURE 9.14 Building locations and severity of surge: (a) categories A to C; (b) severity of surge and upset or damage probability.

building locations, the probabilistic nature of surge severity, and the possibility of upset or damage to systems due to surge. The reader is expected to refer to the set of C62.XX series standards such as C62.41.2 for details [18].

9.4.6 Underwriters Laboratories UL 1449

About two decades ago, Underwriters Laboratories (UL) established a uniform TVSS rating system by creating UL Standard 1449-1987. Although somewhat limited, it was the first step toward establishing benchmarks by which to compare TVSS products. UL established specific test criteria to determine the ability of a TVSS product to stop the travel of a transient voltage surge into protected equipment. TVSS devices were divided into two categories: plug-in or cord-connected devices, and hardwired or direct-connected devices. The plug-in and cord-connected category is subjected to a maximum transient surge impulse of 6000 V with a short-circuit current of 500A available. The voltage available is in a 1.2 × 50 µs waveform, and the current is available in an 8 × 20 µs waveform. Devices included in the hardwired category are subjected to an impulse of 6000 V with

3000 A available in the same type of waveform [19]. In the latest revision of the standard, published in September 2006, the title is changed to "UL Standard for Safety for Surge Protective Devices, UL 1449." This third edition has become a standard by ANSI. In this new edition, the commonly used term TVSS is replaced by surge protective devices (SPDs), and the new test and currents are revised upwards by a factor of six, bringing the test conditions in the let-through voltage tests to 3 kA from 0.5 kA with the same open-circuit voltage of 6 kV. Also, it assigns Type 1 to Type 3 related to C62.41 categories, corresponding to category C to category A. The right-hand side of Figure 9.14(a) indicates this relationship in a practical situation. Type 4 applies to individual components used in all location categories. More details can be found in [20, 21].

9.4.7 Electrostatic Discharge and Circuit Protection

Electrostatic discharge (ESD) is a serious issue in the assembly and testing of systems containing semiconductor devices. There are various standards developed for testing the ESD capability of semiconductor products. These standards have been generated with regard to a specific need related to the electromagnetic compatibility of the system environment. They include the human body model (HBM), machine model (MM), and charged device model (CDM). Each such standard relates to the nature of electrostatic discharge generated within a system application and the potential for damage to the IC. Among the better known standards are:

- Human body model using the MIL-STD-883, Method 3015.7
- Machine model using EIAJIC121
- Human body model using the IEC 1000-4-2 standard

Testing for ESD immunity is more broadly defined to include a device, equipment, or system. Both direct-contact and air discharge methods of testing are used with four discrete steps in the severity level ranging up to 8 KV and 15 kV, respectively. In its simplest form, the Figure 9.15(a) test circuit provides a means of charging the 150 pF capacitor, R_C, through the charge switch and discharging ESD pulses through the 330 Ω resistor, R_D, and discharge switch to the equipment or device under test. The test equipment for the IEC 1000-4-2 standard is constructed to provide the equivalent of an actual human body ESD discharge and has the waveform shown in Figure 9.15(b).

9.4.8 Protection of Printed Circuit Board Components

As we discussed earlier, any remaining components of a transient or an ESD can ultimately end up on a PCB, physically destroying its components or temporarily corrupting any stored data. In order to safeguard physical components in PCBs, semiconductor manufacturers such as Littelfuse and ON Semiconductor have introduced components suitable for board-level protection. Two basic component solutions exist. They are TVS diode arrays and SCR-diode arrays, which come in common IC packages. These devices cater to ESD protection requirements in general.

The SP720, SP721, and SP723 are protection ICs from Littelfuse with an array of SCR-diode bipolar structures for ESD and overvoltage protection of sensitive input circuits.

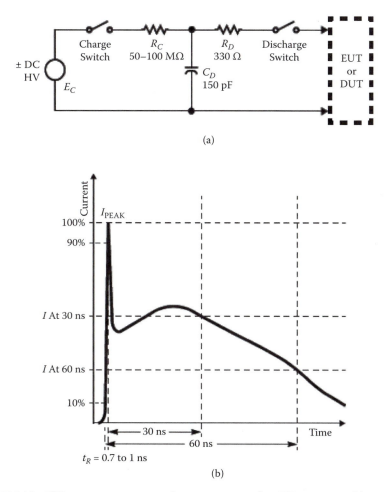

(a)

(b)

FIGURE 9.15 ESD test energy source and output current of an ESD generator. (a) test system (b) waveform.

They have two protection SCR-diode device structures per input. See Figure 9.16(a). The SCR structures are designed for fast triggering at a threshold of one $+V_{BE}$ diode threshold above +V (positive supply terminal) or a $-V_{BE}$ diode threshold below V– (negative or ground). A clamp to V+ is activated at each protection input if a transient pulse causes the input to be increased to a voltage level greater than one V_{BE} above V+. A similar clamp to V– is activated if a negative pulse, one V_{BE} less than V–, is applied to an input. For further details, [24, 25] are suggested.

Similar to the case of SCR-type devices, the TVS diode is also an effective low-cost option to protect sensitive devices from ESD and EMI. In these cases the devices are connected between signal lines and a ground connection or a ground plane on a PCB. Typical examples are the NUP 2105 from ON Semiconductor [26], which is typically used in automotive CAN bus environments and the like. Usually the device connection

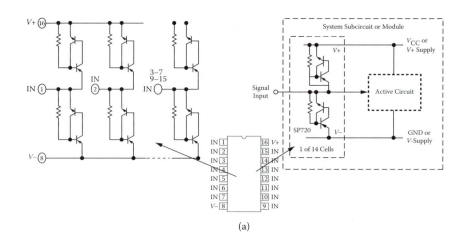

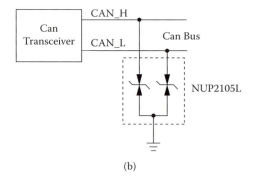

(b)

FIGURE 9.16 ESD protection and techniques: (a) use of SCR-type device arrays; (b) use of TVS diodes. (Adapted from Austin, W. 1999. Littelfuse Application Note AN 9612.2, 1999; and ON Semiconductor Data Sheet, rev. 4, July 2005.)

is as shown on Figure 9.16(b), where the diodes provide common-mode paths between the signal lines and the ground.

9.5 Practical Design Considerations

Having discussed a background on surge protection devices and practices, it is also important to discuss the PCB layout and assembly issues, since these could enhance or degrade the level of protection provided by the devices, particularly in the case of board-level protection. In developing board-level protection stages, we have to keep in mind that capacitive and inductive parasitic elements in a PCB can compromise the level of protection offered by the device such as a TVS diodes, or a TVS thyristor structure, which comes in compact IC packages or as individual elements.

In general TVS devices should be used on all data and power lines that enter or exit a PCB at the I/O connector. Locating the TVS devices as close as possible to the noise source ensures that a surge voltage will be clamped before the pulse can be coupled into any adjacent PCB traces. Figure 9.17(a) depicts this scenario, indicating the good and bad practices. In general the TVS device should be very close to the I/O connector

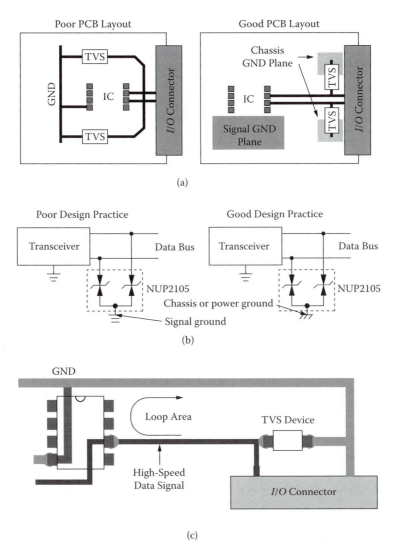

FIGURE 9.17 PCB Layout issues: (a) good and bad practices in placing the TVS device closer to I/O connector with minimum inductive traces; (b) good and bad practices in using the correct ground return, when separate signal and chassis grounds exist; (c) data and ground traces forming a loop, which could act as an antenna with increased RF susceptibility; (d) use of resistor to absorb the majority of surge energy by the external diodes. (Source: Lepkowski, J., *Power Electronics Technology*, March 2006, 48–49. Reproduced with permission.)

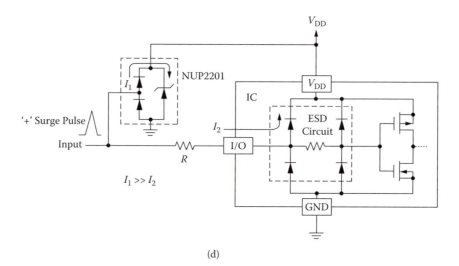

(d)

FIGURE 9.17 (Continued).

and minimizing the length of the TVS traces. In the figure on the left-hand side, due to inductances of the traces part of the energy in the surge pulse can be entering the IC. Unless this component can be absorbed by the ESD protection stages inside the IC, it can have disastrous consequences. The figure on the right-hand side indicates a multilayer PCB, where a separate GND plane is used where return end of the TVS diode is connected, to minimize any inductive impedance. It is simply important to remember that a short trace length equates to a low impedance, which ensures that the surge energy will be dissipated by the TVS device instead of the IC's internal ESD protection stage. Locating sensitive traces in the center rather than near the edge of the PCB is a simple method to protect against any ESD issues during handling [27].

It is important to consider that when a transient is diverted by a TVS device, it can cause a significant current to flow through the device into ground connection or the ground plane. This high current can cause a momentary voltage spike on the ground paths, which can have complex connections within the PCB. In cases where the signal and chassis grounds are separately laid out, if the surge transient-related currents are returned to chassis ground, it may not create a significant transient voltage rise on the signal path ground. Figure 9.17(b) depicts the good and bad practices, noting the two independent GND connections in the right-hand figure. In the left-hand figure, shunting the surge voltage directly to the IC's signal ground can cause ground bounce.

As shown in Figure 9.17(c), another possible issue is the radio frequency susceptibility issue due to loop area formed by high-speed signal lines and the ground return paths. One effective method to minimize this loop problem is to incorporate a separate ground plane. In Figure 9.17(c), we can see that the maximizing the separation distance from the TVS device in IC provides isolation, however, at the expense of forming a larger loop area.

Figure 9.17(d) indicates a case of a diode set such as NUP 2201 from ON Semiconductor used with a semiconductor chip where ESD protection is built in. In such cases, placing a series resistor, such as R in Figure 9.17(d), ensures that the majority of surge energy is absorbed by the external TVS diode. More details on device selection guidelines and placement options are discussed in [27]. ESD failure modes and mechanisms are discussed in [28].

References

1. Kularatna, N. 1998. *Power electronics design handbook*. Burlington, MA: Elsevier-Newnes, chapter 9.
2. Kularatna, N. 2003. *Digital and analogue instrumentation: Testing and measurement*. London: Institute of Engineering and Technology, chapter 2.
3. Duety, S., et al. 2000. The key for processor performance beyond a GHz is clean power. *Proceedings of PCIM 2000*, 68–74.
4. Kularatna, N. 1998. *Power electronics design handbook*. Burlington, MA: Elsevier-Newnes, chapter 4.
5. van Beneden, B. 2003. Varistors: Ideal solution for surge protection. *Power Electronics Technology*, May, 26–32.
6. Littelfuse. 1999. The ABCs of MOVs. Application Note AN 9311-6, July.
7. Littelfuse. 1999. The ABCs of Littelfuse multilayer suppressors. Application Note AN 9671.2, July.
8. 1500 watt Morsorb zener transient voltage surge suppressors. ON Semiconductor data sheet. 1N6267A series.
9. MMT08B310T3 thyristor surge protectors. 2005. ON Semiconductor data sheet, rev. 3, August.
10. Lepkowski, J. 2006. Evaluating TVS protection circuits with SPICE. *Power Electronics Technology*, January, 44–49.
11. Harris/Littelfuse Inc. 1999. Littelfuse varistors—basic properties, terminology and theory. Application Note AN 9767.1, July.
12. Walaszczyk, B. 2002. Multiple protection devices guard against transients. *Power Electronics Technology*, September, 55–64.
13. Dunlap, D. 2001. Thermally protected MOVs resist overvoltage failures. *Power Electronics Technology*, April, 64–68.
14. Clark, M. 2000. Expand the data sheet on silicon transient voltage protection devices. *Power Electronics Technology*, July, 42–73.
15. Clark, O. M. 1990. Transient voltage suppressor types and applications. *PCIM*, November 1990, 19–26.
16. Russell, W. 1996. Transients versus electronic circuits: Survival of the fittest—part I—the need for transient protection. *PCIM*, April, 66–71.
17. Winters, R. 1976. Transients in power supply systems. *Solid State Power Conversion*, September/October, 10–15.
18. IEEE recommended practice on characterization of surges in low-voltage (1000V and less) AC power circuits. 2003. IEEE Standard C62.41.2-2002. IEEE Power Engineering Society, April.
19. Lewis, P. 1995. Transient voltage surge suppression response time. *Power Quality Assurance*, September/October, 56–63.
20. Eaton Corporation. 2008. UL° 1449 third edition: Key changes, May.

21. Siemens Industry Inc. 2009. UL 1449 third edition: SPD/TVSS changes effective September 29, 2009. White paper by Siemens.

22. Levinson, L. M., and H. R. Philip. 1975. The physics of metal oxide varistors. *Journal of Applied Physics* 46 (3, March): 1332–41.

23. Mahan, G. D., L. M. Levinson, and H. R. Philip. 1979. Theory of conduction of ZnO varistors. *Journal of Applied Physics* 50 (4, April): 2799–2812.

24. Austin, W., and R. Sheatler. 1996. Transients vs. electronic circuits: Survival of the fittest, the need for transient protection (part II)—ESD immunity and transient current capability for electronic protection array circuits. *PCIM*, May, 46–53.

25. Austin, W. 1999. IEC 1000-4-2 ESD immunity and transient capability for the SPX72X series protection arrays. Littelfuse Application Note AN 9612.2.

26. NUP2105 dual line CAN bus protector. 2005. ON Semiconductor Data Sheet, rev. 4, July.

27. Lepkowski, J. 2006. Designers maximize TVS diode performance. *Power Electronics Technology*, March, 48–49.

28. Amerasekera, A., W. van den Ableen, L. van Roozendal, M. Hannemann, and P. Schofield. 1992. ESD failure modes: Characteristics, mechanisms, and process influences. *IEEE Transactions on Electronic Devices* 39 (2, February): 430–36.

Appendix A: The "XFET" Reference

D. F. Bowers

The "XFET" reference relies on the principle that the difference in pinch-off voltage (threshold) of two junction field-effect transistors (JFETs) having different channel doping is substantially constant with time and temperature. Figure A1.1 shows the concept.

Two p-channel JFETs, J1 and J2, are run at constant source currents and biased into the saturation region (p-channel JFETs are easier to fabricate than n-channel types on most IC processes). J2 is processed identically to J1 except that it receives an additional channel implant to raise its pinch-off voltage by approximately 0.5 V [1]. The pinch-off voltage for a JFET can be written as [2]:

$$Vp = \frac{a^2 q N_A \left(1 + \frac{N_A}{N_D}\right)}{2\varepsilon} - \Psi \tag{A1.1}$$

where N_A and N_D are the doping densities in the channel and gate respectively, a is the channel thickness, q is the electronic charge (1.6 e^{-19}C), ε is the dielectric constant of silicon ($\approx$1.04 E12), and Ψ is the junction built-in potential, which can further be expressed as:

$$\Psi = \frac{kT}{q} \ln \frac{N_A N_D}{Ni^2} \tag{A1.2}$$

where k is the Boltzmann's constant (1.38e^{-23} J/K), T is absolute temperature, and Ni^2 is the intrinsic carrier concentration.

So the difference in pinch-off voltage between two JFETs differing only in N_A can be written as:

$$\Delta Vp = \frac{a^2 q}{2\varepsilon} (N_{A1} - N_{A2}) \left(1 + \frac{N_{A1} - N_{A2}}{N_D}\right) - \frac{kT}{q} \ln \frac{N_{A1}}{N_{A2}} \tag{A1.3}$$

Appendix A-1

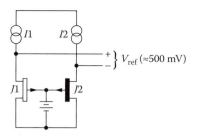

FIGURE A1.1   JFET reference concept.

The last term, the built-in potential difference, is quite small. The channel doping in practice is around 1.1 E12 atoms/cc for the lower pinch-off ($\approx$ 1 V) JFET and around 1.25 E12 atoms/cc for the higher pinch-off ($\approx$ 1.5 V) one. The last term is thus about –2m V and thus contributes about –6.6μ V/°C of overall drift. The measured overall TC runs close to –60 μV/°C (–120 ppm/°C) with about 6 μV/°C of curvature. The excess –53 μV/°C or so of TC is believed to be due to the temperature coefficient of ε, the dielectric constant of silicon, but no accepted authoritative figure for this has been found in the literature. This residual drift is small and predictable enough to be easily compensated for in practice.

Figure A1.2 shows a more practical implementation of the reference (dubbed the XFET reference for eXtra implanted jFET) [3, 4].

Amplifier A1 in conjunction with resistors R1 and R2 amplify the pinch-off difference to provide convenient voltages at the output (R1 and R2 have to be internally trimmed, since the tolerance on the intrinsic reference voltage is quite poor). A network

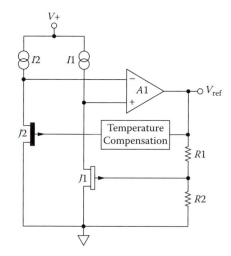

FIGURE A1.2   Practical XFET reference.

of temperature-dependent resistors and current sources placed in series with the gate of J2 provide temperature compensation to better than 3 ppm/°C.

The XFET reference is competitive with buried zeners as far as noise is concerned; the ADR445 5 V version has an output noise of 90 nV/√Hz with 3 mA of supply current. It has the big advantage over the buried zener in that it can work from a supply as low as 2.7 V, whereas zeners are practically limited to about 7 V. The major drawback is the need for very nonstandard processing with a tricky low-current implant to define the important characteristics. Extensive postfabrication trimming is also required to provide the close tolerances needed with practical precision references.

References

1. Bowers, D. F., and L. Tippie. Junction field effect voltage reference. U.S. Patents 5,838,192 and 5,973,550.
2. Gray, Paul R., and Robert G. Meyer. 1977. *Analysis and design of analog integrated circuits*. New York: Wiley, 1977, 52.
3. Goodenough, F. 1997. Sub-5V voltage reference mimics the performance of buried zeners. *Electronic Design*, October 1997, 37–38.
4. Bowers, D. F. 2008. A 37nV/sqrtHz reference based on dual-threshold JFET technology. *BCTM Digest of Technical Papers*, October.

Appendix B

B.1 Buck Converter

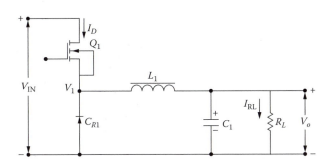

FIGURE B1   Basic configuration.

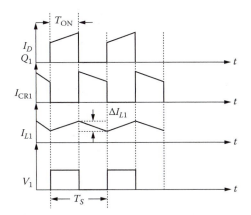

FIGURE B2   Waveforms.

APPENDIX B1   Buck Converter Details

Ideal Transfer Function	Peak Drain Current (I_{DMAX})	Peak Drain Voltage (V_{DS})	Average Diode Currents	Diode Reverse Voltage (V_{RM})	Advantages	Disadvantages	Typical Efficiency
D	$I_{DMAX} =$ $I_{RL} + \dfrac{\Delta I_{L1}}{2}$	$V_{DS} =$ $V_{IN} + V_D$	$I_{CR1} =$ $I_{RL}(1-D)$	$V_{RM} = V_{IN}$	Simple High efficiency No transformer Low switch stress Small output filter Low ripple	No isolation Potential overvoltage if switch shorts High-side switch drive required High-input ripple current	78%

B.2 Boost Converter

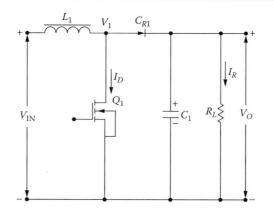

FIGURE B.3   Basic configuration.

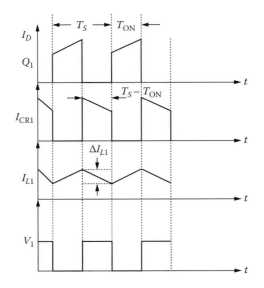

FIGURE B.4 Waveforms. (For the case of both inductor current being positive only)

APPENDIX B2   Boost Converter Details

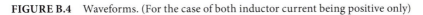

Ideal Transfer Function	Peak Drain Current (I_{DMAX})	Peak Drain Voltage (V_{DS})	Average Diode Currents	Diode Reverse voltage (V_{RM})	Advantages	Disadvantages	Typical Efficiency
$\dfrac{1}{1-D}$	$I_{DMAX} =$ $I_{RL}\left(\dfrac{1}{1-D}\right)+\dfrac{\Delta I_{L1}}{2}$	$V_{DS} =$ $V_O + V_D$	$I_{CR1} = I_{RL}$	$V_{RM} = V_o$	High efficiency Simple No transformer Low-input ripple current	No isolation High peak drain current Regulator loop hard to stabilize High-output ripple Unable to control short-circuit current	80%

B.3 Buck-Boost Converter

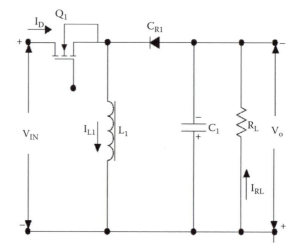

FIGURE B.5   Basic configuration.

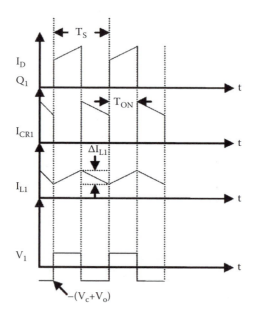

FIGURE B.6   Waveforms.

APPENDIX B3   Buck-Boost converter details

Ideal Transfer Function	Peak Drain Current (I_{DMAX})	Peak Drain Voltage (V_{DS})	Average Diode Currents	Diode Reverse Voltage (V_{RM})	Advantages	Disadvantages	Typical Efficiency
$\dfrac{-D}{1-D}$	$I_{DMAX} =$ $I_{RL}\left(\dfrac{1}{1-D}\right)$ $+\dfrac{\Delta I_{L1}}{2}$	$V_{DS} =$ $V_{IN} + V_O + V_D$	$I_{CR1} = I_{RL}$	$V_{RM} =$ $V_o + V_{IN}$	• Voltage inversion • Simple	• No isolation • High side switch required • Regulator loop hard to stabilize • High output ripple • High input ripple current	80%

B.4 SEPIC Converter

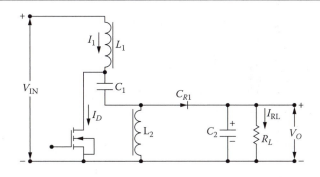

FIGURE B.7   Basic configuration.

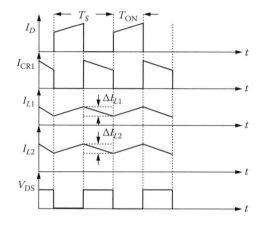

FIGURE B.8   Waveforms.

APPENDIX B3   SEPIC Converter Details

Ideal Transfer Function	Peak Drain Current (I_{DMAX})	Peak Drain Voltage(V_{DS})	Average Diode Currents	Diode Reverse Voltage (V_{RM})	Advantages	Disadvantages	Typical Efficiency
$\dfrac{D}{1-D}$	$I_{DMAX} =$ $I_1 + I_{RL} +$ $\dfrac{\Delta I_{L1} + \Delta I_{L1}}{2}$	$V_{DS} =$ $V_{IN} + V_O + V_D$	$I_{CR1} = I_{RL}$	$V_{RM} =$ $V_o + V_{IN}$	Low-input ripple current Buck and boost with no inversion No transformer Capacitive isolation against a switch failure	No isolation Switch has high RMS/ peak currents (limits power) Capacitors have high ripple currents (low ESR needed) High-output ripple Loop stabilization difficult	

B.5 Forward Convertor

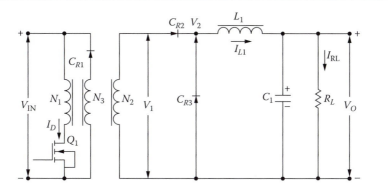

FIGURE B.9   Basic configuration.

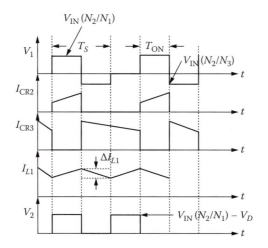

FIGURE B.10   Waveforms.

APPENDIX B5   Forward Converter Details

Ideal Transfer Function	Peak DrainCurrent (I_{DMAX})	Peak Drain Voltage (V_{DS})	Average Diode Currents	Diode Reverse Voltage (V_{RM})	Advantages	Disadvantages	Typical Efficiency
$\dfrac{N_2}{N_1}D$	$I_{DMAX} =$ $\dfrac{N_2}{N_1}\left(I_{RL}+\dfrac{\Delta I_{L1}}{2}\right)$ $+\hat{I}_{MAG}$ $\hat{I}_{MAG}\cdots$ Peak magnetizing current	$V_{DS} =$ $V_{IN}\left(1+\dfrac{N_1}{N_3}\right)$	$I_{CR1} =$ $\hat{I}_{MAG}\left(\dfrac{D}{2}\right)$ $I_{CR2} =$ $I_{RL}(D)$ $I_{CR3} =$ $I_{RL}(1-D)$	$V_{CR1} =$ $V_{IN}\left(1+\dfrac{N_3}{N_1}\right)$ $V_{CR2} =$ $V_{IN}\left(\dfrac{N_1}{N_3}\right)$ $V_{CR3} =$ $V_{IN}\left(\dfrac{N_2}{N_1}\right)$	Drain current reduced by ratio N_2/N_1 Low-output ripple	Poor transformer utilization Poor transient response Transformer design critical Transformer reset limits D Switch requires high-voltage capability High-input ripple current	

B.6 Fly Back Converter

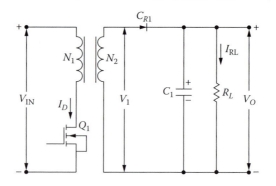

FIGURE B.11   Basic configuration.

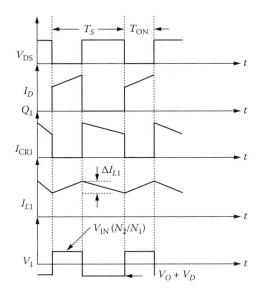

FIGURE B.12 Waveforms.

APPENDIX B6 Flyback Converter Details

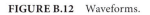

Ideal Transfer Function	Peak Drain Current (I_{DMAX})	Peak Drain Voltage(V_{DS})	Average Diode Currents	Diode Reverse Voltage (V_{RM})	Advantages	Disadvantages	Typical Efficiency
$\dfrac{N_2}{N_1}\left(\dfrac{D}{1-D}\right)$	$I_{DMAX} =$ $I_{RL}\left(\dfrac{N_2}{N_1}\right)\left(\dfrac{1}{1-D}\right)$ $+\dfrac{\Delta I_{L1}}{2}$	$V_{DS} = V_{IN} +$ $\left(\dfrac{N_2}{N_1}\right)(V_{OUT}+V_D)$	$I_{CR1} = I_{RL}$	$V_{CR1} =$ $V_{IN}\left(\dfrac{N_2}{N_1}\right)$	Drain current reduced by ratio N_2/N_1 Low parts count Isolated output No secondary inductors	Poor transformer utilization Transformer stores energy High-output ripple current CR1 requires fast recovery	80%

B.7 Push-Pull Topology

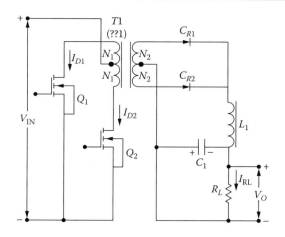

FIGURE B.13   Basic configuration.

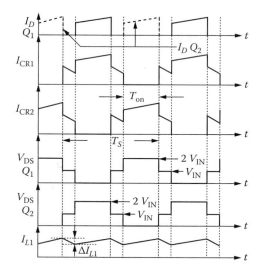

FIGURE B.14   Waveforms.

APPENDIX B7 Phush-Pull Converter Details

Ideal Transfer Function	Peak Drain Current (I_{DMAX})	Peak Drain Voltage (V_{DS})	Average Diode Currents	Diode Reverse Voltage (V_{RM})	Advantages	Disadvantages	Typical Efficiency
$2\dfrac{N_2}{N_1}D$	$I_{DMAX} =$ $\dfrac{N_2}{N_1}\left(I_{RL}+\dfrac{\Delta I_{L1}}{2}\right)$ $+\hat{I}_{MAG}$	$V_{DS}=2V_{IN}$	$I_{CR1}=\dfrac{I_{RL}}{2}$ $I_{CR2}=\dfrac{I_{RL}}{2}$	$V_{CR1}=$ $V_{CR2}=$ $2V_{IN}\left(\dfrac{N_2}{N_1}\right)$	Good transformer utilization Drain current reduced as function of N_2/N_1 Good at low V_{IN} values Low-output ripple current	Cross conduction of Q1/Q2 possible High parts count Transformer design critical Q1/Q2 should be high-voltage capable High-input ripple current	75%

B.8 Two Switch Forward Converter

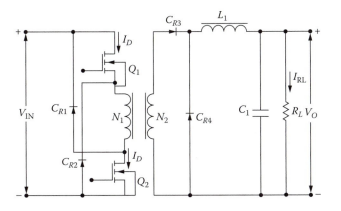

FIGURE B.15 Basic configuration.

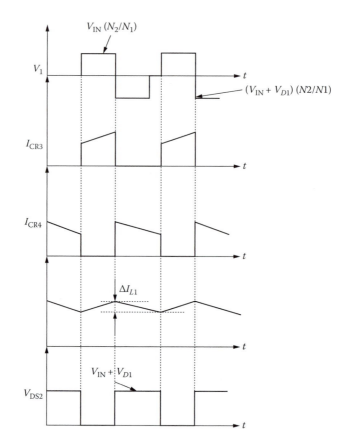

FIGURE B.16   Waveforms.

APPENDIX B8   Two-Switch Forward Converter Details

Ideal Transfer Function	Peak Drain Current (I_{DMAX})	Peak Drain Voltage (V_{DS})	Average Diode Currents	Diode Reverse Voltage (V_{RM})	Advantages	Disadvantages	Typical Efficiency
$\dfrac{N_2}{N_1}D$	$I_{DMAX} =$ $\dfrac{N_2}{N_1}\left(I_{RL} + \dfrac{\Delta I_{L1}}{2}\right)$ $+\hat{I}_{MAG}$	$V_{DS} =$ $V_{IN} + V_{D1}$ (for both transistors)	$I_{CR1} =$ $I_{CR2} =$ $\hat{I}_{MAG}\left(\dfrac{D}{2}\right)$ $I_{CR3} = I_{RL}(D)$ $I_{CR4} = I_{RL}(1-D)$	$V_{CR1} =$ $V_{CR2} =$ V_{IN} $V_{CR3} = V_{CR4} =$ $V_{IN}\left(\dfrac{N_2}{N_1}\right)$	Drain current reduced by turns ratio Lossless snubber recovers energy Drain voltage is half that of single-switch forward converter Low-output ripple	Poor transformer utilization High parts count High-side switch required Transformer reset limits D High-input ripple current	

B.9 Half Bridge Converter

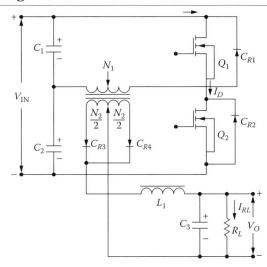

FIGURE B.17 Basic configuration.

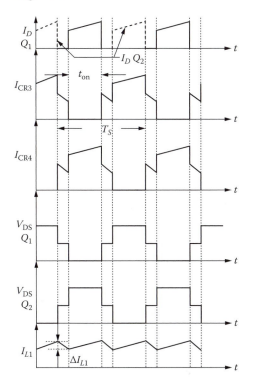

FIGURE B.18 Waveforms.

APPENDIX B9   Half-Bridge Converter Details

Ideal Transfer Function	Peak Drain Current (I_{DMAX})	Peak Drain Voltage (V_{DS})	Average Diode Currents	Diode Reverse Voltage (V_{RM})	Advantages	Disadvantages	Typical Efficiency
$\dfrac{N_2}{N_1}D$	$I_{DMAX} =$ $\dfrac{N_2}{N_1}\left(I_{RL}+\dfrac{\Delta I_{L1}}{2}\right)$ $+\hat{I}_{MAG}$	$V_{DS}=V_{IN}$	$I_{CR3}=\dfrac{I_{RL}}{2}$ $I_{CR4}=\dfrac{I_{RL}}{2}$	$V_{CR3}=V_{CR4}$ $=V_{IN}\left(\dfrac{N_2}{N_1}\right)$	Good transformer utilization Transistors rated at V_{IN} Isolated output I_D reduced as a ratio of N_2/N_1 High power output Low-output ripple current Zero voltage switching possible near $D=1$	Poor transient response High parts count High-side switch required High-input current ripple C1/C2 has high ripple current Cross conduction of Q1/Q2 possible	75%

B.10 Full Bridge Converter

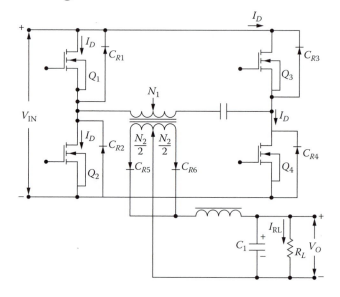

FIGURE B.19   Basic configuration.

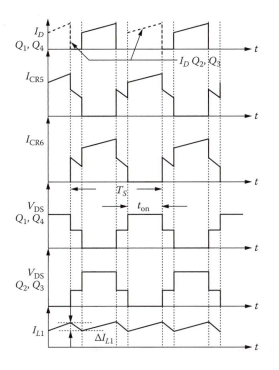

FIGURE B.20   Waveforms.

APPENDIX B10   Full-Bridge Converter Details

Ideal Transfer Function	Peak Drain Current (I_{DMAX})	Peak Drain Voltage (V_{DS})	Average Diode Currents	Diode Reverse Voltage(V_{RM})	Advantages	Disadvantages	Typical Efficiency
$2\dfrac{N_2}{N_1}D$	$I_{DMAX} =$ $\dfrac{N_2}{N_1}\left(I_{RL}+\dfrac{\Delta I_{L1}}{2}\right)$ $+\hat{I}_{MAG}$	$V_{DS}=V_{IN}$	$I_{CR5}=I_{RL}$ $I_{CR6}=I_{RL}$	$V_{CR1}=V_{CR2}=V_{IN}$ $V_{CR5}=V_{CR6}$ $=2V_{IN}\left(\dfrac{N_2}{N_1}\right)$	Nearly same as half bridge	High parts count High-side switch required High-input current ripple C1 has high ripple current Cross conduction of Q1/Q2 or Q3/Q 4 possible	73%

Index